S0-EKK-230

Exploring the Universe

W. M. Protheroe
E. R. Capriotti
G. H. Newsom
all, The Ohio State University

Charles E. Merrill Publishing Company
A Bell & Howell Company
Columbus Toronto London Sydney

Cover photograph by U.S. Naval Observatory

Published by
Charles E. Merrill Publishing Company
A Bell and Howell Company
Columbus, Ohio 43216

This book was set in Times Roman and Eras.
The production editor was Jo Ellen Gohr.
The cover was prepared by Will Chenoweth.

Library of Congress Catalog Number: 78–60353
International Standard Book Number: 0–675–08313–3
2 3 4 5 6 7 8 9 10—85 84 83 82 81 80 79

Printed in the United States of America

Preface

The study of astronomy is an excellent vehicle by which to introduce a student to the study of science. Being one of the oldest sciences, astronomy has a rich history; but, with the modern observational techniques now available, it carries with it the excitement of discovery and confrontation of the unknown. As a discipline, astronomy draws upon all other sciences, particularly the physical sciences.

Attempts to find answers to problems posed by astronomical observations have strongly influenced the intellectual development of the human race. The desire to understand and to predict the motions of the planets gave rise, in large part, to modern mechanics. The growth of mathematics was spurred by astronomical applications. Arguments over the heliocentric and geocentric models of the solar system had profound philosophical implications. Astronomy has supplied the testing grounds for new ideas, such as general relativity, or the impetus for others, such as thermonuclear reactions to explain the sun's energy.

This text is designed to be used in a one-quarter or a one-semester introductory course for nonscience majors, although such a course may well serve a unifying purpose for science majors. Because many students who take a course of this nature do not have an extensive background in science or mathematics, the course must supply the concepts of physics needed to interpret astronomical observations and forego a mathematical development. Based upon our experience in teaching introductory astronomy to nonscience majors, we have incorporated the necessary background at a level which we feel is comprehensible to most students. We have also chosen to emphasize the historical development of astronomy—not as a historian would, but rather to aid the student in seeing how our understanding of the universe has evolved and is still developing. Our experience also indicates that this approach is the most effective one with students having backgrounds in the humanities or the social sciences. Furthermore, our approach does not do a disservice to those from the physical or biological sciences, who quite often have missed how a discipline develops because they are, of necessity, too concerned with the specifics of their particular fields.

We have emphasized the observed, how observations are interpreted, and how those interpretations have influenced our perceptions of the universe. To show that the study of astronomy is a human process, we have introduced each chapter by recounting a story or two appropriate to the material of the chapter. We feel this will make the subject come alive for the student.

There is much from which to choose in developing a course in astronomy, but the nature and length of the course often require that there be some limitation in material. We have tried to emphasize what appear to be the most significant concepts and ideas in the development of astronomy. Our aim is to stimulate the imagination of students, to transfer some of the fascination of the subject, and, hopefully, to leave students with enough background and knowledge that they may follow new developments in the field long after they complete the course.

We are indebted to a large number of individuals and organizations who have supplied us with many of the materials with which we have illustrated the text. They are too numerous to name here, and we can only refer to the list of credits. We are particularly indebted to Delores Chambers for her careful preparation of our manuscript.

W.M.P.
E.R.C.
G.H.N.

Credits

Chapter 1

William Protheroe—Introductory photograph

Chapter 2

Smithsonian Institution Photo No. 55, 385—Figure 2–8
Yerkes Observatory Photographs—Introductory photograph; Figure 2–9

Chapter 3

National Aeronautics and Space Administration—Introductory photograph
Smithsonian Institution Photo No. 37, 117-I—Figure 3–15

Chapter 4

Gerald Newsom—Introductory photograph

Chapter 5

Center of Astrogeology, U.S. Geological Survey—moon map
U.S. Geological Survey—Figure 5–21 (paintings by Don Davis and Don Wilhelm reproduced from *Icarus* 15 (1971): 368, by permission of Academic Press, Inc.)
*The High Altitude Observatory—Figure 5–10c
Lick Observatory Photographs—Figures 5–2, 5–4, 5–5, 5–15
National Aeronautics and Space Administration—Introductory photograph; Figures 5–16, 5–17, 5–19, 5–20
National Space Science Data Center—Figure 5–18
Gerald Newsom—Figures 5–10a, 5–10b, 5–14

---

*A division of the National Center for Atmospheric Research, operated by the University Corporation for Atmospheric Research under sponsorship of the National Science Foundation.

Chapter 6

Education Development Center—Introductory photograph; Figure 6–2
Hale Observatories—Figure 6–10
Gerald Newsom—Figures 6–1b, 6–3, 6–8, 6–9, 6–12, 6–13
Shock Tube Laboratory, Department of Astronomy, Ohio State University—Figure 6–6
J. J. Stoker, from *Water Waves,* copyright 1957 by John Wiley & Sons, Inc.—Figure 6–1a

Chapter 7

Hale Observatories—Introductory photograph; Figures 7–9, 7–10
Kitt Peak National Observatory—Figure 7–11
National Radio Astronomy Observatory—Figure 7–15
Gerald Newsom—Figures 7–2, 7–6
Douglas Wereb—Figure 7–5
Yerkes Observatory Photograph—Figure 7–8

Chapter 8

The Port Authority of New York and New Jersey—Figure 8–5a
Lowell Observatory Photograph—Figure 8–22
JPL/NASA—Introductory photograph; Figures 8–2, 8–3, 8–4, 8–6, 8–8, 8–11, 8–12, 8–14, 8–16, 8–17, 8–18, 8–19, 8–21
National Aeronautics and Space Administration—Figures 8–9, 8–10, 8–13, 8–20
NASA photograph by the Lunar and Planetary Laboratory, University of Arizona—Figure 8–7
National Space Science Data Center—Figure 8–15
U.S. Geological Survey—Mars map

Chapter 9

Center of Astrogeology, U.S. Geological Survey—Mars map
George H. East, Jr.—Figure 9–2
Hale Observatories—Figure 9–4
Harvard College Observatory—Figure 9–8
Lick Observatory Photograph—Figure 9–11
David R. McLean—Figure 9–10
University of Michigan Department of Astronomy—Figure 9–3
Smithsonian Institution—Figure 9–9
Tokyo Astronomical Observatory—Figure 9–6
Yerkes Observatory Photograph—Introductory photograph

Chapter 10

Aerospace Corporation San Fernando Observatory—Introductory photograph; Figures 10–1, 10–16a
*Courtesy of the High Altitude Observatory—Figure 10–10
Kitt Peak National Observatory—Figure 10–8a
E. W. Maunder, from the *Monthly Notices of the Royal Astronomical Society* 82 (1922): 534, with permission from the Royal Astronomical Society—Figure 10–12
McMath-Hulbert Observatory of the University of Michigan—Figures 10–5a, 10–5b, 10–8b, 10–16b, 10–18a, 10–18b

---

*A division of the National Center for Atmospheric Research, operated by the University Corporation for Atmospheric Research under sponsorship of the National Science Foundation.

**Project Stratoscope of Princeton University—Figures 10–3, 10–13
William Protheroe—Figure 10–4
Sacramento Peak Observatory, Air Force Geophysics Laboratory—Figure 10–6
Sacramento Peak Observatory, Association of Universities for Research in Astronomy, Inc.—Figure 10–7
H. Zirin, Big Bear Solar Observatory—Figures 10–14a, 10–17, 10–19

Chapter 11

Copyright by the National Geographic Society—Palomar Observatory Sky Survey. Reproduced by permission from the Hale Observatories—Introductory photograph

Chapter 12

Paul L. Byard—Introductory photograph
Dominion Astrophysical Observatory, Victoria, B.C.—Figure 12–11
Perkins Observatory—Figures 12–1, 12–2, 12–5, 12–9

Chapter 13

Hale Observatories—Introductory photograph
Lick Observatory Photograph—Figure 13–16
From James Wanner, "The Visual Binary Kruger 60" in *Sky and Telescope* 33, no. 1, January 1967. Reproduced by permission of Sky Publishing Corp., Sproul Observatory, and Leander McCormick Observatory—Figure 13–1

Chapter 14

Akadamie der Wissenschaften der DDR; S. Marx—Introductory photograph
Hale Observatories—Figure 14–8

Chapter 15

Akadamie der Wissenschaften der DDR; S. Marx—Introductory photograph
Hale Observatories—Figures 15–4, 15–8

Chapter 16

Lick Observatory Photograph—Figure 16–5
Observatorium Lund, Sweden—Introductory photograph
Sterrewacht, Leiden—Figure 16–8

Chapter 17

Hale Observatories—Introductory photograph; Figures 17–1, 17–2, 17–8, 17–11, 17–12, 17–16, 17–17
By permission of the Harvard College Observatory—Figure 17–4
Paul W. Hodge—Figure 17–10
Kitt Peak National Observatory—Figure 17–3
Lick Observatory Photograph—Figures 17–15a, 17–15b
François Schweizer, Cerro Tololo Inter-American Observatory Photograph—Figure 17–14a
Alar Toomre with permission from D. Reidel Publishing Co., Holland—Figure 17–14b

---

**Supported by the National Aeronautics and Space Administration, the National Science Foundation, and the Office of Naval Research.

Chapter 18

Ivan King, Lick Observatory Photograph—Figure 18–10c
T. D. Kinman, Lick Observatory Photograph—Figure 18–11
Kitt Peak National Observatory—Introductory photograph
Maarten Schmidt, Hale Observatories Photographs—Figures 18–9, 18–10a
Dan Weedman—Figure 18–10b

Color Plates

American Science and Engineering, Inc.—Plate 8C
© by the Association of Universities for Research in Astronomy, Inc., The Cerro Tololo Inter-American Observatory—Plates 13D, 16B
© by the Association of Universities for Research in Astronomy, Inc., The Kitt Peak National Observatory—Plate 13C
James M. Baker—Plate 7A
British Crown Copyright, reproduced with permission of the Controller of Her Britannic Majesty's Stationery Office—Plate 2A
British Crown Copyright, Science Museum, London—Plate 2C
Allan Cook—Plate 9
Dennis di Cicco—Plates 2B, 4B, 7B, 15B
George H. East, Jr.—Plate 8A
Copyright by the California Institute of Technology and the Carnegie Institute of Washington, reproduced by permission from Hale Observatories—Plates 5D, 11C, 12A, 12B, 12C, 14B, 15A
Lick Observatory Photographs—Plates 13A, 13B, 14A
The University of Michigan Department of Astronomy—Plate 10
Muzeum Okregowe, Torun, Poland—Plate 1A
National Aeronautics and Space Administration—Plates 5A, 5B, 5C, 6A, 6B, 8B
National Portrait Gallery, London—Plate 1C
Gerald Newsom—Plates 3A, 3B, 11A
Sacramento Peak Observatory, Air Force Cambridge Research Laboratories—Plate 3C
Scala/EPA, Inc., portrait by Oltario Leoni, Maurcelliana Library, Florence—Plate 1B
Smithsonian Astrophysical Observatory—Plate 4A
U.S. Naval Observatory—Plates 11B, 15C, 16A, 16C

Contents

The Beginning of Astronomy

The Halys River rises in the mountains of central Turkey, flowing on a southwesterly path which carries it through the foothills of the Taurus Mountains. Eventually it curves to the north past Ankara and empties into the Black Sea. In the winter of 585 B.C., the Median hordes, under the command of Cyraxerses, were encamped to the west of the Halys. To the east lay Lydia, at that time ruled by Alyattes. Between Lydia and the army of Medes stood the Lydian army, which had engaged the Medes in many battles during the previous five years.

By any standards, these battles were fought with fantastic ferocity; for the Medes were tough, hard, and merciless, and the Lydians were the most sophisticated soldiers of their time. They were well trained, well disciplined, well equipped. As mounted lancers, they had no peers.

In the spring of 585 B.C., another ferocious battle was fought. Time after time, the Median mounted archers and the Lydian lancers charged each other. The battle raged until the day suddenly became like night, as a solar eclipse occurred. The moon's shadow had swept across the earth, directly over the battlefield. Frenzied warriors who had been trying their very best to hack each other to pieces laid down their arms and milled around in quiet confusion and fear. The tough, fierce Medes and the sophisticated, disciplined Lydians acted together and cowered under the shadow of the moon.

Not only did the eclipse of the sun end the battle, but it finished the war. Alyattes and Cyraxerses both agreed that the eclipse was evidence that the gods had been angered by the fighting, so a peace was promptly negotiated. Such was the power of the astral religion—the worship of sun, moon, planets, and stars—which prevailed in the Middle East from at least 3000 B.C. until near the end of the Roman Empire.

In ancient times, people from all parts of the world had some knowledge of the motions of the sun, moon, and planets. As early as 4000 B.C. in Egypt, and at later

Stonehenge

times in Mesopotamia, Britain, China, and America, people recognized the motions of the sun and moon to be regular and periodic.

Every day the sun rises in the east and sets in the west. The moon passes through its phases about once every thirty days. During each year, the sun appears to move eastward, to travel completely around a great circle relative to the stars. The people of antiquity used the day, month, and year as the basic units of time to construct calendars.

Although various people in various places at various times developed ideas about the universe, the development of ancient astronomy was influenced most strongly by the people who lived in Mesopotamia, the region that is located in modern Iraq. In particular, the people of the Second Babylonian Empire, which flourished during the sixth century B.C., developed a very sophisticated astronomy.

The astronomy of the Babylonians was religiously motivated, for they as well as most other peoples of Mesopotamia worshipped the sun, moon, planets, and stars. It was the job of the astronomer-priests of Babylon, called Chaldeans by the Greeks, to know the future positions of the sun, moon, and planets so that they could in turn know what the future held in store. A remnant of the religion of the Babylonians exists even today in the form of astrology.

The Greeks, from the sixth century B.C. through the second century A.D., were strongly influenced by the Chaldeans. They borrowed heavily from the mathematical methods which the Chaldeans had developed in their astronomical studies. In fact, the Chaldeans were so renowned for their mathematics that the very word **mathematics** derives from the Latin word for Chaldeans, that is, Mathematici.

However, the Greeks developed concepts of the universe which were far less naïve than those of the Chaldeans. For the most part, they were more interested in learning about the nature of the universe for the pure sake of knowing than for the purpose of determining their own fates. The Greeks became so sophisticated that, until the sixteenth century, very little was added to the knowledge of the universe which they had gained over the eight centuries between 600 B.C. and A.D. 200.

Primitive Astronomy

The study of astronomy began when early people tried to explain what they saw when they looked toward the sky. They had no instruments such as telescopes to help them, only their unaided eyes. Their knowledge of the true nature of the sun, the moon, and stars was nonexistent. They began, however, by using a method used in modern science—devising a mental model which would describe what is seen. Sometimes the model approaches reality, and sometimes it deviates far from reality. As long as it helps to describe and to predict what is seen, the model is helpful.

Early records show that the first model of the universe consisted of a flat earth, upon which human beings lived, surrounded by a sphere to which was attached the sun, the moon, and the stars. Early peoples were true to their senses. The earth, except for distractions such as mountains and valleys, does appear quite flat to the eye. The sky overhead looks equally far away in all directions and therefore appears to have the properties of a sphere. They did not sense themselves to be in motion, so early peoples concluded that the rising and setting of the sun and stars were due to an actual westward movement of the spherical sky above them. The problem of night and day is easily explained by the rotating sphere. As shown in Figure 1–1, when the rotation of the sphere carries the sun above the edge of the earth, it is daytime. When the dark part of the sky, with the stars attached, rotates above the earth, it is nighttime; the daylight sky is hidden from view. Early peoples' sense of distance, we now know, was completely wrong, because some thought that the stars could be touched by climbing a mountain of sufficient height. Although their model was incorrect, a modified model of a spherical sky is useful, as will be seen later in this book where it will be a helpful device for visualizing the sky and for locating objects for study.

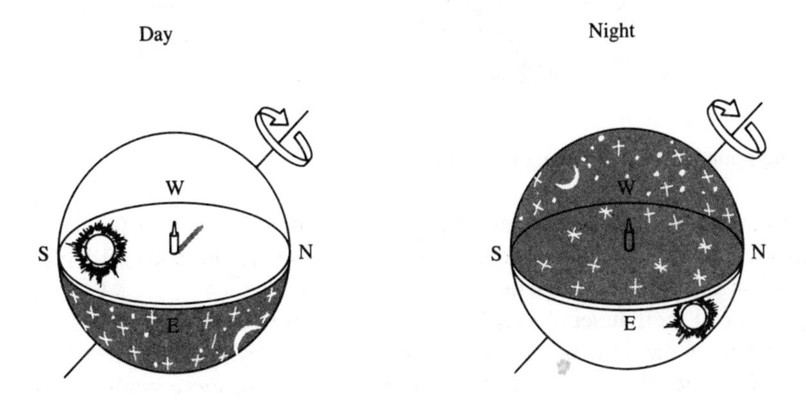

Figure 1–1 The early concept of the sky explaining night and day.

The primitive concept of the sky may have first been questioned when the change in the rising and setting points of the sun was noticed and later correlated with the seasonal changes in the weather. Figure 1–2 shows schematically how the sunrise points change throughout the year in the northern temperate zones. In the summer, the sun rises at the most northerly point; while, in the winter, it rises at the most southerly point. At two times, once in spring and once in autumn, the sunrise point lies due east. By counting the number of days for the sunrise points to go through a full cycle, the interval which we call the year was measured. A number of ancient structures, such as Stonehenge in England (Plate 2A), were used in the observation of the sunrise points and in the determination of the beginning of the seasons.

Babylonian Astronomy

We are particularly indebted to the ancient civilizations of Mesopotamia, in what is now called Asia Minor, because they began the systematic study of the sky which developed into modern astronomy. As early as 3000 B.C., they had developed a solar calendar using the sun and moon to keep time and mark the seasons. Marking the seasons was particularly important because theirs was an agrarian culture, so planting and harvest times were very important. They had an aristocracy of astronomer-priests who were charged with keeping the calendar. These astronomer-priests developed a complicated sky-lore based upon the motions of the sun, moon, and the five visible planets—Mercury, Venus, Mars,

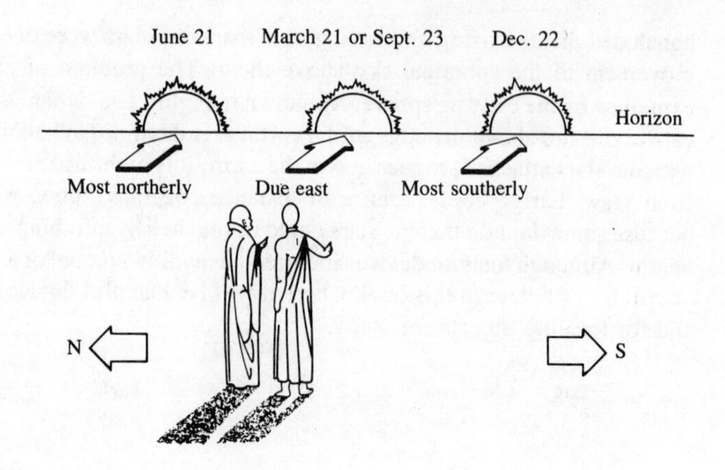

Figure 1–2 The variation in the rising points of the sun during the year.

Jupiter, and Saturn. They deified these celestial objects and, along with observing them for practical reasons, watched them because they came to believe that the celestial deities determined events on the earth by their motions in the sky.

What we would call their superstitious beliefs drove them to develop methods for following the motions of the sun, moon, and planets, which they observed to travel through the stars. They developed star maps upon which to record these motions and discovered and recorded the apparent path of the sun through the sky. This path, particularly important because it located their most important god, is called the **ecliptic.** It is shown in a flat map in Figure 1–3, with the sun's position indicated at various times of the year according to our current calendar. The sun moves eastward along the ecliptic, passing alternately north and south of the celestial equator. Note that this map of the sky is drawn with east to the left rather than to the right, as is common for maps of the earth; this is because the sky is viewed from the inside rather than from the outside as the earth is viewed. The early astronomers correlated both the rising points of the sun and its position

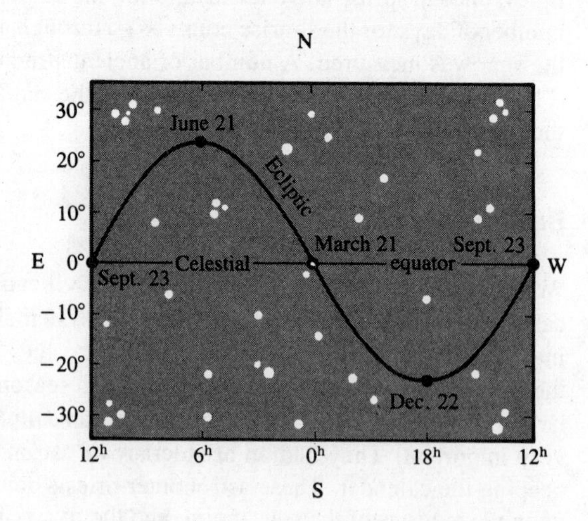

Figure 1–3 The annual path of the sun through the stars.

along the ecliptic with the seasons. They therefore devised a calendar based both upon the seasons and upon the sun's position relative to the stars.

They also carefully studied another powerful god, the moon. Its path (*see* Figure 1–4) was always found to lie close to the ecliptic; and its **phases,** or changing appearance, were related to its position on its path relative to the sun. There are roughly twelve complete changes of the moon in a year. The period of one complete phase change is a month. Both the month, based on the moon, and the year, based on the sun, were used in setting up the ancient Mesopotamian calendar.

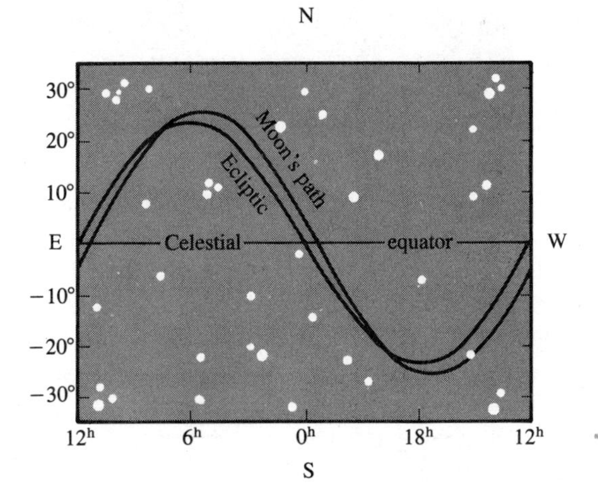

Figure 1–4 The relation of the moon's monthly path to the ecliptic.

The five starlike objects which also moved through the stars, that is, the **planets** or wanderers, were also considered important gods. The planets were particularly puzzling because every so often as they moved through the stars, they interrupted their eastward motion to move westward for a short time. This motion, shown in the reversal in the path of Mars in Figure 1–5, is referred to as a **retrograde** or backward motion. Being relatively uncommon, retrograde motion of a planet was given particular importance by the ancient astronomer-priests. They also found that the paths of the five planets always fell close to the ecliptic.

In addition, they ascribed special meaning to groupings of the fixed stars, some of which we still recognize today as constellations. The band of twelve constellations along the ecliptic was considered to be special because the important gods—the sun, moon, and planets—are always located within these constellations. By 600 B.C. the Chaldeans, the astronomer-priests of Babylon, were able to predict the future positions of the moon and planets within this band of constellations, which they called the **Zodiac** (as depicted in Figure 1–6). It was, therefore, not difficult for them to convince themselves that they could predict future events which would be determined by the future positions of the gods. They formalized this belief into what is best called an *astral religion,* which we know as **astrology.**

During the early period of their astral religion, the Babylonians believed that the universe had a definite beginning. All that existed before its creation were one god and one goddess. Eventually other gods were born to them and their descendants. One of these,

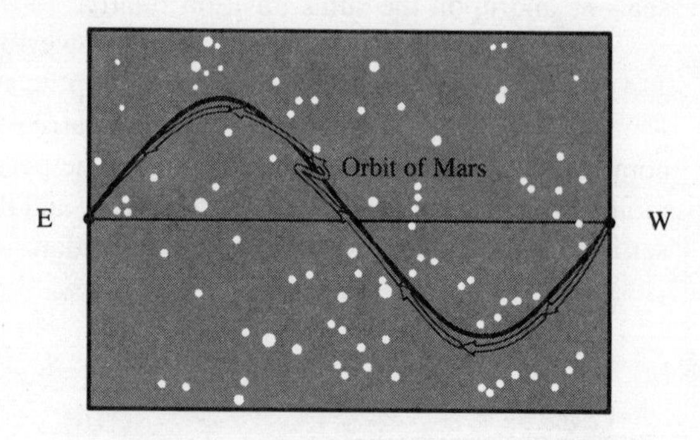

Figure 1–5 Retrograde motion of Mars.

Marduk, the great-grandson, created the earth and man. However, through their discovery of the repetitious cycles of the sun, moon, and planets, the Chaldeans were led to the concept of eternity. They invoked a concept of the heavens which had neither beginning nor end and which was ruled by celestial objects which they took to be ageless gods. Their philosophy contrasted strongly with that of the contemporary Jews, who believed in a material world created at a particular time in the past by a single, universal god. In contrast to the Babylonian belief that each action on earth results from the multiple actions of their many gods, the Jews believed that their god made people free to guide their own lives and interceded only when they needed to be corrected.

Greek Astronomy

The next people to enter the development of astronomy were the Greeks. They began an age of inquiry, always seeking reasons why things occurred rather than assuming that there were no reasons or that the gods did it, as the Babylonians believed. The Greeks

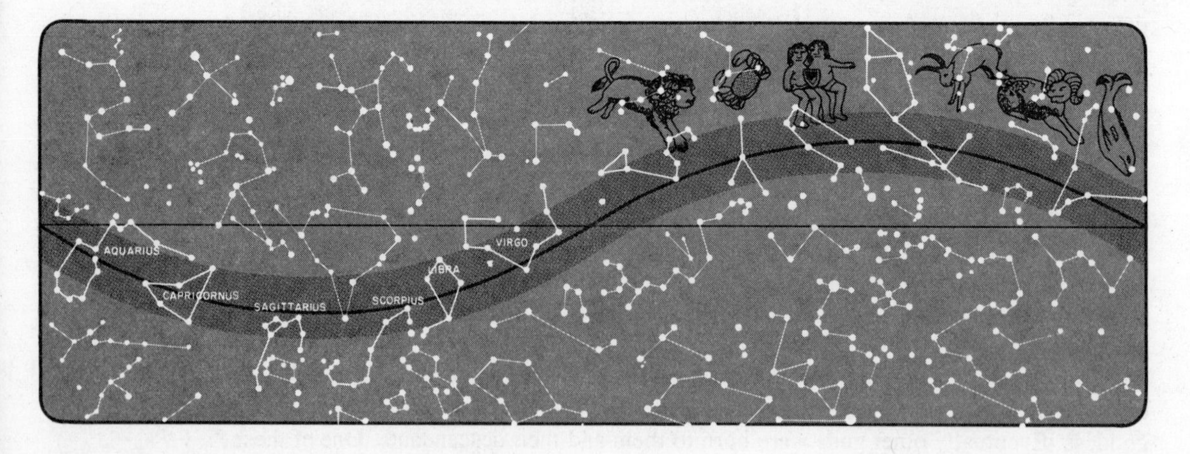

Figure 1–6 The constellations of the Zodiac.

knew of the astronomical work of the Chaldeans and were impressed by it. Some of them were taken by the Babylonian astrology, while others considered it hardly worthy of consideration. They began, however, to try to explain the celestial phenomena which the Babylonians had recorded and which they themselves observed. One early Greek, Anaxagorous, gave the correct explanation for the phases of the moon. Instead of seeing the phases as some mysterious property of the moon, he proposed that they are caused by the moon's motion about the earth. In addition, since the moon is a dark, nonluminous sphere, it could only shine by reflecting light which it receives from the luminous sun. Since only one-half of the moon can be illuminated by the sun at any time, the phases arise because different amounts of the illuminated half of the moon are visible as it circles the earth. The outer ring of drawings in Figure 1–7 illustrates these phases as seen from the earth. This conclusion quite naturally led the Greeks to understand the phenomenon of eclipses. **Solar eclipses** occur when the moon's shadow strikes the earth. **Lunar eclipses** result when the moon passes through the shadow cast by the earth. The Greeks realized that the earth, moon, and sun form a three-dimensional system, with the sun more distant than the moon. This contrasts strongly with the earlier proposal that the sun and moon are attached, like all other objects, to the spherical sky.

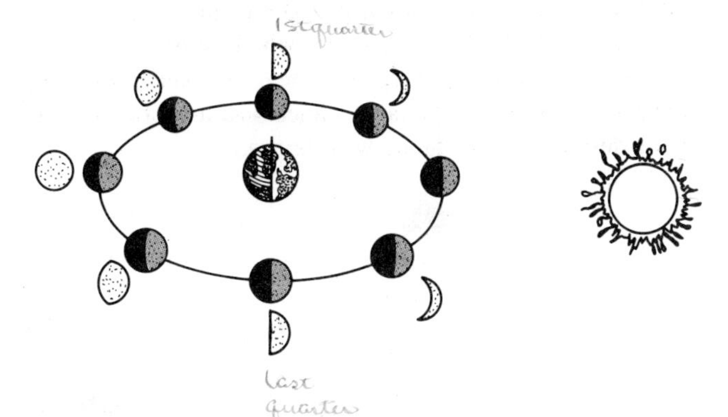

Figure 1–7 The phases of the moon.

Eclipses also gave the Greeks one of their strongest arguments in favor of the earth being spherical rather than flat. The shadow of the earth upon the moon during an eclipse always appears to be a section of a circle. Aristotle argued that the earth consequently must be spherical. He reasoned that, as illustrated in Figure 1–8, if the earth were a flat disc, it would cast a circular shadow on the moon when an eclipse occurred at midnight. But the shadow cast on the moon would have to appear as a straight line for an eclipse viewed at sunrise. Since this did not correspond to his observations, the earth could not be a flat disc. The only shape for the earth which satisfied all observed eclipses was a spherical shape.

In the fourth century B.C., another Greek astronomer, Aristarchus, developed a method to find the distance to the moon. He constructed a diagram like that shown in Figure 1–9. He knew that when the moon is in quarter-phase, the line from the moon to the sun must make a right angle to the line from the moon to the earth. The triangle formed by sides *A*, *B*, and *C* would therefore be a right triangle. The Greek mathematicians had already developed geometry and had studied the properties of the right triangle. By measuring the angle between sides *A* and *B*, Aristarchus could apply these properties to find comparative values for sides *A* and *B*, the distances to the sun and the moon. While he could not find the actual distances, he did conclude that the sun is many times more distant

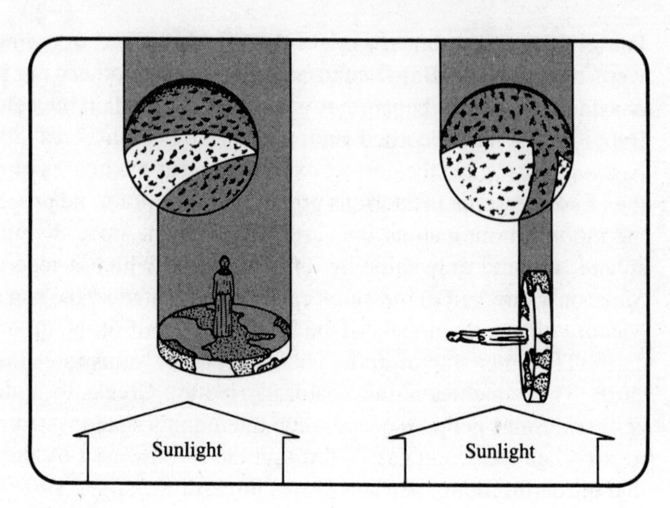

Figure 1–8 Aristotle's proof of the earth's spherical shape.

than the moon. We know today that his result was numerically incorrect and that the sun is much more distant than he concluded, because the method by which he measured the angle between sides *A* and *B* was itself incorrect. Nevertheless, the basic idea was sound; and the value he obtained for the relative solar and lunar distance was used for several centuries until a better method for determining distance was devised.

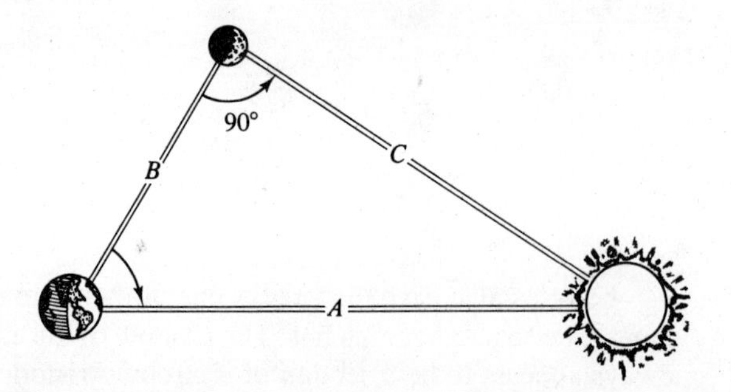

Figure 1–9 Aristarchus' method for estimating the relative distances of the sun and moon.

Since the sun and the moon both have the same apparent size (one-half degree) as viewed from the earth, Aristarchus knew that the linear sizes of these bodies must have the same relationship as their distances. The sun was therefore larger than the moon because it was more distant. He, of course, underestimated the sun's size, because the distance to the sun which he used was twenty times too small.

He also estimated the size of the earth relative to the sun and moon. Observations of lunar eclipses showed that the moon took about two and two-thirds times as long to cross the earth's shadow as it did to move its own apparent diameter on the sky. The diameter of the earth's shadow at the moon's distance was therefore two and two-thirds as large as the moon's diameter. By drawing a diagram similar to Figure 1–10, he could deduce the size of the earth. First he drew the sun and the moon at distances *A* and *B*, maintaining the same ratio as the one he had determined. Then he could draw in *C* and *D* at the end points of *A* and B, with the same ratio as A and B. At the moon he could construct a line, indicated by the arrows, representing the slice through the shadow of the earth two and two-thirds times larger than the moon. Connecting the ends of this line with the ends of the line representing the sun's diameter, he could find the diameter of the earth by measuring across the converging lines at the point from which A and B were drawn. Figure 1–10 has been exaggerated for clarity. In a true drawing, the angles are all much smaller, and the drawing becomes difficult to interpret. Even though his results were in error because of the mistake in the distances, Aristarchus did place the three bodies in the proper order of size. The sun is larger than the earth which, in turn, is larger than the moon.

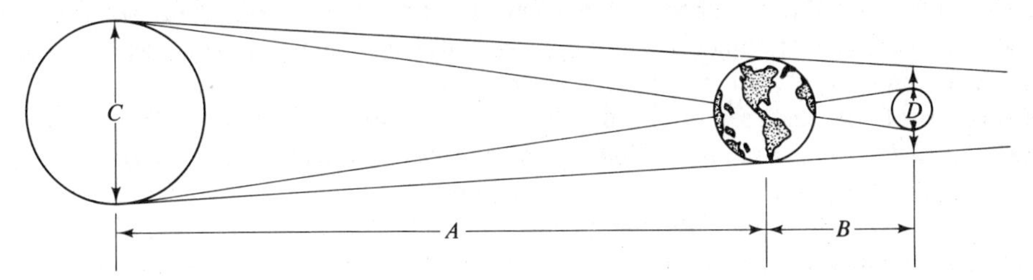

Figure 1–10 Determination of the relative size of the earth from lunar eclipse observations.

Later, Eratosthenes, a Greek living in Egypt, was able to measure the circumference of the earth by a method closely related to the methods used by surveyors today. He had heard that on the first day of summer the sun passed directly over the ancient city of Syene; no shadow was cast by the walls of a deep well which had been dug there. On that same day in Alexandria, 800 kilometers north of Syene, a vertical stick did cast a shadow. The length of the shadow compared to the height of the stick indicated that the angle between the stick and incoming sunlight was about seven degrees, or one-fiftieth of a circle. Assuming the earth to be round, he could then (as indicated in Figure 1–11) find the full circumference of the earth by multiplying the distance between Alexandria and Syene by fifty. His result was surprisingly accurate and differs by only a few percent from the modern value for the earth's circumference. Two thousand years before Columbus sailed the Atlantic Ocean, some Greek astronomers had already concluded that the earth is a sphere. They even knew its size.

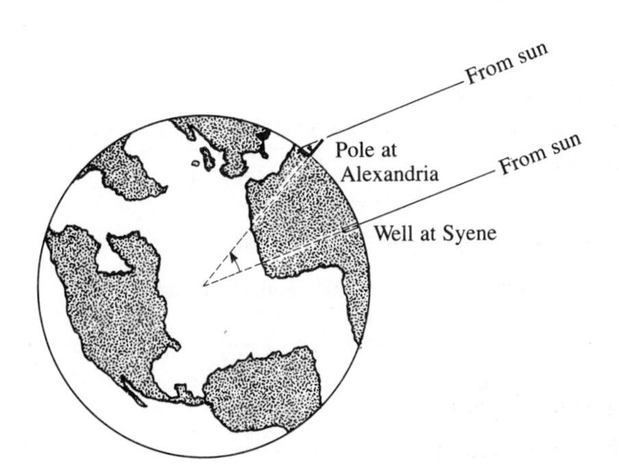

Figure 1–11 Eratosthenes' method for determining the circumference of the earth.

They were still troubled by one obvious sensation, or rather the lack of one—the earth seemed perfectly motionless. Most Greek astronomers, therefore, concluded that the earth is the motionless center of the universe. However, a few people, including Aristarchus, wanted to place the sun at the center and let the earth, with its companion the moon, move about the luminous sun. A very convincing argument was proposed against this **heliocentric,** or sun-centered, model. If the earth moves about the sun, the stars should show an apparent change in position, or **parallax,** as they are viewed at different times of the year. No such effect was observed, and no small wonder—it was many centuries later, after the telescope was invented, that the first stellar parallax was detected. The **geocentric,** or earth-centered, model consequently held sway. Even though the Greeks succeeded in removing the sun, moon, and later the planets from the sphere of the stars, they could not place them in their correct spatial order because they lacked the confirmation of the earth's motion. By the second century B.C., probably the most accomplished Greek astronomer, Hipparchus, had devised a sophisticated geocentric model of the universe capable of describing and predicting most of what could then be observed in the sky. He retained the sphere of stars which bounded the universe, at least in his mind. At the center of this sphere he placed the earth. The sun, moon, and the planets moved about the earth in a complicated system of circles. This geocentric model, with some modification and improvements, was to be the picture of the universe that was accepted until the sixteenth century A.D., and in some circles, at least, for a hundred or so years beyond that.

KEY TERMS

- planet
- ecliptic
- lunar phases
- retrograde motion
- Zodiac
- parallax
- astrology
- lunar eclipse
- solar eclipse
- geocentric
- heliocentric

Review Questions

1. In the primitive conception that the earth is stationary and the spherical sky rotates around it, in which direction does the sky rotate?
2. What is the ecliptic?
3. When does the sun reach the northernmost point on the ecliptic?
4. How did ancient people tell which objects in the sky were stars and which were planets?
5. What is retrograde motion?
6. Do the sun and moon show retrograde motion?
7. What were two early explanations of the nature of planets?

8. What are constellations?
9. What distinguishes the constellations of the Zodiac from the rest of the constellations?
10. What causes the phases of the moon?
11. How long does it take the moon to go through a complete cycle of phases?
12. Which of the ancient people were most noted for their efforts to understand the natural causes of observed phenomena?
13. How can an understanding of eclipses be used to show that not all objects in the sky are attached to the inside of a sphere surrounding the earth?
14. How did ancient Greeks deduce that the earth is a sphere?
15. No conclusive observations were made in ancient times to demonstrate whether the earth goes around the sun or the sun goes around the earth. What observation was made which was interpreted to indicate that the sun goes around the earth?
16. On what basis might Aristarchus have concluded that the earth travels around the sun?
17. What is a geocentric universe?

Discussion Questions

1. If the sun moves from west to east relative to the stars, why doesn't it rise in the west and set in the east?
2. From about March 21 to September 23, people in the Northern Hemisphere see the sun rise north of due east. Where do people in the Southern Hemisphere see the sun rise during this same time?
3. How did Aristarchus try to estimate the distance to the moon, compared to the distance to the sun?
4. How did Aristarchus discover that the sun is much bigger than the moon?
5. How did Aristarchus discover that the earth is bigger than the moon?
6. How was the size of the earth measured in ancient times?
7. If the rising and setting of the sun, moon, planets, and stars is a result of the rotation of the earth, in what direction does the earth rotate?
8. What advantages in observing the skies did ancient people have over the average person today?

2

EFFIGIES TYCHONIS BRAHE O.F.
ÆDIFICII ET INSTRUMENTORUM
ASTRONOMICORUM STRUCTORIS
A° DOMINI 1587 ÆTATIS SUÆ 40

The Astronomical Revolution

Copernicus, Tycho, Kepler, Galileo, Newton: In a period spanning barely 150 years, their efforts led to the conclusion that the universe is a well-ordered system which works according to the dictates of a few, simple laws.

When Copernicus was born in 1473, Ptolemy's **Almagest,** a book that contained practically all existent information about astronomy, was almost fourteen hundred years old. Its main thesis was that the earth is the immobile center of the universe. The apparent motions of the sun, moon, planets, and stars were taken to be real.

Copernicus decided that things were far simpler if you assume that the earth rotates on its axis and revolves around the sun. Then, in part, the apparent motions of the sun, planets, and stars result from the fact that the earth moves.

The later work of Tycho, Kepler, and Galileo supported Copernicus' hypothesis and ultimately, in the late sixteenth century, inspired Isaac Newton to develop a few simple laws which explain the motions of the objects in the solar system.

Certainly, of the five people who influenced the revolution in science, Tycho Brahe is the most neglected. Copernicus, Kepler, Galileo, and Newton are renowned historical figures who have been immortalized by literature, art, and even by postal stamps and commemorative festivals. Tycho, on the other hand, is known for the most part only to professional astronomers.

Tycho was at times lacking in patience and reservation, qualities which are seemingly necessary for one who would spend years in the refinement of methods for measuring the positions of celestial objects. As a university student, he quarreled with a colleague over an issue involving mathematics. The argument lasted about a week and ended at a Christmas party where each of the two young Danish noblemen, their good senses having been deteriorated by huge quantities of beer and wine, drew their usually only decorative swords and began to hack away at each other.

Tycho Brahe and his mural quadrant

Tycho and his adversary must have been better mathematicians than they were swordsmen, for neither was able to get his weapon much closer than a nose's length to the body of the other. Tycho's opponent did just that, however, and with one swipe ended the fight by lopping off Tycho's nose. Not only was Tycho forever deprived of one of the pleasures of Copenhagen's gentry, the sniffing of snuff, but, as a legend has it, he was no longer considered to be an eligible bachelor by the debutantes of Denmark. After all, a husband without a nose? Indeed!

Tycho did have a nose in a manner of speaking. He had a metalsmith fashion him an artificial one which he attached to the remnant of his real nose with a paste adhesive. The artificial nose was good enough for one particular woman of common origin, and Tycho married her.

The taking of a commoner as a wife was to cause Tycho continued troubles with the Danish aristocracy, for a marriage between a noble and a commoner was not considered to be a marriage at all. He was, in fact, living in sin and, in the process, spawning one bastard child after another.

Luckily, Tycho lived most of his married life a safe distance from the hostile nobility. He built a magnificent observatory called Uraniborg on the island of Hveen, which had been given to him by King Frederick. Over a period of twenty years at Uraniborg, he measured precisely the positions of celestial objects, obtaining an accuracy never before reached.

By the end of the sixteenth century, even though he had spent a great deal of time at his sanctuary on Hveen, Tycho had managed to alienate not only the nobility but the church and the new king of Denmark. He left Denmark in a fit of temper and eventually, in 1599, took a position at the University of Prague in Bohemia. He died there a little more than a year later, but not before he had hired Johannes Kepler as an assistant and instructed him in the use of his data. The accuracy of Tycho's data enabled Kepler to solve the problem of the motions of the planets.

Not only was Tycho's life bizarre in many respects, but so was his death. In his later years he developed bladder trouble, a most reasonable development in view of the tremendous amounts of beer that he had consumed in his lifetime. Living in Prague could only aggravate his problem, since the Bohemian brew-masters were and are reputed to be the world's very finest.

In October of 1601, he dined at the castle of a Bohemian baron; and during the meal, he drank his usual huge volume. The etiquette of the day demanded that no one leave the table until led by the host, who in this case decided to remain seated for what to Tycho must have been a laboriously long time. One can only admire the fantastic control of Tycho, who politely and steadfastly remained seated until his bladder exploded. A few days later, he was dead.

His body today lies at rest in a position of importance in Tyne Church on Old Town Square in Prague. His metal nose rests elsewhere, however, having been stolen by grave robbers centuries ago.

The Ptolemaic System

In the first century A.D., Claudius Ptolemy summarized the then-current knowledge of astronomy, along with his own contributions, in a book titled the *Almagest*. By then the geocentric model of the solar system, a main feature in Ptolemy's book, had more or less reached its final form. There were two important assumptions in this model: The observed motions of the sun, moon, and planets had to be accounted for by circular motions, and the basic motions of the objects had to be uniform.

Both the sun and the moon periodically appear to speed up and slow down on their paths through the stars, although always continuing their eastward motions. To account for this, the centers of the circles that were supposed to describe their motions were displaced from the earth. Figure 2–1 shows the scheme as it applies to the moon. An off-center circle like this is said to be *eccentric*. The earth is located to one side of the *center of eccentric,* and another point equally displaced to the other side is called the **equant.** The uniform *angular* motion of the moon on the circle, which came to be called the **deferent,** takes place about the equant. That is, the moon travels through equal angles measured at the equant in equal time intervals. From the earth, then, the moon would appear to move faster in the sky when its distance becomes smaller and, on the other hand, slower when its distance becomes greater.

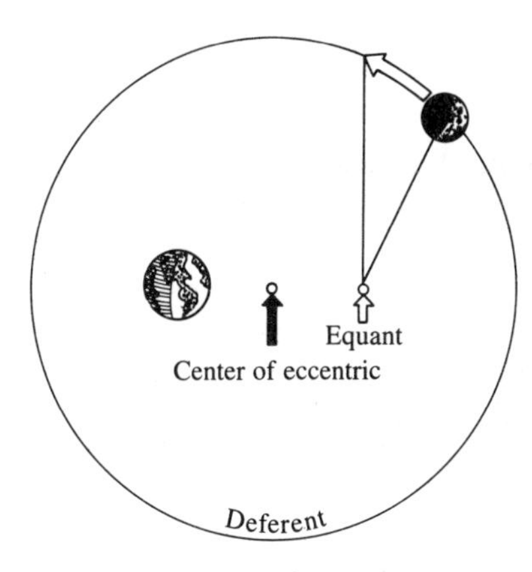

Figure 2–1 The Ptolemaic model of lunar motion.

The planets imposed a greater difficulty, namely, retrograde motion. Not only do they speed up and slow down, but they occasionally reverse direction and move westward for a while. To explain this, the concept of **epicycles** was introduced. Figure 2–2 shows the eccentric circle, again called the *deferent,* which accounts for the general slowing down and speeding up of the planet in its course through the stars. Retrograde motion was explained by having the planet itself move on the small circle, or epicycle, while the center of the epicycle moves uniformly on the deferent about the equant of the planet. By choosing a faster motion of the epicycle than on the deferent, the planet will appear to be traveling retrograde whenever it reaches the closer side of its epicycle. Notice that even though the basic motions on both the deferent and the epicycle are eastward or counterclockwise, they still explain the occasional westward digressions of the planets. In order to gain better agreement with observations, even more epicycles were added to the original one, giving epicycles moving on epicycles; eventually the planet itself is moving on the last and smallest epicycle. This model, even though incorrect, was an ingenious one, capable of predicting the observed motions of the planets within the accuracy of observation possible at that time.

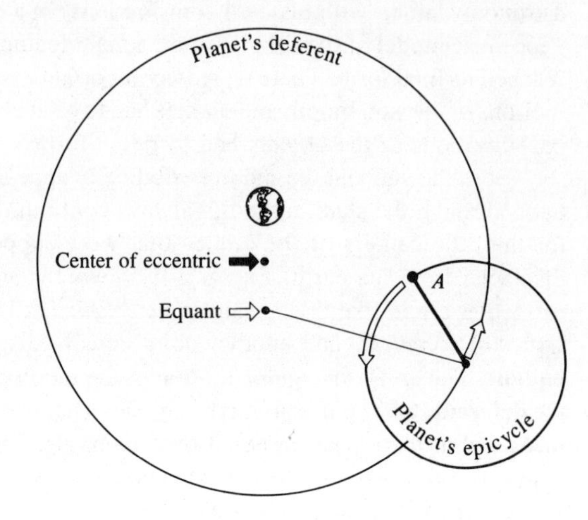

Figure 2–2 The epicyclic motion required in the geocentric system to explain retrograde motion.

The Copernican System

The first serious break with what came to be known as the Ptolemaic system occurred with the work of Nicolas Copernicus (Plate 1A) which was published in 1543. The Copernican system is heliocentric: The earth is just another planet moving around the sun. Copernicus placed the known planets in order of distance from the sun as Mercury, Venus, Earth, Mars, Jupiter, and Saturn, as shown in Figure 2–3, reproduced from one of his own drawings. The closer the planet is to the sun, the faster it moves about the sun. He referred to the planets inside the earth's distance, Mercury and Venus, as **inferior planets.** The planets Mars, Jupiter, and Saturn were labeled **superior planets** because of their greater distances from the sun.

There are a number of useful terms applied to the positions of the planets in the sky relative to the position of the sun, as they are seen from the earth. Figure 2–4 illustrates these positions, or **configurations,** for an inferior planet. When the planet lies in the same direction as the sun, it is said to be in **conjunction.** There are two conjunctions possible for an inferior planet—inferior conjunction when it is closest to the earth and superior conjunction when it is farthest from the earth. The angular distance of the planet from the sun is called its **elongation.** Because the inferior planets always lie closer to the sun than the earth does, they can never appear in the opposite part of the sky from the sun. Their extreme elongations from the sun are called **greatest eastern** and **greatest western elongation,** depending upon whether they appear to the east or the west of the sun.

The superior planets have three more configurations. Being more distant than the earth from the sun, a superior planet can lie in the opposite part of the sky from the sun. When the planet is on the opposite side of the earth from the sun, it is said to be in **opposition.** It can, of course, appear in the direction of the sun, that is, in conjunction. As

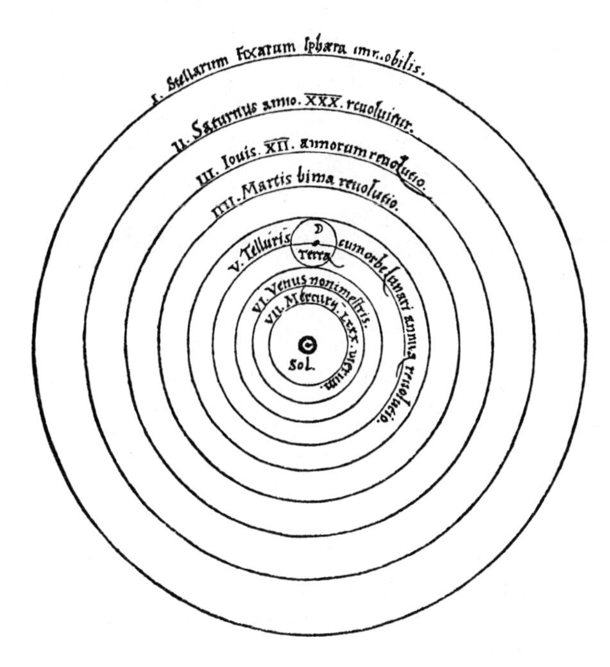

Figure 2–3 Copernicus' original drawing for the heliocentric solar system.

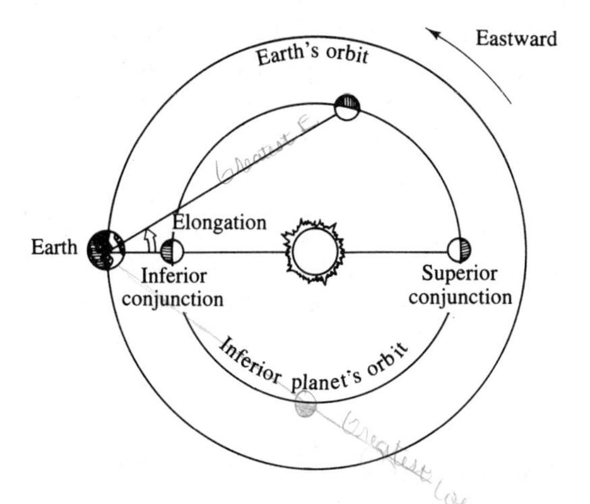

Figure 2–4 Configurations of an inferior planet.

shown in Figure 2–5, there are two positions relative to the earth and sun where a superior planet has an elongation of exactly 90°. These two cases are called **eastern** and **western quadrature.**

Copernicus wanted to find how long it actually takes a planet to travel one time around the sun. Unfortunately, he could only observe how long it appears to take as seen from the moving earth. The length of time for a planet to travel once about the sky relative to the stars, as it would be seen from the sun, is known as its **sidereal period,** the period Copernicus wanted. The length of time for a planet to go through one complete cycle of configurations, as seen from the earth, is called its **synodic period,** the period Copernicus

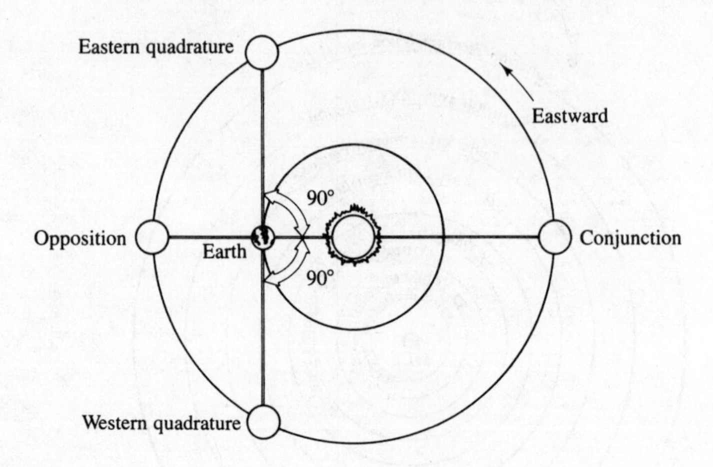

Figure 2–5 Configuration of a superior planet.

could observe. The method by which the sidereal period can be calculated from the synodic period can be understood from Figure 2–6, which represents the sky as it might be seen from the sun. Assume that the earth and Mars, a superior planet, happen to lie in the same direction. (Note that this means Mars would be in opposition to the sun as seen from earth.) One day later, the earth would have moved 360° divided by its sidereal period, P_E, expressed in days. The earth's sidereal period can, of course, be directly observed from the earth; it is one year. Mars, on the other hand, would have moved 360° divided by its sidereal period, P_M, expressed in days. Being a superior planet, Mars moves more slowly and will not proceed as far to the east as the earth will in the same day, as seen from the sun. The difference between the two amounts of travel is the amount the earth appears to gain on Mars each day, as viewed from the sun.

When the earth has gained a full 360° on Mars, the two planets will again be lined up; that is, Mars will again be in opposition as seen from the earth. But the length of time from opposition to opposition is the synodic period of Mars. Therefore, 360° divided by the synodic period, S, in days must equal the difference in the daily rates of the earth and Mars, as seen from the sun. Writing this mathematically gives

$$\frac{360}{S} = \frac{360}{P_E} - \frac{360}{P_M}$$

Since 360 is common to all terms, it can be cancelled out and

$$\frac{1}{S} = \frac{1}{P_E} - \frac{1}{P_M}$$

This expression, which can also be used for Jupiter and Saturn, was used by Copernicus to calculate the sidereal periods of the superior planets. By remembering that the inferior planets move faster than the earth, it is easy to show that the expression to use to calculate their sidereal periods is

$$\frac{1}{S} = \frac{1}{P_I} - \frac{1}{P_E}$$

where P_I equals the sidereal period of the inferior planet. The calculated sidereal periods of the planets and their synodic periods are shown in Table 2–1.

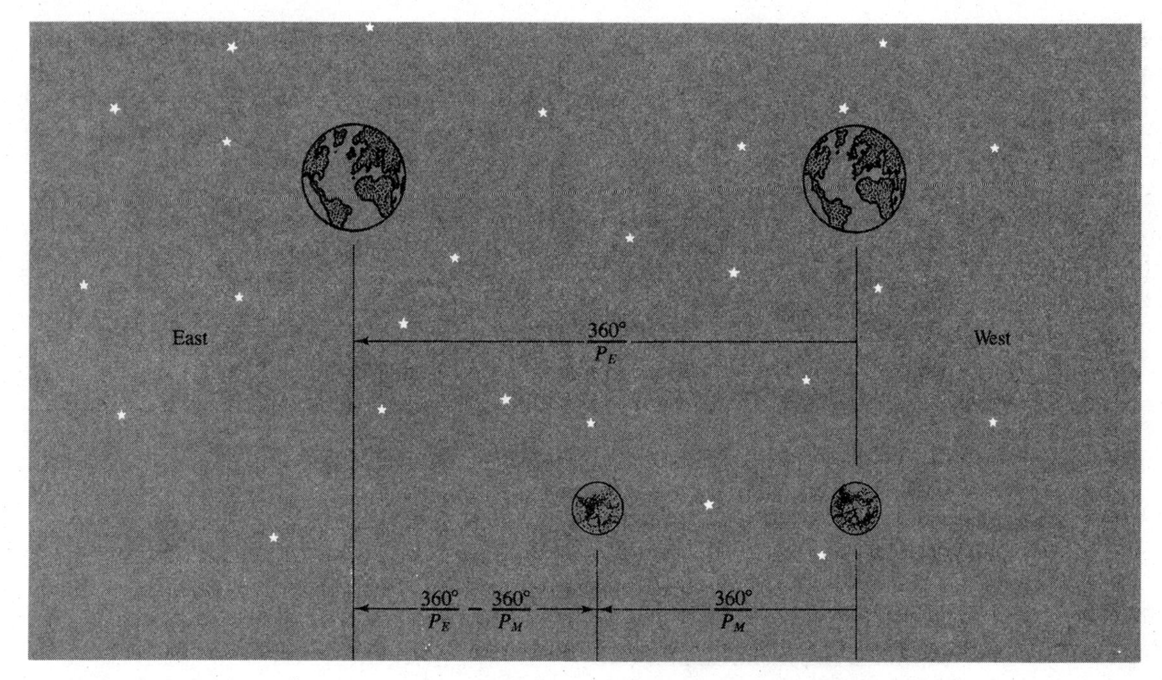

Figure 2–6 Daily motions of Earth and Mars as seen from the sun.

Table 2–1

Sidereal and synodic periods of the planets

Planet	Synodic period	Sidereal period
Mercury	115.88 days	87.97 days
Venus	583.92 days	224.70 days
Earth		365.26 days = 1 year
Mars	779.94 days	1.88 years
Jupiter	398.88 days	11.86 years
Saturn	378.09 days	29.46 years

Copernicus used the sidereal periods of the superior planets to help him determine the distance of these planets from the sun. Because the earth moves about the sun faster to the east than the superior planets do, a superior planet, lagging behind, appears to move westward relative to the earth and sun. Starting from opposition, the superior planet appears to drift toward eastern quadrature. The diagram in Figure 2–7 shows the relative motion which has occurred during the time the planet moves from opposition to quadrature. The angle between the sides labeled A and B in the diagram is simply the full relative movement in a synodic period, 360°, times the fraction of the synodic period which has expired between opposition and eastern quadrature. Timing opposition to quadrature therefore determines the angle. The angle between sides A and C, on the other hand, is, by the definition of quadrature, a right angle. Applying the properties of a right triangle, Copernicus found the value of B, the solar distance from the planet, relative to the value of

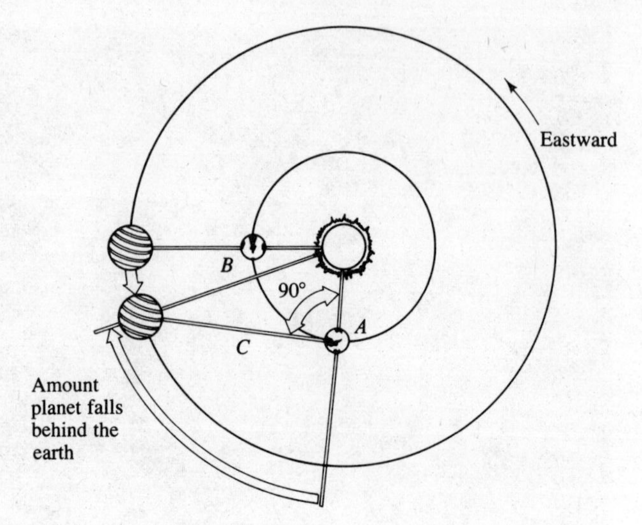

Figure 2–7 Copernicus' method for determining the relative distances of the superior planets.

A, the solar distance from the earth. This distance between the earth and the sun is usually called the **astronomical unit** (AU).

The distances of the inferior planets are even easier to determine. At greatest elongation, the angle between the line joining the planet and the sun makes a right angle to the line from the planet to the earth. The planet, the earth, and the sun, therefore, form a right triangle. Measuring the value of the angle of elongation at greatest elongation thus permits the inferior planet's distance to be expressed as a fraction of an astronomical unit.

Table 2–2 compares the values which Copernicus determined for the planets with the modern values. Even though his distances were found nearly 500 years ago, his largest error was only 4 percent, in the case of Saturn.

Table 2–2

The distances of the planets from the sun expressed in astronomical units

Planet	Copernicus	Modern
Mercury	0.38	0.387
Venus	0.72	0.723
Earth	1.00	1.000
Mars	1.52	1.52
Jupiter	5.22	5.20
Saturn	9.17	9.54

The Laws of Kepler

Before the heliocentric theory of Copernicus was to gain acceptance, another sixty years would transpire. During the intervening years, the Danish astronomer Tycho Brahe

(Figure 2–8), claimed by some to be one of the greatest observers who ever lived, carried out a long series of observations of the sky which would serve as the basis for the acceptance of the heliocentric theory. Tycho himself never accepted the Copernican universe because he could never detect the stellar parallax which must accompany it. He had measured a parallax for the moon by observing its apparent change in position amongst the stars at moonrise and moonset. But, if Copernicus were corrcct, Tycho felt he should have been able to detect a stellar parallax; because now he would be observing from two greatly separated points, one on either side of the earth's orbit, instead of from two points even less than a diameter of the earth apart when he found the lunar parallax. Although Tycho probably achieved the highest accuracy possible with the unaided eye, the eye, without a powerful telescope, is unfortunately incapable of seeing the minute parallax of a star.

Figure 2–8 Tycho Brahe (1546–1601).

In 1600 Tycho hired a young assistant, Johannes Kepler (Figure 2–9), to analyze the long series of planetary observations which he had accumulated. The Ptolemaic theory, with its epicycles on epicycles, did not quite agree well enough with his observations. It probably needed only a minor correction or two—add another epicycle here or there—to come into agreement with his excellent observations. No matter what he tried, however, Kepler could not reconcile Tycho's observations of the position of Mars with the geocentric model within the observational accuracy of the accumulated data. Tycho had already died before Kepler made significant progress in the interpretation of Tycho's observations. No longer restricted by Tycho's insistence that the earth is the immobile center of the universe, Kepler attempted to fit the observational data for the planets to a model of the Copernican universe. He started to understand the motions of the planets when he discovered that the orbit of Mars about the sun is an ellipse, a discovery resulting from the arduous fitting of various curves to the data. This discovery was the first serious break with the idea that all heavenly bodies move only on paths consisting of combinations of perfect circular motions.

Figure 2–9 Johannes Kepler (1571–1630).

An **ellipse** is a curve such that any point on it has the same total distance from two certain fixed points as any other point on the ellipse. These two fixed points are called the **foci of the ellipse.** Thus, in Figure 2–10, the sum of the distances from point 1 to the two foci F and F' is equal to the sum of the distance from point 2 to F and F'. The size of an ellipse is characterized by its **major axis,** which is the longest line across the ellipse. This line passes through the two foci. Astronomers prefer to give the size of an ellipse in terms of the **semi-major axis,** or the distance from the center of the ellipse, through a focus, to the ellipse itself. The shape of the ellipse depends upon the spacing of the two foci. It is characterized by the **eccentricity,** which is defined to be the distance between the foci divided by the length of the major axis. Note that the circle is a special case of an ellipse. The eccentricity is zero, the major axis is the diameter of the circle, and the semi-major axis is its radius. The eccentricities of the orbits of the planets are all quite small. For Mars, the planet which was most useful to Kepler in his analysis of planetary motions, the eccentricity of the orbit is only one-tenth. Its deviation from a circle is therefore not very great, and it was only the great precision of Tycho's observations that allowed Kepler to make progress.

To describe the motion of a planet, Kepler needed two laws: one to describe the shape of the planet's orbit and the other to specify how the speed of the planet changes along its orbit. Kepler's first law of planetary motion states that the planets move on elliptical orbits with the sun at one focus, as indicated in Figure 2–11. His second law, the law of equal areas, states that the line joining the planet and the sun sweeps out equal areas in equal intervals of time. Therefore, as shown in Figure 2–11, if it takes as long to go from point 1

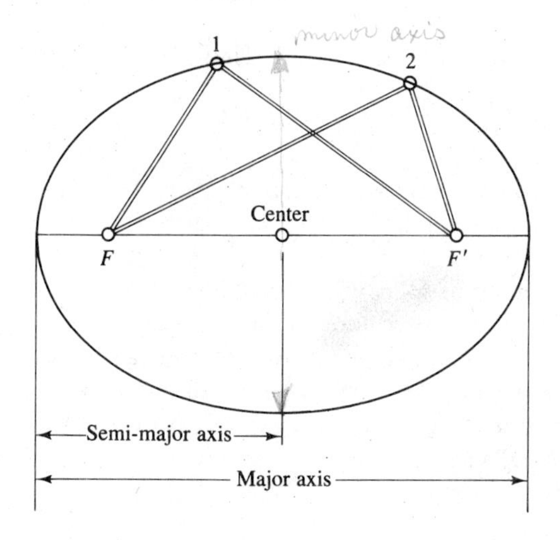

Figure 2–10 Properties of an ellipse.

to 2 as it takes to go from 3 to 4, the areas of the sectors of the ellipse between the first pair of lines and the second pair of lines must be exactly the same. The two end points of the major axis are given special names. **Perihelion** means "close to the sun"; by the law of equal areas, the planet must have its greatest orbital speed when at the perihelion point. **Aphelion** means "far from the sun," and at this point in its orbit the planet must move with its slowest orbital speed.

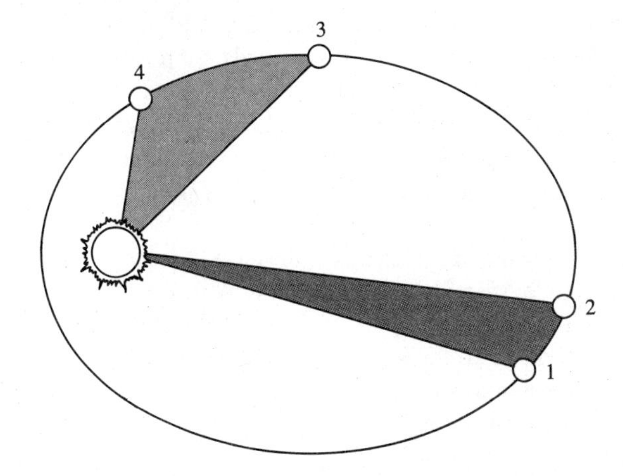

Figure 2–11 The law of equal areas.

Kepler continued working with the observations of Tycho. Ten years after the announcement of his first two laws he put forth his third law, often called the harmonic law, which allows a comparison of the orbital motion of the different planets. Although Kepler was an astute mathematician, he was also a mystic. His harmonic law explained the music or harmony of the celestial spheres—music, by the way, which is heard only by angels and the most perceptive of mortals. This law, which relates the sidereal period of a planet to the size of its orbit, states that the square of the sidereal period of a planet is proportional to the cube of the semi-major axis of its orbit. If the period is measured in units of the earth's sidereal period, called the *sidereal year,* and the semi-major axis in astronomical units, the coefficient of proportionality is equal to one. Thus, Kepler's third law is summarized by the equation

$$P^2 = a^3$$

where P is the sidereal period in sidereal years and a is the semi-major axis in AUs. This equation predicts with amazing accuracy the known distances to the planets from their known sidereal periods.

Although Kepler did not observe the stellar parallax which would have proved his theory, the simplicity, beauty, and accuracy of his description of the planetary motions began winning converts to the heliocentric philosophy. Kepler's laws represent a particularly good example of a scientific discovery which describes an observed phenomenon with such precision that the basic concept becomes accepted even in advance of the positive proof—the discovery of stellar parallax in this case.

Galileo's Observations

Galileo Galilei (Plate 1B), an Italian scholar who was a contemporary of Kepler, did much to advance the heliocentric theory. He was one of the first persons to make telescopic observations. Even though his telescopes, which still exist (Plate 2C), were inferior to many modern binoculars, they still gave him the ability to see things which had previously been hidden from view.

One of his startling discoveries was that Jupiter had four moons (although today a total of thirteen moons are known). These moons continually changed their positions as he viewed them in his telescope. If Jupiter could have moons circulating about it like our moon circulates about the earth, why could the sun not have the planets circulating about it? Galileo communicated his finding to Kepler. Kepler, in turn, found that the so-called Galilean moons of Jupiter obey his harmonic law, a magnificent confirmation of Kepler's results.

Galileo was also the first person to see the irregularities of the moon's surface. While not in itself a disproof of the Ptolemaic system, his observation flew in the face of the contemporary belief that the celestial bodies were all perfect spheres. His observation of dark marks on the sun was equally disturbing to the idea of perfect spheres. In addition, when he noticed the regular motion of these sunspots across the disc, he concluded that the sun must be in rotation. Again, this was not proof of the Copernican viewpoint, but it certainly made it easier to accept.

Along with the moons of Jupiter, Galileo's observation of the phases of Venus must certainly rank as one of his most important observations in gaining the acceptance of the heliocentric model. Through his telescope Galileo saw Venus in crescent, quarter, and gibbous phases. He could not see the new or full phase of Venus, because at these times Venus is in conjunction with the sun and is lost in its glare. As shown in Figure 2–12, the only way to explain these phases is to place Venus on an orbit about the sun inside the orbit of the earth, exactly as Copernicus had predicted.

Today we find it easy to accept Galileo's observations as supporting, if not proving, the validity of the heliocentric theory. In his own day, after thousands of years of acceptance of the geocentric model, this acceptance was not so easily accomplished. The Ptolemaic system had become so ingrained that it had even become built into the dogma of the Catholic Church and other churches. Galileo had a long argument with the authorities

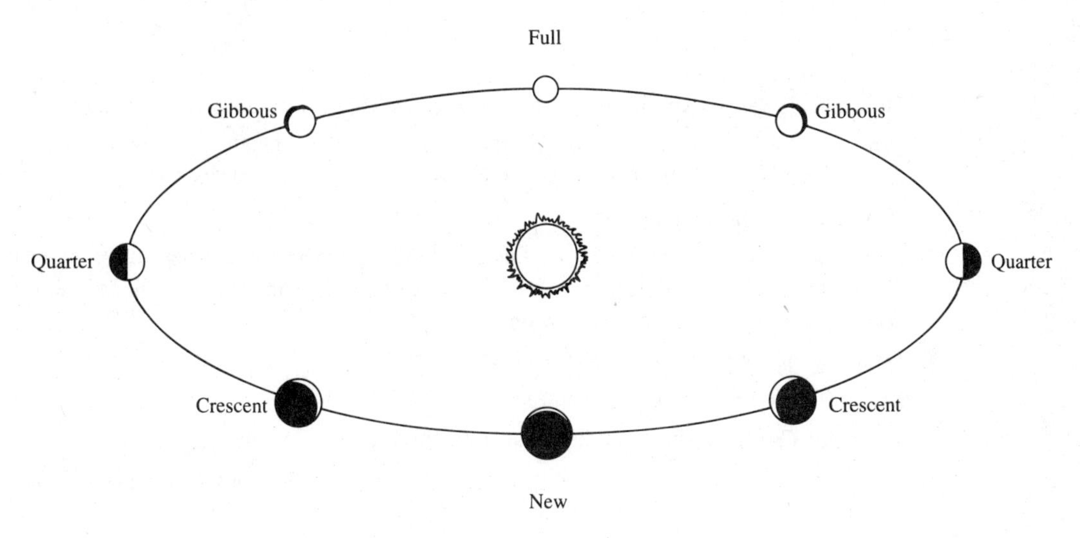

Figure 2–12 Phases of Venus.

and eventually was tried and convicted of heresy. In 1632 he was forced to recant and was sentenced to house arrest for the remainder of his days.

Galileo is also known to us as one of the founders of the branch of physics known as *mechanics,* the study of moving bodies. In his work with falling objects, he gathered evidence which he used for formulating his law of inertia. In this law he opposed the old Greek ideas handed down from the time of Aristotle. The Greeks had thought that rest is the natural condition for a material body. Galileo proposed that rest is only a special condition, because any body, once set in motion, would continue to move unless subjected to a force. He recognized friction as the force which causes a sliding body eventually to come to rest.

The work of Galileo, along with that of Kepler and Copernicus, has set the stage for the dynamic study of the solar system. The introduction of *dynamics,* or the prediction of the motions of bodies from a knowledge of the forces applied to them, was left to Isaac Newton. The next chapter will be devoted to his findings.

KEY TERMS

deferent
equant
epicycle
inferior planet
superior planet
planetary configuration
conjunction
elongation
opposition
quadrature
sidereal period
synodic period
astronomical unit
ellipse
focus of an ellipse
major axis
semi-major axis
eccentricity
perihelion
aphelion

Review Questions

1. How did Ptolemy explain the motion of the sun and moon amongst the stars?
2. Why did Ptolemy's explanation of the motion of the planets have to be more complicated than his explanation of the motion of the sun and moon?
3. How did Ptolemy explain retrograde motion?
4. In Ptolemy's geocentric model of the solar system, where epicycles move along deferents, how fast do the epicycles of Venus and Mercury move along their deferents compared to the speed of the sun on its deferent?
5. What do *inferior planet* and *superior planet* mean?
6. List the inferior planets.
7. What did Copernicus conclude about the orbital speeds of planets?
8. If Mars goes around the sun from west to east, why does it sometimes appear in the sky to go from east to west, relative to the stars?
9. What are inferior conjunction and superior conjunction?
10. What would the telescopic appearance of Venus be when it is at greatest elongation?
11. An inferior planet is at greatest eastern elongation. What is the configuration of the earth viewed from that planet?
12. When Jupiter is in opposition, where would you look to see it?
13. What is the distinction between sidereal and synodic periods?
14. Many of the discoveries in modern astronomy depend on one astronomer to make observations of a phenomenon and another to develop a theory to explain the causes of the phenomenon. Which two astronomers have been discussed who, a few centuries ago, showed this division of labor?
15. Suppose you tie the ends of a string around two thumbtacks and move a pencil around, keeping the string taut. What curve is drawn?
16. If the figure is drawn again with the same string but with the thumbtacks closer together, what happens to the shape of the figure?
17. Saturn's distance from the sun is a little less than ten times the earth's distance. How many earth years should Saturn take to complete one orbit around the sun?
18. Who was the first person known to have looked at the heavens through a telescope?

    (1) Galileo

    (2) Tycho

    (3) Kepler

    (4) Copernicus

    (5) Newton

19. List some of the astronomical discoveries made by Galileo.

Discussion Questions

1. How did Copernicus demonstrate that Mercury's orbit is smaller than the orbit of Venus?
2. When Venus is at eastern elongation, is it referred to as a morning star or an evening star?
3. Which is longer, the sidereal or the synodic period of Mercury?
4. If a planet were exceedingly far from the sun, taking thousands of years to orbit the sun, what would its synodic period be?
5. At greatest elongation, Mercury's distance from the sun in the sky can be anywhere from 18° to 28°. Why aren't all the greatest elongations the same?
6. When closest to the sun, at perihelion, Mercury is 46 million kilometers from the sun. When farthest from the sun, at aphelion, its distance increases to 70 million kilometers. How fast does Mercury travel in its orbit at perihelion compared to aphelion?
7. How were Galileo's telescopic observations received by other learned men of that time?
8. If you slide a brick across the floor, it slows down and stops. How would Aristotle and Galileo each describe the reason why it stopped?

3

Newton's Laws and the Earth's Motions

Our everyday experience leads us to associate movement or motion with activity. Birds move through the air by beating their wings. We walk or run by contracting and extending our leg muscles. A baseball soars through the air upon the impact of a bat. We travel in cars powered by gasoline vapor exploding in engines.

In time, motion stops if it is not sustained by some force. If the bird stops beating its wings, it falls to the ground and comes to rest, unless it is kept aloft by air currents. If we do not extend and contract our muscles, we do not move. The well-hit baseball comes to earth, rolls a bit, and then stops.

It appears that when there is no force, motion stops. Look around you at the chair, books, walls. Nothing moves unless you push it or throw it; even then, it eventually comes to a stop. It is small wonder that for thousands of years, scholars charged with the job of describing nature came to consider the motion of an object as dependent on force and consider rest or stillness as the normal state of an object—the state at which it would ultimately arrive if deprived of a sustaining force. However, for thousands of years scholars charged with the job of describing nature were wrong.

It was not until the sixteenth century that Galileo began to uncover the truth about motion. Through his experiments, he discovered that a moving object which lacks a sustaining force will not come to rest unless, in fact, an impeding force stops it. For example, the rolling baseball comes to rest because of the frictional forces between it and the ground. Galileo argued that motion is as natural as rest. In short, if an object is at rest, it remains at rest unless forced into motion; if an object is in motion, it continues in motion unless forced to stop.

The main character in the story of motion is not Galileo, however, for Isaac Newton gave a complete description of motion and the way in which forces affect it. This description is beautifully contained in three separate laws.

Newton applied his laws of motion to the planets. Those scholars who considered rest to be the only natural state of objects thought that the planets were rather

Newton's third law of motion: Apollo 11 at launch

peculiar because they move without the influence of obvious sustaining forces. Those scholars proposed that the planets move by virtue of divine will.

According to Newton's laws of motion, however, a planet in motion would move along a straight line with uniform speed and would eventually recede beyond our range of view unless there were some force constraining its motion to an orbit around the sun. Through application of his laws of motion to Kepler's laws of planetary motion, Newton was able to show that all planets and the sun mutually attract one another with a force which we call **gravity.** He correctly deduced that the same force which influences the motion of the planets around the sun also influences the motion of the moon around the earth and the motion of objects at the surface of the earth. Just as Mars orbits the sun because of the sun's gravity, an apple falls to the ground and the moon orbits the earth because of the earth's gravity.

The formulation of the law of gravitation by Isaac Newton ranks as one of the major achievements in recorded history. This law, one short statement, beautifully describes the gravitational force—one of only four forces known to us at this time. More importantly, the success with which this law describes the gravitational force supports scientists' conviction that there is order in the universe, order that can be described clearly and in a straightforward manner.

Newton's Laws of Motion

Sir Isaac Newton (Plate 1C) formulated three laws of motion which allow us to understand the motions of bodies in the solar system. These laws relate the motions of the bodies to the forces acting upon them. They also define the basic concepts of **force** and **mass.**

Newton's first law, which had already been found by Galileo, is called the law of inertia and states that *a body continues in its present state of motion unless it is acted upon by a net external force*. This law requires a considerable change in thinking from that of the early Greeks. A body at rest is simply a special case—one in which the velocity of the body is zero—and there is no unbalanced force upon it which would cause it to move. The Greeks had thought that rest was the natural state of objects, but Galileo and Newton concluded otherwise. Constant straight-line motion is the natural state as long as no unbalanced forces act upon the object being studied. Thus a baseball, given a particular velocity by a ball player's bat, would continue in a straight line, with the same speed, indefinitely. We are assuming, of course, that the force applied to the baseball by the earth and the influences of air friction are eliminated. A little further thought, however, quickly leads to the conclusion that this analysis is incomplete. If the earth can influence the ball, then the sun, the moon, the planets, and, eventually, even the distant stars will influence it. Thus, to keep the ball moving uniformly along a straight line forever requires a universe in which nothing but the ball exists. The law of inertia is still useful, however, because it allows us to recognize that any body not moving in a straight line with constant speed must be subject to some unbalanced force. The earth, for example, must be subject to a force in order to move upon its curved path about the sun.

Newton's second law, the law of force, relates the change in the velocity of a body to the net, or unbalanced, force applied to the body. *The net force applied to a body is equal to the mass of the body times the acceleration caused in the body by that force. In addition, the acceleration is in the same direction as the force.* This law is beautifully embodied in the simple equation

$$F = ma$$

The **acceleration** of a body measures how rapidly the velocity of that body is changing. Remember that to specify the **velocity** of an object, it is necessary to state both how fast it is moving—that is, its speed—and also the direction of its motion. Acceleration can, therefore, result in a change in the speed, in the direction of motion, or both. Acceleration is expressed as the amount of change in the velocity divided by the interval of time over which that change occurred. If Δv represents the change in velocity and Δt represents the corresponding interval of time, then the acceleration is

$$a = \frac{\Delta v}{\Delta t}$$

As an example, if a body continually moving in the same direction has its velocity changed from five meters per second to eleven meters per second, Δv is eleven minus five, or six meters per second. If this change occurred in the span of one second, Δt is one second. The acceleration in this case is six divided by one, or six meters per second per second. If the velocity change occurred over two seconds, Δt would have been two, and the acceleration would only have been three meters per second per second. Note that Newton's force law would require twice the force to accomplish the first acceleration as it would to accomplish the second one.

Newton's force law can also be used to measure mass. The mass of a body, that is, the amount of its matter, is a measure of the body's resistance to change in motion. By Newton's law, $F = ma$, two bodies have the same mass if the same force applied to both bodies causes the same acceleration in both. If, on the other hand, it takes four times as much force to cause the same acceleration in one body as in another, that body has four times the mass of the other.

The final velocity of a body depends upon the direction of the force applied to the body relative to its original direction of motion. Figure 3–1 shows three examples of what can happen. In the first case, the force is applied in the same direction as the original motion of the body. The body's acceleration will be such that its final velocity is increased, but the direction of its motion will be unchanged. In the second case, the force is applied opposite to the body's original motion. In this situation, the final motion of the body will be diminished, even to zero, or its motion may actually be reversed, depending upon the magnitude of the force and how long the force is applied to that body. In the third case, since the force is neither with nor against the beginning velocity, the final velocity will always be in a new direction; but its magnitude, or speed, may be either greater than, the same as, or less than the original speed. It will depend upon the precise value and direction of the force and the length of time for which it is applied.

Remember that Newton's force law refers to the *net force* applied to a body, that is, the combination of all forces acting upon it at any instant. In Figure 3–2, the upper diagram shows two equal but opposite forces applied to a body. The body may become deformed by the forces, but it will not be accelerated because the two forces will cancel each other. In the middle diagram, the acceleration will be the sum of that which would result from either force acting alone. In this particular case, since the two forces are equal

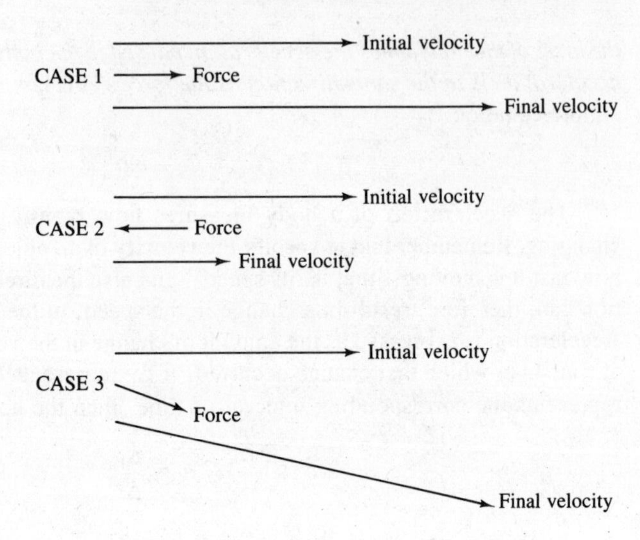

Figure 3–1 The effect of a force upon the velocity of a body.

and in the same direction, the net force on the body is twice the value of one of the forces by itself. The third diagram illustrates what would happen if the two forces were applied at right angles to each other. The acceleration caused by the forces would lie diagonally between them. The single force—that is, the net force—which could be used to replace the two forces can be found by completing the box formed by the two forces (as indicated in the diagram). The diagonal of the box represents the net force.

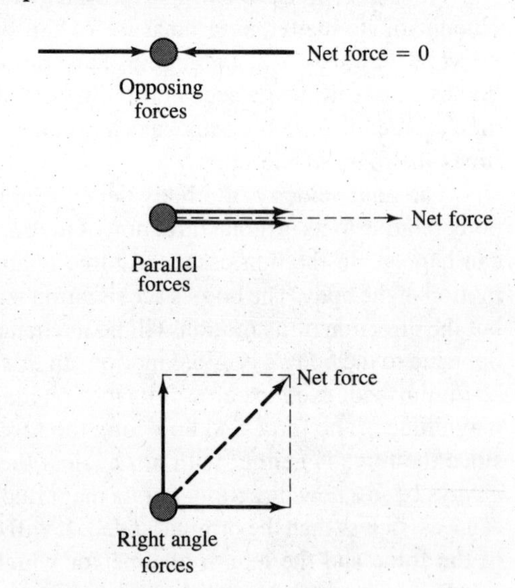

Figure 3–2 Net force: the combination of all forces acting upon a body.

Newton's third law, called the reaction law, states that *if one body exerts a force upon another body, the second body exerts a force just equal and opposite to that force on the first body.* This law is sometimes summarized by saying that for every force, there is an equal and opposite force.

This law implies that if a body exerts a force on another body, causing it to move, the first body also moves, due to the equal and opposite force exerted upon it by the second body. The amount of acceleration induced in each body can be calculated by Newton's law of force. The less massive body will suffer a greater acceleration; the more massive, a lesser acceleration. Thus, in the earth-moon system, the moon does most of the moving caused by the force of attraction of the earth and the moon upon each other, since the moon contains the smaller amount of mass.

Newton could explain the motion of the planets about the sun by proposing a force directed toward the sun which gave the correct accelerations to cause the planets' velocities to change direction continually. Thus, as illustrated in Figure 3–3, the earth continues to circulate indefinitely about the sun, its direction of motion continually changing, because of the sun's force of attraction on it.

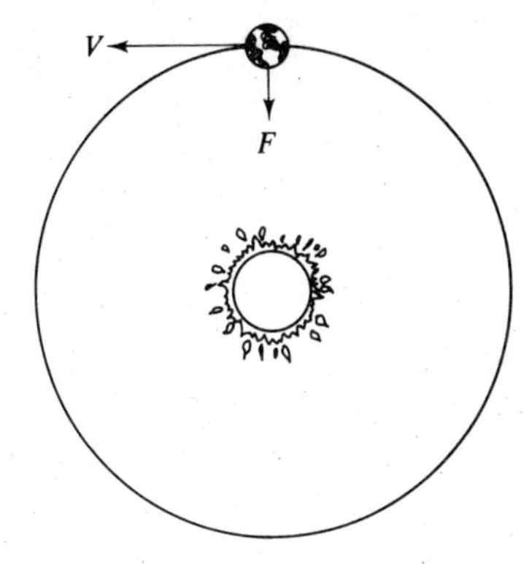

Figure 3–3 The earth's orbital velocity and its relation to the sun's gravitational force.

Newton's Law of Gravitation

Newton invoked the concept of **gravity,** the same force that makes an object fall to the earth's surface, to explain the planetary motions. He began by first considering the moon's apparent motion about the earth. Since he knew the size of the moon's orbit about the earth and the period of its orbital motion, he could calculate how fast it travels along its orbit. It moves roughly one kilometer per second along its orbit. In that same time, it must be accelerated by an amount which will cause it to fall 1.3 millimeters toward the earth to remain on its observed orbit, as schematically depicted in Figure 3–4.

Newton then had to find a law of force which would supply the correct acceleration to the moon. Obviously, whatever he proposed had to give a description of planetary or satellite motion which would agree with Kepler's laws of planetary motion. Working backwards from Kepler's laws, Newton was able to find the form of the force law required to explain the observed planetary motions. To do this, he had to develop a new form of mathematics, which we now call *calculus*.

The description of the force between the sun and the planets is called the *law of universal gravitation,* because Newton proposed that it would hold true throughout the universe. The force of gravity between two bodies is given by the equation

$$F = G\frac{M_1M_2}{D^2}$$

where G = Newton's constant of gravitation
M_1 and M_2 = the masses of the attracting bodies
D = the distance between the centers of the bodies

The force of gravity is always directed along the line joining the two bodies.

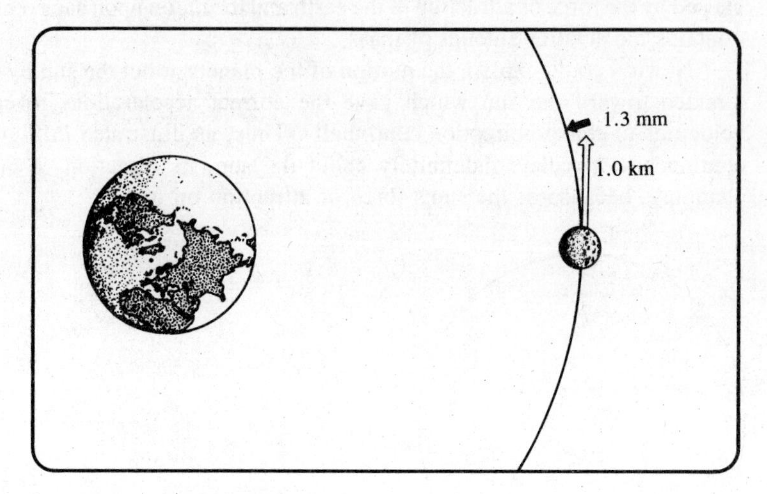

Figure 3–4 The acceleration required to keep the moon in its orbit about the earth.

This law applies to any two bodies. Therefore, it applies to a body at the earth's surface and the earth itself. Using the gravitation law and Newton's law of force, it is possible to calculate the acceleration of a body at the earth's surface.

Thus, if m = the mass of a body, the force law gives

$$ma = F$$

But the value of the gravitational force is given by the law of universal gravitation,

$$F = G\frac{mM_{Earth}}{R^2_{Earth}}$$

where M_{Earth} = the mass of the earth
and R_{Earth} = the radius of the earth

Since the force causing the acceleration is gravity,

$$ma = G\frac{mM_{Earth}}{R^2_{Earth}}$$

Dividing both sides of this equation by m, the mass of the body, gives

$$a = \frac{GM_{Earth}}{R^2_{Earth}} = g$$

The acceleration at the surface of the earth is usually symbolized by g and has a value of roughly 980 centimeters per second per second. Notice that g depends only upon Newton's gravitational constant, the mass of the earth, and the radius of the earth; it does not depend upon the mass of the body being attracted by the earth. This is a theoretical confirmation of Galileo's earlier experimental result—namely, bodies of different masses all undergo the same acceleration at the surface of the earth.

There is also a qualitative explanation for this result. Imagine that you hold a bowling ball in your left hand and a golf ball in your right hand. You obviously would feel the bowling ball pulled much more strongly by the earth's gravity than the golf ball. If you now were to strike the golf ball and the bowling ball each in turn with a golf club, the golf ball would soar into the air, but the bowling ball would barely move; you probably would be left with a badly bent or broken club. The bowling ball, with its greater mass, has much more inertia and is much harder to accelerate—hence the damaged golf club. The stronger pull of gravity on the bowling ball is balanced by its greater inertia, and its acceleration ends up being just the same as that of the golf ball as they both fall to the ground. You may have noticed that in all of our examples, we have chosen objects which will suffer a relatively small influence from air friction. A feather, for example, falls more slowly than a lead weight, but this result is due to air friction. In a vacuum, with the air removed, the feather will accelerate toward the earth at exactly the same rate as the lead weight.

To test his gravitational law, Newton applied it to the moon. The gravitational force decreases as the square of the distance between attracting bodies. Consequently, since the distance to the moon is sixty times the earth's radius, the acceleration caused by the earth's gravity at the moon's distance must be one-3600th that caused at the earth's surface. Dividing g by 3600 gives the same value for the moon's acceleration that Newton had determined from a knowledge of the moon's orbit.

Newton's Formulation of Kepler's Laws

Using Newton's laws of motion and gravitation, it is possible to generalize Kepler's laws of planetary motion. It thus became possible for Newton to predict a number of phenomena which subsequently were verified by observation.

First, by the reaction law, we know that if two bodies mutually attract each other, both must be accelerated; and, hence, both are in motion. Since the force of gravity always lies along the line joining the two bodies, the common point about which they move must also be on that line. The forces on the two bodies, although oppositely directed, are equal. Therefore, the accelerations of the bodies will be inversely related to their masses. The more massive body will show the lesser motion, while the less massive body will show the greater motion. The common point about which the bodies move is called the **center of mass.** It is located such that the product of the mass of the first body and its distance from the center of mass equals the same product for the second body. Thus, in Figure 3–5, $m_1r_1 = m_2r_2$.

If Newton's calculus is applied to his laws of motion and gravity, it is possible to extend Kepler's first law of planetary motion. Two bodies, under the force of mutual gravitation, move on similar orbits about their common center of mass. The orbits will always be one of three kinds of curves: **ellipses, parabolas,** or **hyperbolas.** An ellipse is a closed curve, in which case the bodies repeat their paths about each other indefinitely. Remember that a circle is simply a special case of an ellipse. Both the hyperbola and parabola are open curves. They represent orbits which result from a single gravitational encounter between two bodies. Figure 3–6 shows possible motions of two objects on

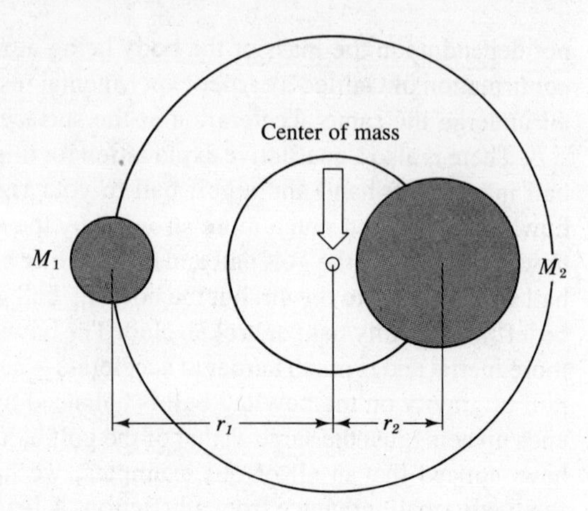

Figure 3–5 The individual orbits of two bodies about their center of mass.

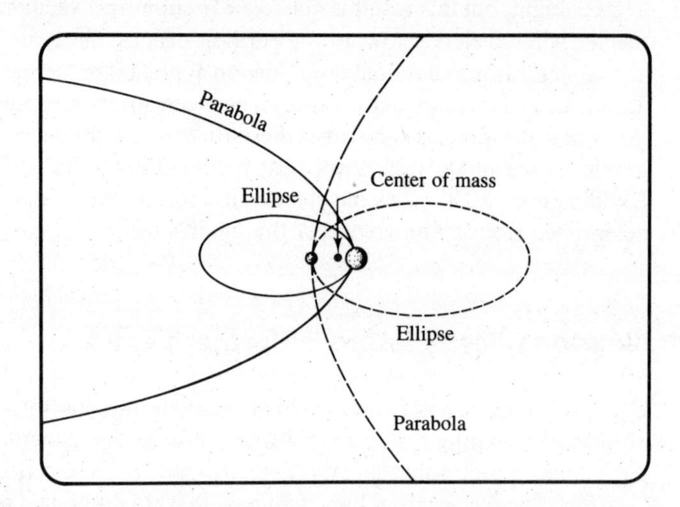

Figure 3–6 Orbital shapes: Each body moves on a similar ellipse, parabola, or hyperbola about the center of mass.

mutual elliptical orbits and mutual parabolic orbits. Hyperbolic orbits would be even more open than the parabolas.

Kepler's law of equal areas becomes a very general law when analyzed using Newton's theory. The line joining two bodies will sweep out equal areas in equal units of time if the force between them always lies along that line. Thus, in Figure 3–7, the dotted area equals the cross-hatched area. It will take the same length of time for the line joining the orbiting bodies to generate the two areas. This will hold true for any force law, as long as it has this special directional property. Therefore, for the case of the law of universal gravitation, the law of equal areas holds whether the orbits are ellipses, parabolas, or hyperbolas. Thus, bodies must always move more rapidly in their orbits when they are at smaller separations, and they must move with slower orbital speeds when they reach greater separation.

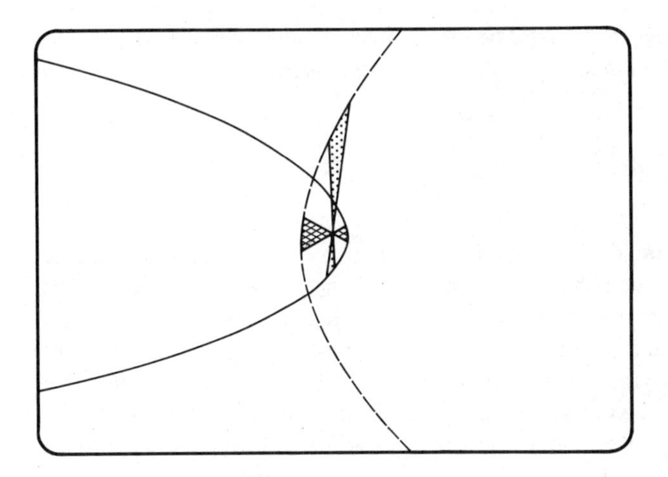

Figure 3–7 The law of equal areas: The line joining the bodies sweeps out equal areas in equal intervals of time.

Kepler's harmonic law applies only to elliptical orbits and requires that the relative orbit of one body about the other be used. The **relative orbit** is the sum of the two similar individual orbits about the center of the mass. It has a shape exactly the same as either individual orbit. The semi-major axis of the relative orbit equals the sum of the semi-major axes of the two individual orbits.

The analysis of elliptical motion, using Newtonian mechanics, introduces a new term into Kepler's harmonic law: the sum of the masses. In Newton's formulation, it is written

$$(M_1 + M_2)\, P^2 = a^3$$

where M_1 and M_2 = the masses of the bodies expressed in solar masses
P = the orbital period in years
a = the semi-major axis of the relative orbit expressed in astronomical units

The sum-of-the-masses term escaped Kepler because the masses of the planets are so small compared to the sun's mass that they can be taken as zero, at least to the accuracy of the data which he was studying. Even the most extreme case in the solar system, the planet Jupiter, has a mass which is only one-thousandth that of the sun. Ignoring Jupiter's mass causes an error in the calculated size of its orbit of only 0.03 percent, much below Kepler's limit of detection.

The sum-of-the-masses term becomes extremely important in astronomy. In fact, it allows the masses of stars to be determined. In a later chapter we will discuss a group of stars called *double stars*. In these systems, two stars orbit each other under their mutual gravitation. They therefore obey Kepler's harmonic law. If the period of their revolution and the semi-major axis of their relative orbit can be measured, Newton's formulation of the harmonic law can be used to calculate the mass of the system. As an example, consider a double-star system having a period of two years and relative semi-major axis of three

AU. The combined mass of the stars would be three cubed, or twenty-seven, divided by two squared, or four. The system mass is thus 6.75 times the mass of the sun. This can be symbolized as

$$(M_1 + M_2) = \frac{a^3}{P^2} = \frac{(3)^3}{(2)^2} = \frac{27}{4} = 6.75$$

Newton described the principle involved in launching an artificial satellite when he published his laws of motion in 1686 in a book entitled the *Principia*. Figure 3–8 is adapted from one of his drawings in the *Principia*. He gave his discussions in terms of a heavy lead ball placed in motion by means of a charge of gunpowder. Newton's laws of motion and his law of gravity predict the value of the orbital speed which an object in a circular orbit just skimming the earth's surface would require. This speed is 7.9 kilometers per second. The velocity in a parabola, an open orbit, at the point of closest approach to the earth's surface can also be calculated. This speed is 11.2 kilometers per second. An even larger velocity would result in a hyperbolic orbit.

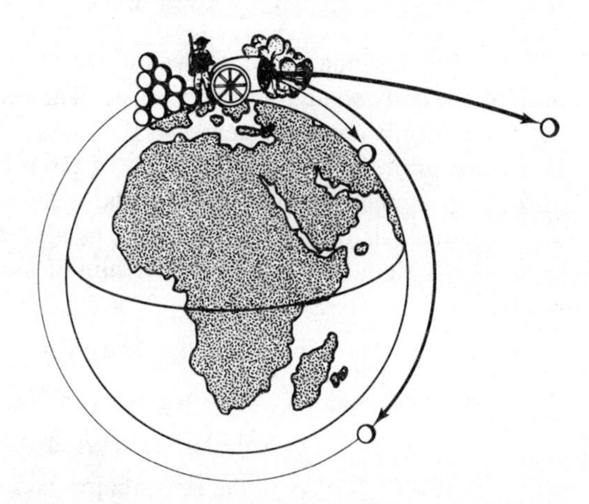

Figure 3–8 Newton's scheme for the launch of artificial satellites.

We can simulate Newton's suggestion by imagining a large cannon—the only thing available for giving a projectile a high enough velocity in Newton's day—being placed on a mountaintop and the barrel of the cannon leveled. If a powerful enough charge could be used to eject the cannon ball at 7.9 kilometers per second, it would travel about the earth in a circular orbit. If a weaker charge were used, the projectile would not move fast enough along the circumference of the earth to remain on a circular orbit as gravity continually accelerated it toward the surface. It would ultimately strike the earth. On the other hand, if the charge were more powerful, the velocity would be greater and the orbit would be elliptical, reaching its greatest distance from the earth on the opposite side of the earth from the cannon. Eventually, with even more powerful charges, when the muzzle velocity reached 11.2 kilometers per second or greater, the cannon ball would enter an open orbit and escape from the earth forever. Newton even discussed the effect of air friction upon the motion of such a projectile. Between air friction and irregularities in the surface of the earth, the minimum size determined for a circular orbit just skimming the earth's surface is only hypothetical. Had Newton tried what we have just proposed, he would have had an even greater difficulty, however, since cannons capable of the launching were not available in the first place.

What Newton proposed as a mental exercise has now become almost routine in practice. Instead of a launching cannon, rockets are used to give artificial satellites the appropriate velocities to place them in orbit; but Newton's laws are still used to calculate the forces necessary to obtain a desired orbit. Modern technology permits us to go him one better. We now launch artificial satellites of the sun and have even placed a few on orbits which will permit them to escape from the solar system.

Newton also recognized that Kepler's laws were strictly correct only for the case of two isolated interacting bodies. In the many-bodied case of the solar system, the attraction of the planets upon each other, as shown schematically in Figure 3–9, will cause deviations from the orbit which any particular planet would follow if only that planet and the sun were present. Since the sun contains most of the mass in the solar system (99.86 percent), it will still control the planetary motions. The relatively small forces of the planets on each other are called *perturbing forces*. These forces cause small deviations in the orbits of the planets called **perturbations.** It was considered a great triumph of Newtonian mechanics when the position of Neptune was predicted by studying the perturbations in the motion of Uranus. In 1846, the hitherto unseen Neptune was found within one degree of the position predicted by perturbation theory.

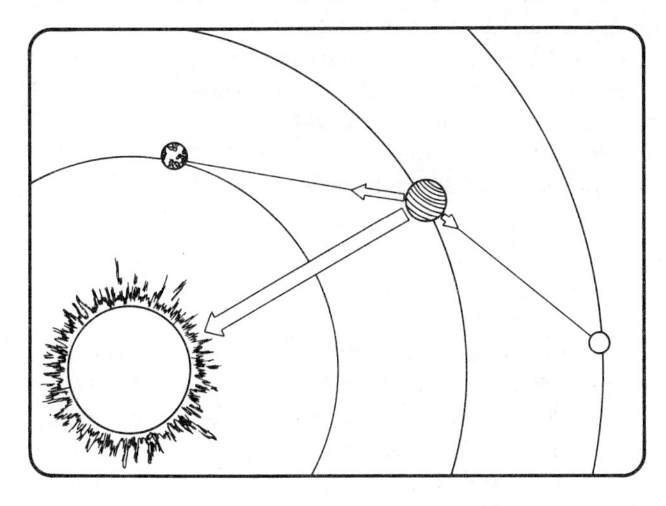

Figure 3–9 Planetary perturbations.

The Tides

The phenomenon of the ocean tides, also given a physical explanation by Newton, is a gravitational phenomenon. The **tides** arise because of the difference in the gravitational force due to both the moon and the sun across the diameter of the earth. Figure 3–10 illustrates the principle of the tidal force. Imagine three small masses of the same size connected by springs and placed in the gravitational influence of the large mass, M. The small mass closest to M suffers the greatest gravitational force due to M. Each succeeding mass will suffer less force, since the gravitational force decreases with increasing distance. The difference in force relative to the middle mass may be found by subtracting the force on the middle mass from the force on each of the other two masses. As the arrows in the diagram show, the most distant mass has a net force away from M, and the closest mass has a net force toward M when measured relative to the mass in the middle. Exactly the same effect would occur if the outer masses were pulled apart by two opposing forces. When the waters of the oceans are subjected to this effect, the water reaches its highest level on the two locations, on opposite sides of the earth, in line with the tide-causing body. Due to its rotation, the earth spins beneath the tides, causing an observer on the earth to see two high and two low tides each day.

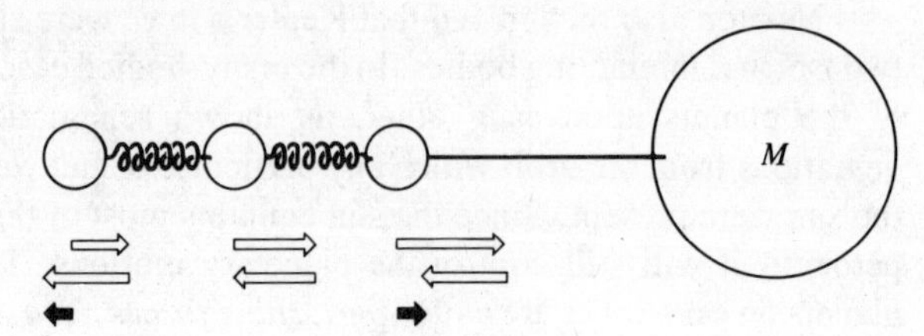

Figure 3–10 Tidal forces.

Both the sun and the moon cause tides. The moon is nearly twice as effective as the sun, since it is much closer to the earth than the sun. The height of a particular high or low tide depends upon the direction to the sun and moon at that moment. When the moon is new or full, the tides due to the sun and moon line up and the maximum effect occurs. At quarter moon, the solar high tides occur at the location of the lunar low tides and vice versa. The high and low tides are therefore reduced in height during this phase of the moon. Tides observed in coastal areas often do not line up exactly with the tide-producing bodies as described above because of irregularities in the ocean floor near the shoreline.

The Major Motions of the Earth

There are three major motions of the earth: rotation on its axis, revolution about the sun, and precession. Apart from observing the rotation of the earth from a space vehicle, the earth can be proven to rotate by means of a **Foucault pendulum,** a massive weight suspended on a steel wire which, in turn, is attached to a pivot. The pendulum is therefore able to swing in any direction. If, as in Figure 3–11, such a pendulum were placed at the North Pole and set in motion toward a distant star, the plane of its swing would remain fixed relative to the star, but it would appear to rotate relative to the ground. In reality, the plane of swing is fixed in space, and it is the earth that turns beneath it. Newtonian mechanics predicts a similar but slower apparent rotation of the plane of swing relative to the ground when the pendulum is placed away from the pole and closer to the equator. At the equator, there is no apparent rotation of the plane of swing. This type of pendulum is frequently seen in museums of science and technology. It is one of the most easily demonstrated proofs that the earth rotates.

The revolution of the earth about the sun can be proven by the observation of stellar parallax or stellar aberration. Parallax is the angular change in the line of sight to a nearby star as it is viewed from different points in the earth's orbit. The effect is very small and was not observed until 1838. More than 100 years earlier, James Bradley, while trying to

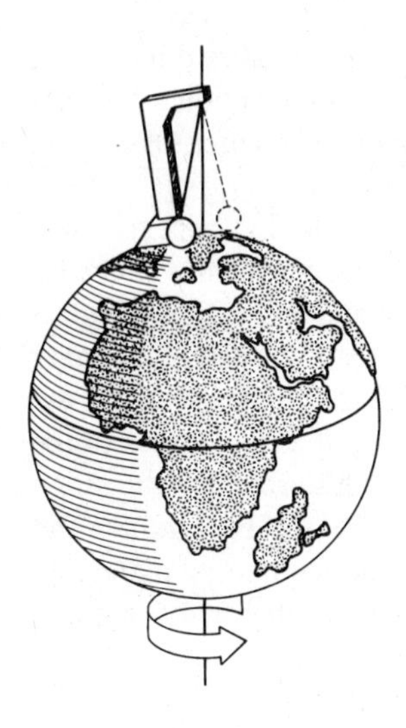

Figure 3–11 A Foucault pendulum at the North Pole.

detect parallax, discovered the phenomenon of **stellar aberration.** He observed that all stars show an annual angular motion about their average positions. The maximum displacement is roughly 20.5 seconds of arc. Aberration is accepted as direct observational proof of the earth's orbital motion. It is best explained by an analogy with falling rain.

Imagine standing in a gentle rainfall on a windless day. For maximum protection, you would hold your umbrella vertically. But if you start to run rapidly through the rain, you will now want to tip the umbrella in the direction in which you are running, because the apparent motion of the raindrops now becomes a combination of your motion and the natural descent of the raindrops.

Since the velocity of light is not infinite, the apparent direction from which it comes will also be influenced by the motion of the observer. As shown in Figure 3–12, the continual change in the direction of orbital motion of the earth causes an observer to

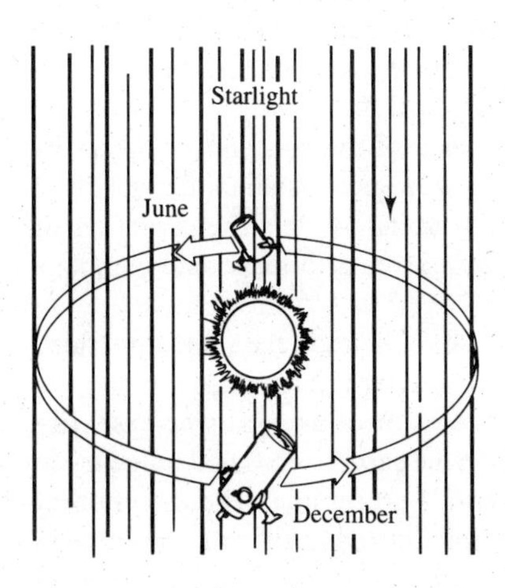

Figure 3–12 Stellar aberration.

continuously adjust the direction of his telescope to agree with the apparent direction of the starlight. The precise correction in the direction depends upon the position of the star in the sky and the direction of motion of the earth at that moment. The latter, of course, depends upon the time of year when the observation is being made.

The third of the earth's major motions, **precession,** is the one that appears most mysterious. It also can be explained by the application of Newtonian mechanics. Because of its rapid rotation, the earth is deformed from a perfect sphere and has a slight bulge at the equator. The deformation is very small, the equatorial diameter of the earth becoming only one-third percent larger than its polar diameter; but it is still enough to cause the observed precession. The plane of the earth's equator is inclined at 23½ degrees to the plane of the earth's orbit. The sun is always in the plane of the earth's orbit, and the moon is never far away from that plane. Consequently, the unequal attraction of both the sun and the moon upon the earth's rotational bulge tries to rotate the earth's equatorial plane into the plane of its orbit. Figure 3–13, with the rotational bulge of the earth greatly exaggerated, shows this effect.

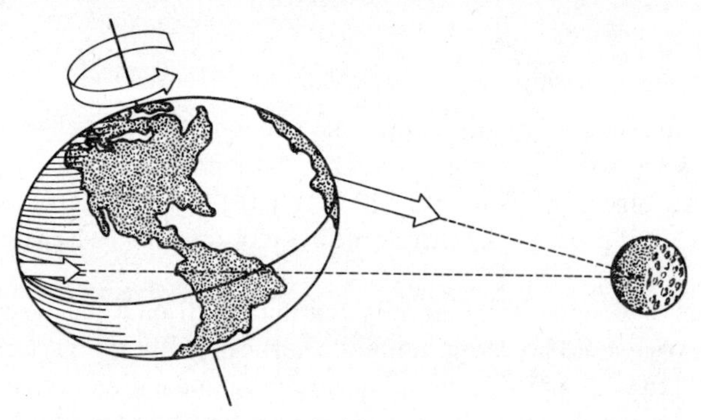

Figure 3–13 The unequal gravitational attraction of the moon upon the rotational bulge of the earth tries to reduce the tilt of the earth's axis.

Newtonian mechanics predicts that when an attempt is made to rotate a rapidly spinning body about an axis different from its axis of spin, the body moves instead about a third axis. This third axis is always perpendicular to both the spin axis and the axis about which the attempted rotation is made. A toy top is a good example of this phenomenon. The force of gravity acts to cause the top to fall over on its side. But instead of falling over, the top's axis of spin precesses in a cone about the vertical direction.

The earth has a similar motion due to the influence of the sun and moon on its rotational bulge. The earth's axis of rotation makes a cone in space, maintaining its tilt of 23½ degrees with the perpendicular direction to its orbital plane. The direction of movement is westward; it takes the earth roughly 26,000 years to make one complete conical motion.

Note in Figure 3–14 that the top and the earth rotate, or spin, in the same direction, but they precess in opposite directions. This is because gravity tries to make the top fall over; that is, it tries to cause the angle of the cone of precession of the top to increase. For the earth, however, the forces of the sun and moon upon the earth's rotational bulge try to decrease the 23½-degree angle of the cone of the earth's precession. The direction of precession of a spinning body depends upon both the direction of spin and the direction in which the forces which give rise to the precession are being applied. Hence, the reversal

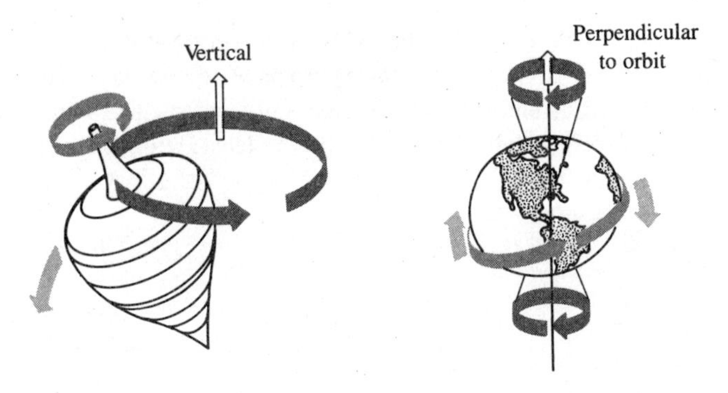

Figure 3–14 A comparison of the precession of a top and the precession of the earth.

in the turning forces between the top and the earth results in the reversal of their precessions.

The application of Newton's laws of motion to moving objects is called *Newtonian mechanics*. Newtonian mechanics gives a very good description of nature, at least as far as the movement of the planets, baseballs, automobiles, spacecraft, and the like are concerned. Early in this century Albert Einstein (Figure 3–15) introduced the theory of relativity as another way of describing nature. He also developed a different way of describing gravity in his general theory of relativity. His relativistic mechanics predicts, with greater precision than Newtonian mechanics, the motions of bodies when their velocity approaches that of light or when in the presence of very intense gravity. Relativity

Figure 3–15 Albert Einstein (1879–1955).

is thought to be more general than Newtonian mechanics, because the laws of Newton can be derived from Einstein's theory when the velocities of the objects are small—that is, closer to those of everyday experience—or the intensity of gravity is small enough. In Appendix I you will find a discussion of relativity and some of the apparently strange predictions it makes. Many of these predictions have, however, been confirmed by observation. Relativistic considerations have an important place in several astronomical applications, such as the explanation of some radiations from space, the behavior of objects in the extremities of the universe, and the rather strange objects such as black holes and neutron stars which are now either being discovered or, at least, being eagerly searched for by astronomers.

KEY TERMS

velocity
acceleration
force
mass
gravity
center of mass
ellipse
precession
parabola
hyperbola
relative orbit
perturbation
tides
Foucault pendulum
stellar aberration

Review Questions

1. What is velocity?
2. What is acceleration?
3. If a rocket increases its speed through space by 100 kilometers per hour by firing its engine for one second, how much will its speed increase if it fires its rocket for two seconds?
4. Is the acceleration greater when the rocket fires its engines for two seconds rather than for just one second?
5. What controls on an automobile are used to accelerate the car?
6. What are Newton's three laws of motion?
7. A certain subcompact car has just half the weight of a full-size car, yet both vehicles have the same maximum acceleration. How do the forces that accelerate these two cars compare?
8. How can we use the law of gravity to find the mass of a body?
9. How can we use Newton's second law to find the mass of a body?
10. The mass of the earth is eighty-one times the mass of the moon. How strong is the force of the earth's gravity pulling on the moon, compared to the moon's pull on earth?

    (1) Eighty-one times stronger

    (2) Nine times stronger

(3) The same

(4) Eighty-one times weaker

11. How does the amount of acceleration of the moon towards the earth compare to the amount of acceleration of the earth towards the moon?

    (1) Eighty-one times greater

    (2) Nine times greater

    (3) The same

    (4) Eighty-one times smaller

12. If the mass of the moon were doubled, how would the force of gravity between the earth and moon change?
13. If the mass of both the earth and moon were doubled, how would the force of gravity between them change?
14. If the distance from the earth to the moon doubled, how would the force of gravity between the earth and moon change?
15. It is easy to see why the pull of the moon's gravity raises a high tide on the side of earth toward the moon. But why should it raise a high tide on the opposite side?
16. The force of the sun's gravity on the earth is nearly 200 times stronger than the moon's, yet the moon raises tides on earth twice as high as tides produced by the sun. Why?
17. What phenomenon does the Foucault pendulum demonstrate?
18. What are two methods observers on earth can use to show the earth goes around the sun?
19. What causes the equatorial bulge of the earth?
20. What causes the precession of the earth's axis?
21. What are three motions of the earth?

Discussion Questions

1. If you stand on the ground, why doesn't the force of the earth's gravity accelerate you toward the earth's center?
2. How does the acceleration of the earth toward the sun compare with the acceleration of the moon toward the sun?
3. If the moon's orbit around the earth were a perfect circle, Kepler's second law would require that its orbital speed be a constant. Since the pull of earth's gravity continuously accelerates the moon, how can its speed remain constant?
4. In the vacuum of outer space, where there isn't any air for the exhaust gases to push against, how does a rocket engine provide any thrust?
5. Precise observations show that the earth's orbit has slight deviations from the elliptical orbit of Kepler's first law. Why?

6. Suppose an object orbits the sun in a long thin ellipse:

The shape of the orbit is exactly the same at the point nearest the sun as it is at the point farthest from the sun. How can this be possible when the force of gravity is so much stronger when the object is closest to the sun?

7. An observer watching a Foucault pendulum at the North Pole would see the pendulum appear slowly to change its direction of swing. Why? The earth rotates from west to east once each day. How fast and in what direction would the plane of swing of the pendulum change?
8. At the present time, the North Star, Polaris, is very close to being directly over the North Pole of the earth. Will it continue as our north star for many thousands of years?

4

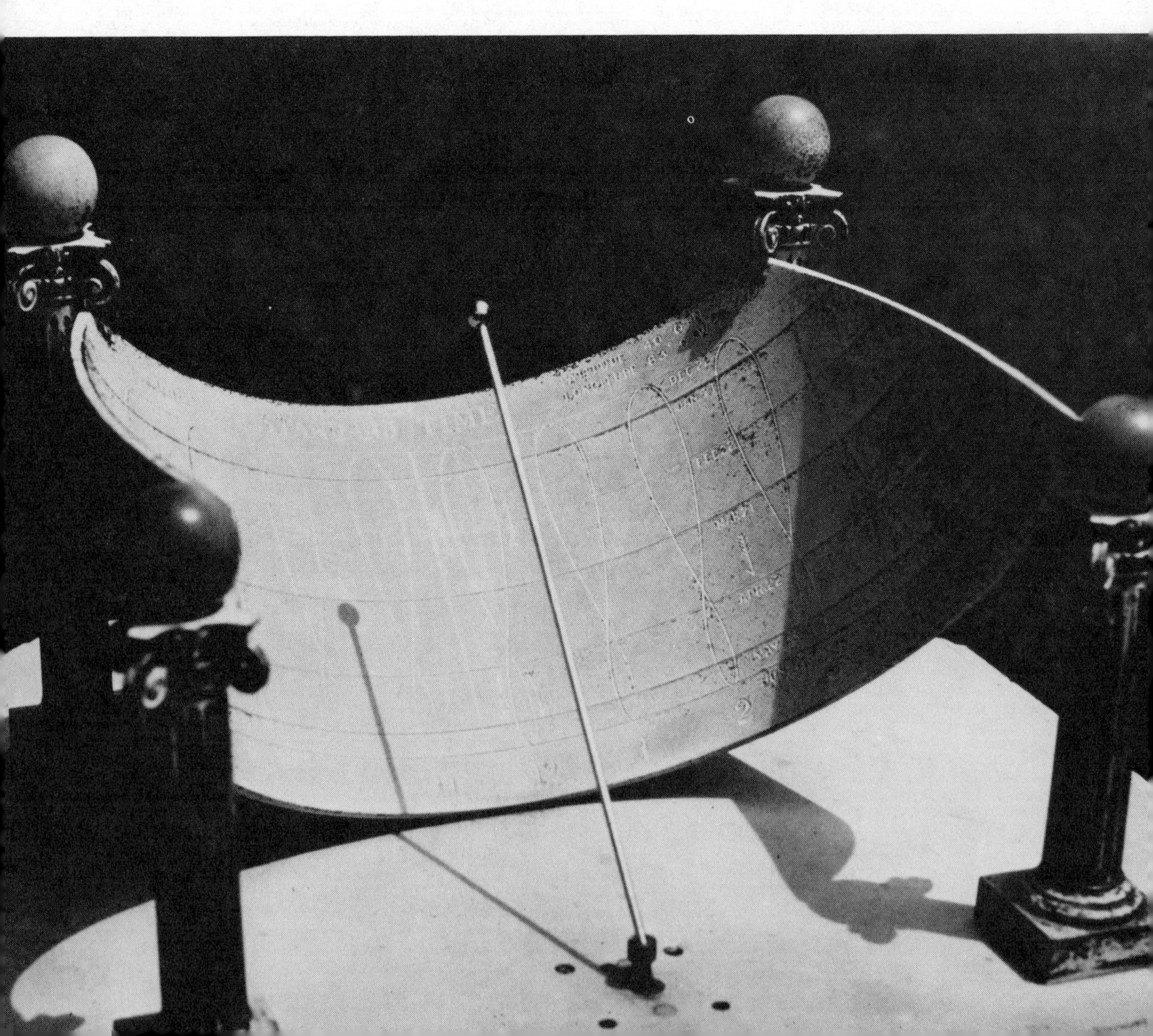

Celestial Coordinate Systems and Timekeeping

Not only has astronomy been humanity's timekeeper or chronologer, but it has also paid its way by providing a system of navigation. Ships at sea, camel caravans on the trackless desert, and aircraft on long journeys all found their way by sighting celestial objects.

Finding your way by the stars—celestial navigation, as it is called—makes use of two systems of coordinates. One is the celestial coordinate system which marks the positions of the stars on the sky. The other is the terrestrial, or geographical, coordinate system, the familiar longitude-latitude system which marks the positions of places on earth.

If the earth did not rotate relative to the stars, celestial navigation would be easy and uncomplicated, for a particular region of the sky would always lie over the same region on the surface of the earth. Thus, by recognizing the region of the sky which is overhead, you could immediately find your position on earth. However, the earth does rotate, and different regions of the sky pass over any particular region on earth at different times.

So the fact that the earth rotates makes it necessary to include time measurements in the linking of the celestial and terrestrial coordinate systems. This necessity was well understood by sailors on long ocean journeys in the days prior to radio communications. Errors in navigation could and did lead to shipwreck and loss of life. One such tragedy, a British naval disaster which occurred near the Scilly Isles in 1707, cost the lives of about 2000 sailors.

Spurred by the tragedy, Sir Issac Newton, then a member of the British Parliament, suggested that prizes with a top award of 30,000 pounds, a considerable fortune at that time, be given to someone who could, by whatever method of timekeeping he chose, determine longitude at sea to within one degree, or 60 miles (96 kilometers). A "Board of Longitude" was appointed to make the awards when a majority of the commissioners were convinced that the proposed method was prac-

A sundial

tical and when the method was proven accurate to within the prescribed limits on an ocean voyage from Britain to the West Indies.

An Englishman, John Harrison, attacked the problem of finding longitude at sea by building very accurate clocks. As an alternate method, a German astronomer, Tobias Mayer, constructed tables of the positions of the moon for various times. His tables were based on the very accurate gravitational analysis of the moon's motion that had been made by the great mathematician Leonhard Euler.

After years of tests and squabbles between the Board of Longitude, astronomers, Parliament, the admiralty, and other mariners, Harrison's method was shown to be the more accurate, within one-half degree of longitude; but Mayer's method was shown to be sufficiently accurate and more practical. In fact, Mayer's method was adopted by most ships' captains.

In 1765 the board decided to make partial awards to Harrison and to Mayer, the latter's award being posthumous. However, Harrison had valid claim to the top prize; and in 1773, upon the intervention of King George III, he received the remaining amount of his prize.

Time, navigation, coordinate systems—these are, in fact, concepts which have been formalized within the framework of astronomy. These concepts deal with the most basic aspects of astronomy, that is, with the shape of the earth and its motions relative to the coordinate system which locates the stars on the sky.

The ancient concept of the **celestial sphere** with the earth placed at its center is still useful for keeping track of celestial objects. In fact, the coordinate systems used to locate stars, the planets, the sun, and the moon are based upon this concept. These coordinate systems are, in turn, used to aid us in both defining and keeping time.

The Appearance of the Celestial Sphere

One of the most obvious features of the nighttime sky is its apparent westward rotation about a fixed point called the **celestial pole.** The stars appear to move across the sky on circular paths called **diurnal circles,** as shown in Plate 2B. These circles are concentric with—that is, centered upon—the celestial pole. This pattern is particularly obvious for the so-called *circumpolar* stars, which are simply stars with diurnal circles which never dip below the horizon. These circumpolar stars never set and would be visible for a full twenty-four-hour period if the bright light from the daytime sky could be removed. The stars at the opposite side of the celestial sphere, about the pole which is below the horizon, are also called circumpolar if their diurnal circles never rise above the horizon. These circumpolar stars, then, are the stars always hidden from view by the horizon for an observer at a particular location on the earth.

We identify a number of points and circles on the imaginary celestial sphere surrounding an observer. Two obvious points are the North Celestial Pole (NCP) and the South Celestial Pole (SCP), that is, the two points on the sky about which the celestial sphere appears to rotate each day. The circle on the sphere exactly midway between these points divides it into two equal halves; it is called the **celestial equator.** Directly

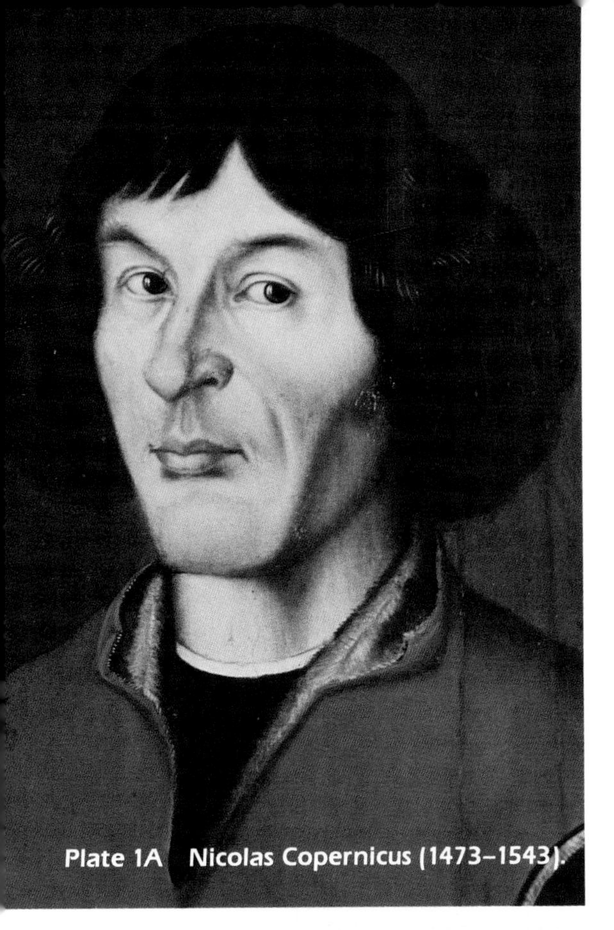

Plate 1A Nicolas Copernicus (1473–1543).

Plate 1B Galileo Galilei (1564–1642).

Plate 1C Isaac Newton (1643–1727).

Plate 2A Aerial view of Stonehenge.

Plate 2C Replicas of two of Galileo's telescopes.

Plate 2B Polar star trails.

overhead, for any observer, is the point known as his **zenith.** Directly below him, and therefore hidden by the earth, is his **nadir.** The circle midway between these two points, and hence at right angles to the upward direction, is the **astronomical horizon.** These circles and points are identified in Figure 4–1. At sea, the astronomical horizon and the visible horizon (where the sky meets the earth) are almost exactly coincident. On land, mountains, trees, and buildings frequently protrude above the astronomical horizon, so the visible horizon is irregular in appearance. The astronomical horizon is thus the imaginary circle on the celestial sphere which would separate the visible half and the invisible half of the sky if the earth were a perfectly smooth sphere.

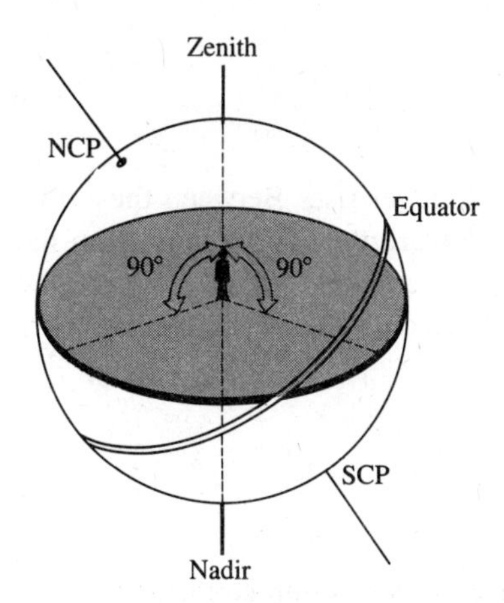

Figure 4–1 The relationship of the celestial equator to the observer's horizon.

Early observers noticed that the **altitude** of the visible celestial pole, that is, its angular height above the horizon, increased as they moved toward it, which implies that the surface of the earth is not flat. As an observer moves across the earth's surface, his zenith will swing through the sky because of the curvature of the earth's surface. Since the astronomical horizon is fixed by the observer's zenith, it will also swing across the sky. Thus, the altitude of the pole must change when the observer moves either northward or southward.

There are two rather unique situations possible. When the observer's zenith coincides with one of the celestial poles, the diurnal circles of the stars will all be parallel to his horizon, as shown in Figure 4–2a. All stars will be circumpolar, half always above the horizon and half always below the horizon. This situation will occur when the observer is at the North or South Pole of the earth. The other unique situation, as shown in Figure 4–2b, occurs when the observer's zenith lies on the celestial equator. In this case, which occurs when the observer is at the earth's equator, the planes of the diurnal circles will be perpendicular to his horizon. Each object in the sky will spend twelve hours above and twelve hours below the horizon each day. From the earth's equator, an observer can see everything which is permanent in the sky, if he waits long enough.

Most persons are more familiar with the intermediate situation, neither at the pole nor at the equator, which is depicted in Figure 4–2c. In this case, only objects on the celestial equator are above the horizon for exactly twelve hours each day and below it for the other twelve hours. About the visible pole and inside the diurnal circle which just touches his horizon are found the circumpolar stars which never set. A similar circle about

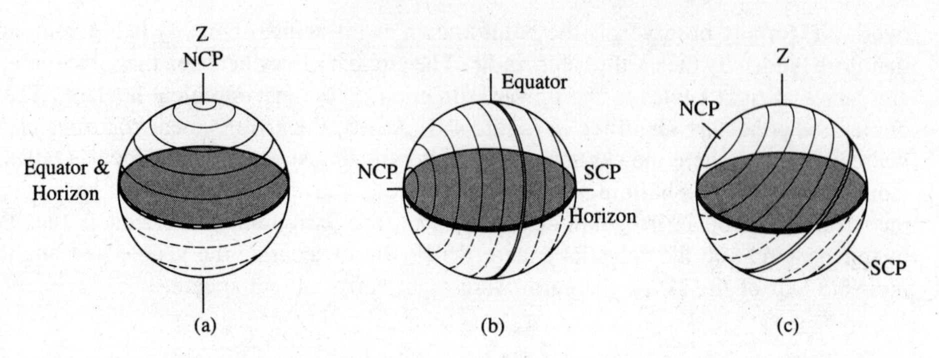

Figure 4–2 The orientation of the diurnal circles to the horizon at (a) the North Pole, (b) the equator, and (c) an intermediate latitude.

the invisible pole encloses the circumpolar stars which never rise. Between these two circles lie the objects which alternately rise and set. Toward the visible pole from the equator, objects are above the horizon for more than twelve hours each day. Toward the invisible pole from the equator, objects are above the horizon for less than twelve hours each day.

Longitude and Latitude on the Earth

The change in the appearance of the sky with the observer's position confirms that the earth is essentially spherical in shape, which leads to a system for locating positions on the surface of the earth. In reality, as we have seen, it is the earth which actually rotates in the eastward direction. The apparent westward rotation of the sky is merely a reflection of the earth's motion. The two points on the earth's surface in the same line joining the celestial poles are called the *poles* of the earth's rotation. The circle on the earth midway between these points is the *equator* of the earth. The distance of a location from the equator toward either of the earth's poles can be measured by finding the angle from the equator to the line joining that location and the center of the earth. This angle is called the **latitude** of the location. All points at the same latitude will lie on a circle parallel to the equator called a **parallel of latitude.**

Because of the obvious difficulties of measuring the angle directly, we pass a circle through the poles of the earth and the location of interest. The latitude of the point is then the fraction of the circle between the point and the equator expressed in degrees. The value of the latitude must lie between 0° and 90°, since the pole, the farthest point from the equator, lies one-quarter of a circle away from the equator. The circle through the point along which its latitude is measured is called a **meridian of longitude.** The latitude of a point is called *North* if it lies closer to the North Pole than to the South Pole and *South* if the reverse is true.

The location of a place along its parallel of latitude is found by comparing its meridian of longitude with the meridian of longitude through Greenwich, England, which is called the **prime meridian.** The **longitude** of a point on the earth's surface is the fraction of a circle which is intercepted on the equator by the prime meridian and the meridian of longitude of the point. Longitude is measured from the prime meridian from 0° to 180°

eastward in the direction of the earth's rotation and 0° to 180° westward in the opposite direction. Thus, the two arcs indicated in Figure 4–3 uniquely locate Washington, D. C., which has coordinates longitude 77°4′ W and latitude 38°55′ N.

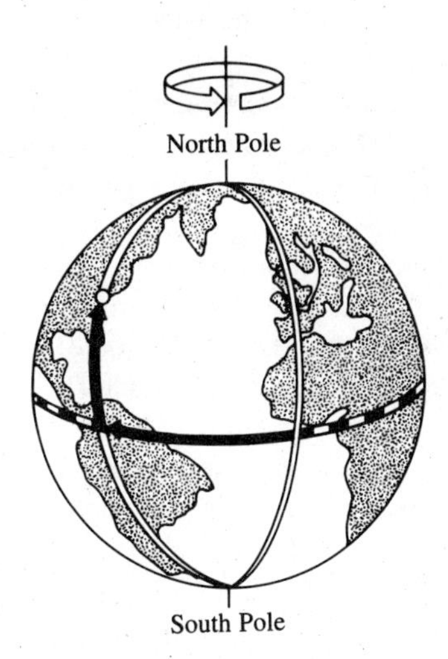

Figure 4–3 The longitude and latitude system.

This description of latitude and longitude is easy to apply to a globe representing the earth. On the actual earth, the determination of latitude and longitude requires the use of the apparent celestial sphere. First, consider latitude. As shown in Figure 4–4a, when an observer is at the equator, his zenith lies on the celestial equator. Therefore, the North Celestial Pole lies on his horizon; or, to put it another way, the altitude of the pole is zero degrees. As the observer moves toward the pole, as in Figure 4–4b, his zenith moves toward the celestial pole by the same angle that he has moved in latitude. His northern

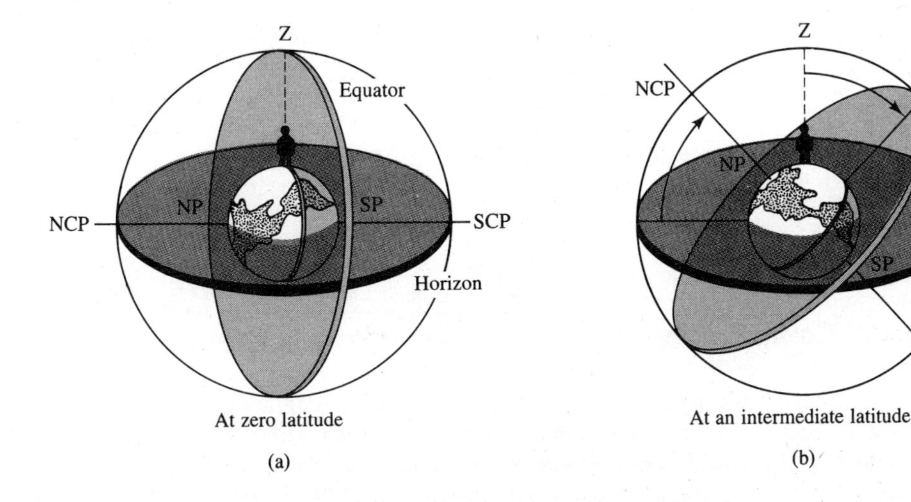

Figure 4–4 The change in the celestial pole and the celestial equator as the observer changes latitude.

horizon, therefore, appears to dip below the pole by this same amount. In other words, the altitude of the pole is equal to the latitude of the observer. The basic task of a surveyor in measuring a latitude is consequently the measurement of the altitude of the celestial pole seen from that location. Of course, some complicated, but small, corrections to this basic measurement are required to find the geographical latitude of a location, because the earth is not perfectly spherical.

Longitude is somewhat more difficult to measure, although the measurement is, in principle, simple. Begin by identifying the observer's **celestial meridian,** the circle passing through the celestial poles and the observer's zenith. The observer's celestial meridian lies directly over his meridian of longitude. Consider another observer at Greenwich, where the celestial meridian of Greenwich corresponds to the prime meridian of longitude. Assume a star to be on the celestial meridian of Greenwich, as shown in Figure 4–5. Then, since the earth rotates eastward once in twenty-four hours, the meridian of an observer to the west of Greenwich will advance toward the same star at a rate of 15° each hour. Timing how long it takes the star to reach the observer's meridian after it has crossed that of Greenwich is, thus, equivalent to measuring the longitude from Greenwich to the observer. The longitude for this situation would be labeled *West*. If he had been in the Eastern Hemisphere, the star, naturally, would cross his meridian before it crossed that of Greenwich. The time interval converted to degrees at the rate of 15° per hour would still be the longitude. The longitude would be labeled *East* in this case, however. Frequently the longitude is expressed as the time difference between when a star crosses the meridian of Greenwich and when it crosses the meridian of the observer, without converting to degrees. Longitude determination requires a precise knowledge of time and so led to the development of accurate clocks to permit navigation at sea.

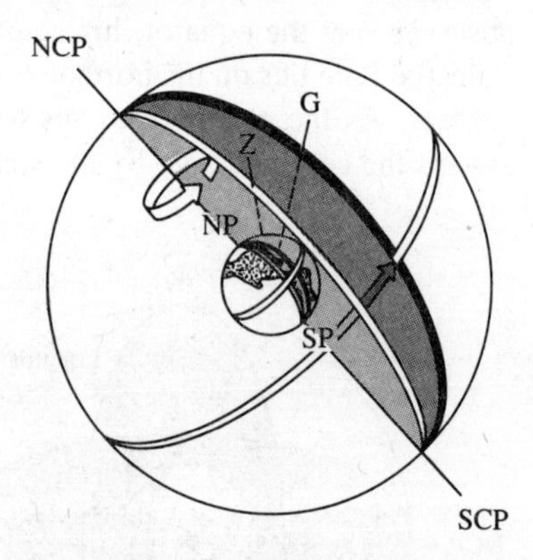

Figure 4–5 The determination of longitude by timing star transits.

Celestial Coordinates

The same technique for locating objects on the earth may be applied to locating stars on the celestial sphere. The celestial equator is now the important circle. The angle between the equator and the direction of a star is called the **declination** of the star; it is usually symbolized by δ or dec. Since a star keeps the same declination throughout a day, its diurnal circle is also a **parallel of declination.** All other stars with the same declination lie on the same parallel. To locate a given star on its parallel, the celestial meridian can be used in much the same way that the meridian of Greenwich is used on the earth. Pass a circle through the celestial poles and the star. This circle is called the **hour circle** of the star. The angle at the equator between the celestial meridian and the hour circle of the star is called the **local hour angle** (LHA) of the star. It is always measured westward from 0° through 360° or 0 through 24 hours. Again, 1 hour of angle equals 15° of angle. When a star has an hour angle of 0, it is high on the meridian and is said to be at *transit*. The precise measuring scheme is shown in Figure 4–6. Notice that since the sky appears to move westward, the local hour angle of the star is continually changing. For some purposes this is a nuisance; but for timekeeping, it will prove to be very helpful.

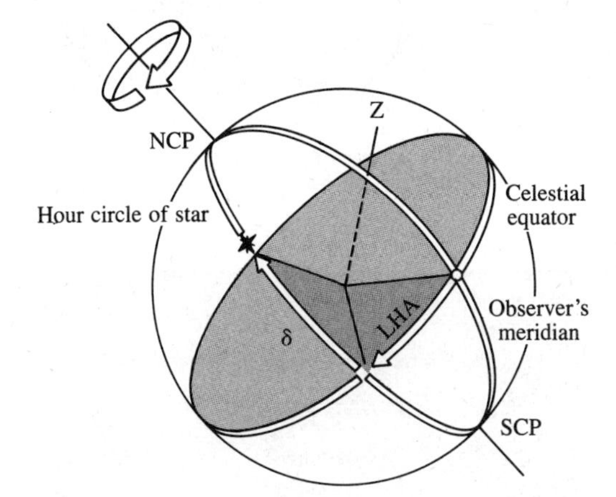

Figure 4–6 The declination–hour angle coordinate system.

For cataloguing the positions of stars it is desirable to have a coordinate system fixed to the stars, where both coordinates are unchanging, rather than a system fixed to the observer, who is actually in motion because of the earth's rotation. Declination can be used as before because a star's declination does not change. The coordinate measured along the equator can also be fixed by using a fixed point on the equator from which to measure. The point of reference which is chosen is the **vernal equinox,** indicated by the sign of Aries the Ram, ♈ . The vernal equinox is the point where the ecliptic (the sun's apparent annual path) cuts the equator such that a body traveling eastward on the ecliptic would pass from the southern to the northern hemisphere of the sky. The **right ascension** of a star is defined as the angle along the equator from the vernal equinox to the circle through the celestial poles which also passes through the star. Figure 4–7 indicates how the right ascension–declination system is used. Right ascension, symbolized by α or RA, is always measured eastward from 0 through 24 hours.

Using the right ascension–declination system, we may draw maps of the sky similar to Figure 4–8. Right ascension is always shown in the horizontal direction, increasing to the left, and declination is always shown in the vertical direction. Remember that the vernal equinox is not permanently fixed relative to the stars because of the precession of the earth. Therefore, the right ascension and declination of a star are continually changing, but very slowly. It is necessary to continually update the position of a star for the precession of the equinoxes. This is a nuisance requiring a rather straightforward calculation when the coordinates of a star are needed for a certain night, but its position has been given on a map or in a catalogue of positions for a different time.

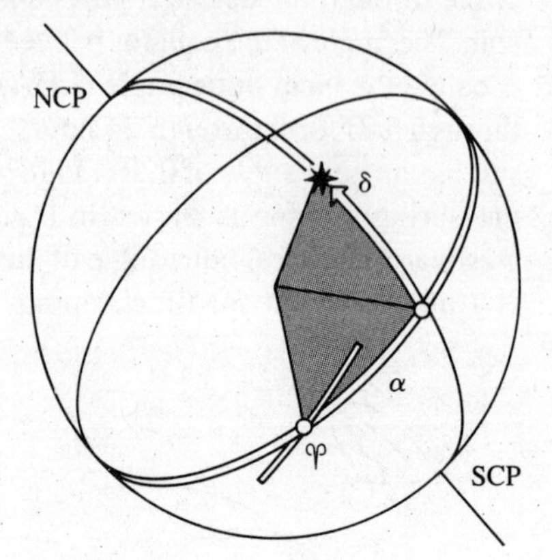

Figure 4–7 The declination–right ascension coordinate system.

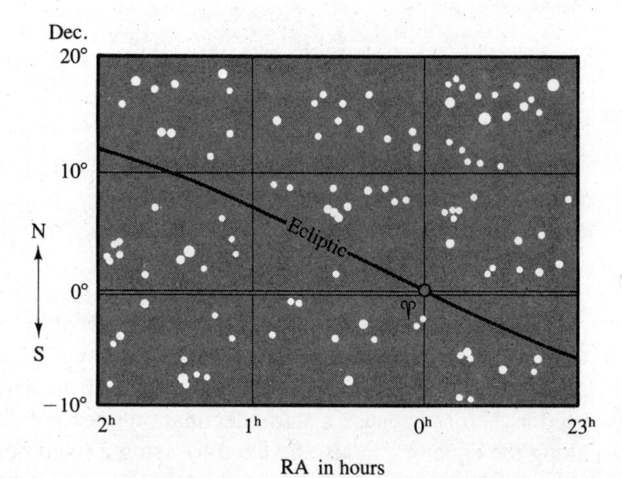

Figure 4–8 A map of the celestial sphere.

Timekeeping

The hour angle system can now be used to help us keep time. **Local sidereal time,** or star time, is defined as being equal to the local hour angle of the vernal equinox. The word *local* is required because this time is measured in reference to the observer's, hence the local, celestial meridian. Unfortunately, the vernal equinox is not visibly marked on the sky. We know that if the local sidereal time has a particular value, the vernal equinox is west of the observer's meridian by that hour angle. Alternately, the observer's meridian is eastward of the vernal equinox by that same angle. But the angle measured eastward from the vernal equinox is right ascension. Therefore, as shown in Figure 4–9, the local sidereal time is equal to the right ascension of a star in transit at the given instant.

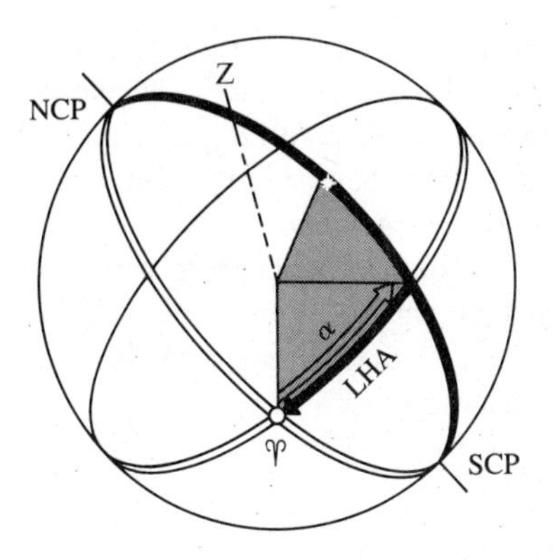

Figure 4–9 Determination of local sidereal time by star transits.

A telescope called a *meridian transit* is used to determine sidereal time. This instrument has a horizontal axis which is aligned due east and west. As the telescope is rotated on its axis, its field of view sweeps along the observer's celestial meridian. If the telescope is pointed at the declination of a particular star, the observatory sidereal clock can be synchronized with the earth's rotation by noting that the clock should read a time equal to the right ascension of the star when the star is precisely in the center of the field of the telescope.

Sidereal time is useful for finding the positions of stars in the sky, but not for everyday use. For this purpose we prefer to tell time by the sun. Because the sun appears to move around the ecliptic once each year, it is continually changing its position relative to the vernal equinox. Sidereal noon, when the vernal equinox transits the meridian, and solar noon, when the sun transits the meridian, therefore generally do not agree. In other words, during the year solar noon occurs at progressively changing times by the sidereal clock.

The time used for our daily living is synchronized with the sun. **Local apparent solar time** equals the local hour angle of the apparent sun plus twelve hours. The twelve hours are added to make the solar day start at midnight rather than at noon. The apparent sun is simply the sun as we see it in the sky. Figure 4–10 illustrates the measurement of apparent solar time.

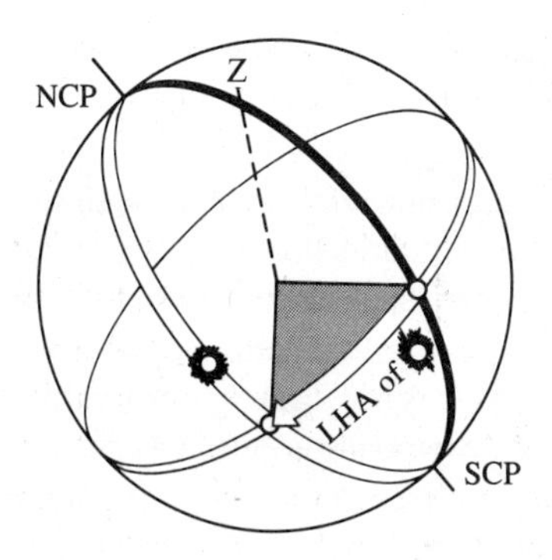

Figure 4–10 Local apparent solar time.

To estimate the difference between the solar and the sidereal day, refer to Figure 4–11. Assume the earth to be at the point in its orbit where sidereal and solar noon happen to occur together. One sidereal day later the earth will have made one full rotation relative to the vernal equinox; but it will also have advanced nearly one degree in its orbit during that time. Therefore, it will have to rotate approximately one degree more to bring the sun back to the observer's meridian. It takes the earth four minutes to rotate through one degree. Therefore, an interval of twenty-four hours of solar time is roughly four minutes longer than twenty-four hours of sidereal time. A solar clock consequently appears to run slower than a sidereal clock, "losing" four minutes each day.

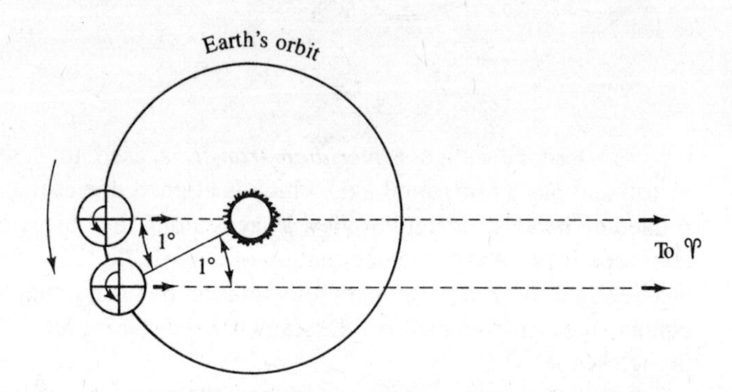

Figure 4–11 Comparison of the solar and the sidereal day.

The best way to tell apparent solar time is with a sundial. Until mechanical clocks were invented, a sundial was the only good way of keeping time. Mechanical clocks do not keep exact pace with the sundial, however—not because the clocks are bad, but rather because the apparent solar day varies in length throughout the year. Figure 4–12 illustrates the two causes of this variation. First, the earth's orbit is eccentric. It therefore speeds up and slows down in its motion in accord with Kepler's law of equal areas. Thus, the apparent solar day is longer when the earth is at perihelion than when it is at aphelion. The second cause of the changing length of the apparent solar day results from the **obliquity of the ecliptic.** Because of this 23½-degree angle between the ecliptic and the equator, a movement of the sun near a solstice causes a greater change in the right ascension of the sun than an equivalent movement of the sun near an equinox. Since time is reckoned relative to the equator, this difference introduces a further variation in the length of the apparent solar day.

To circumvent this problem, the concept of the *mean sun* is introduced. The mean sun is a fictitious sun which moves uniformly along the celestial equator. It goes once around the equator each time the real sun appears to go once around the ecliptic. **Local mean solar time,** schematically depicted in Figure 4–13, is equal to the local hour angle of the mean sun plus twelve hours. Since the mean sun is fictitious, it is not directly observable. However, from a knowledge of the earth's orbital motion, the position of the mean sun relative to the vernal equinox can be calculated. The position of the vernal equinox can be found by observing transits of the stars. Consequently, solar clocks are regulated at

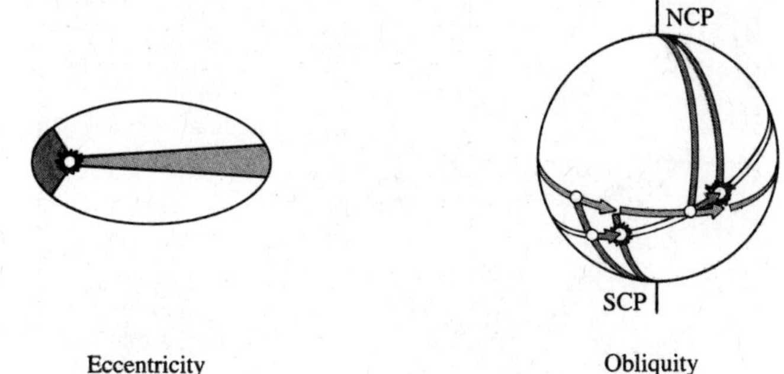

Figure 4–12 Causes of the variations in the length of the apparent solar day.

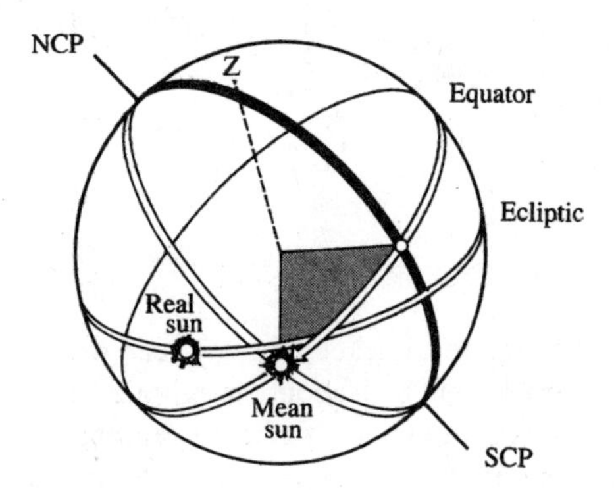

Figure 4–13 Local mean solar time.

observatories by timing transits of stars. The sidereal time so observed is transformed to mean solar time by calculation; this time, in turn, is used to check a mean solar clock.

Local mean solar time changes as you move eastward or westward in longitude. In fact, the difference in reading between two mean solor clocks at two places will be exactly the difference in the longitudes of the two locations. Strictly speaking, whenever you move eastward or westward, you should reset your local clock, no matter how little your longitude has changed. To remove this nuisance, standard time zones have been established. By common agreement all clocks within a time zone are set to read the same, that is, to keep **standard time.** The standard meridians of each zone differ by multiples of fifteen degrees in longitude from the meridian of Greenwich. The standard zones, therefore, always differ by an exact number of hours from the time at Greenwich. Notice in Figure 4–14 how the zone boundaries have been adjusted to conform to geographical and political boundaries. The numbers at the top of the diagram indicate the number of hours that must be added to that particular standard time in order to find Greenwich time. Greenwich time, by the way, is frequently called **universal time;** it is the time used in the tabulation of the occurrence of astronomical events. **Daylight saving time** is simply the time that would be kept at the next standard meridian east of the normal meridian for a particular zone. Clocks are therefore advanced one hour, which makes sunset occur one hour later by the clock and gives the appearance of adding one more hour of daylight in the evening, hence its name. Of course, one hour of daylight is stolen from the morning.

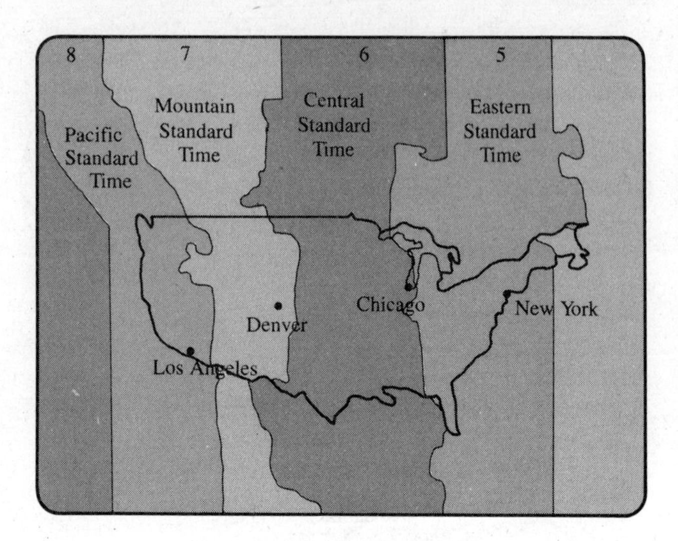

Figure 4–14 Standard time zones in the United States.

The Year and the Calendar

Besides the day, there is a longer period that can be used for timekeeping: the year, or the length of time it takes the sun to make one complete apparent path about the ecliptic. But, just as there are different kinds of days, there are different kinds of years, depending on the reference point by which one trip around the ecliptic is measured. We will be concerned with only two kinds, the sidereal year and the tropical year. The **sidereal year** is the interval between successive conjunctions of the sun with a particular star. Strictly speaking, we should say a fixed direction in space instead of a particular star, because, as will be demonstrated later in the book, the stars are not really fixed. The **tropical year** is the interval between successive passages of the sun through the vernal equinox. Since the vernal equinox is moving slowly westward through the stars due to precession, the tropical year is slightly shorter than the sidereal year, as shown in Figure 4–15. The sidereal year is the true period of revolution of the earth and is 365.2564 days in length. The tropical year is often called the *year of the seasons;* it is 365.242199 days in length.

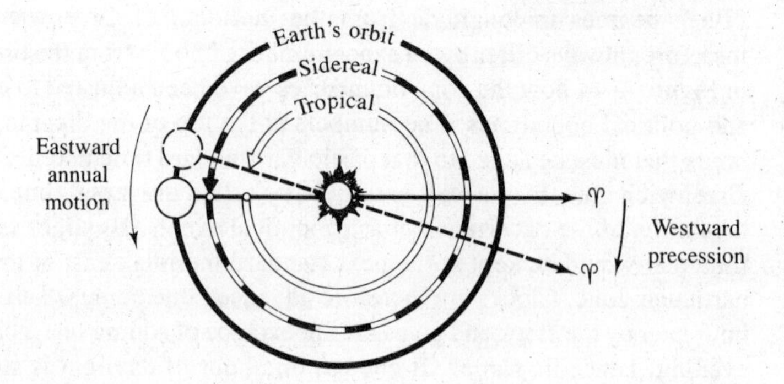

Figure 4–15 Comparison of the sidereal and the tropical years.

Due to the phenomenon of the seasons, the tropical year was the first to be discovered, although it was only crudely determined. The seasons are a direct result of the obliquity of the ecliptic. Figure 4–16 is a map of the sky showing the relationship of the ecliptic to the celestial equator. The two points of intersection of these circles are called the **equinoxes,** which means equal nights. When the sun is at either of these points, its declination is zero, and it therefore is above the horizon for twelve hours and below it for twelve hours. The two points of extreme distance from the equator are called the **solstices,** which means the sun comes to a stop. It does in fact stop its movement in declination since it must reverse its direction of movement in declination at these locations. It has a declination of 23½°N at the summer solstice and 23½°S at the winter solstice. At the solstices, the length of time the sun is above the horizon is different from that for which it is below the horizon. In the Northern Hemisphere there are more daylight hours than nighttime hours during the interval between the vernal and autumnal equinoxes. The reverse is true during the interval between the times of the autumnal and vernal equinoxes.

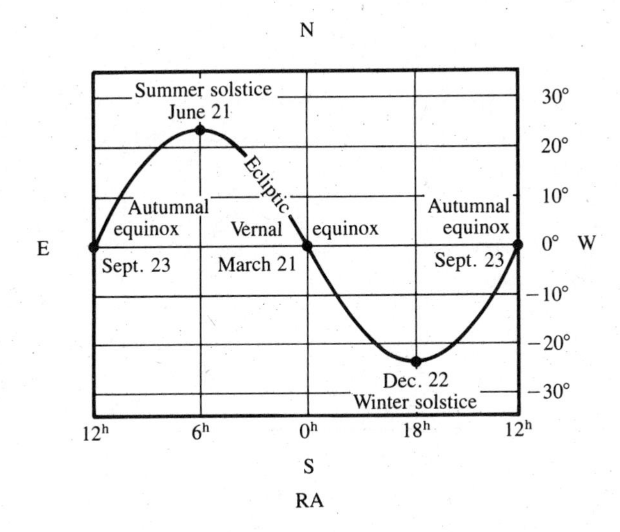

Figure 4–16 Map of the ecliptic on the celestial sphere.

To understand the seasons, consider Figure 4–17, which schematically represents the circumstances at the summer solstice. Since there are more than twelve hours of daylight in the Northern Hemisphere, that hemisphere receives more energy and warms up. The Southern Hemisphere suffers the opposite effect and cools off. When the sun is at the summer solstice, the sun can be seen for twenty-four hours during the day within the cap bounded by the Arctic Circle, which lies 23½ degrees from the North Pole. This is the region of the midnight sun. At the same time, within the Antarctic Circle, around the South Pole, there will be twenty-four hours of darkness during the day. In the winter, at the other solstice, the circumstances in the caps reverse, and the Antarctic Circle defines the region of the midnight sun.

Along with the increased length of the day, the rays of the sun strike the surface of the earth more directly in the Northern Hemisphere from the time of the vernal to the autumnal equinox, which results in more effective heating and adds to the seasonal changes. The reverse is true in the other half of the year. The variation in the daylight

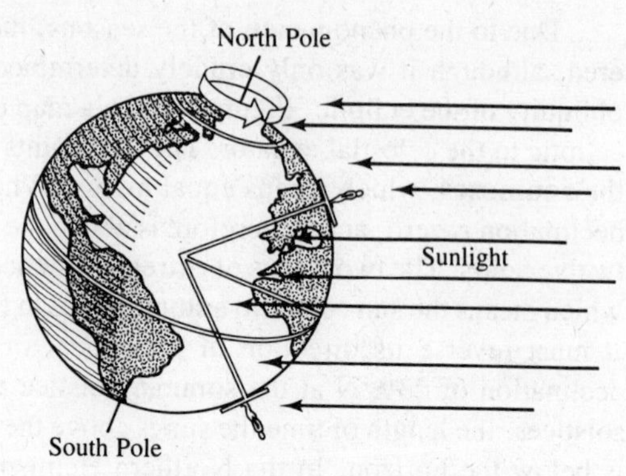

Figure 4–17 Detailed causes of the seasons.

Table 4–1
Comparison of various calendars

Type of calendar	Average length (in days)
Early Roman: 365 days each year	365
Julian: Every year divisible by 4 is a leap year	365.25
Gregorian: Century years not divisible by 400 are ordinary years	365.2425
Modified Gregorian: The years 4000, 8000, 12,000, . . . are ordinary years	365.24225
Tropical year	365.242199

hours and the changing angle at which the sunlight strikes the earth's surface both result from the obliquity of the ecliptic and are the detailed causes of the seasons.

The calendar was devised to count days and identify events according to the year. The early Roman calendar, from which ours is derived, contained 365 days per year. Because the number of days in a tropical year is not an exact number, the day of the beginning of spring continually fell later in the year. In 720 years, it would have moved from what is now March into September. The astronomer Sosigines pointed out that the tropical year is about 365.25 days long. By adding a leap year of 366 days every fourth year, the average calendar year would become 365.25 days. Julius Caesar accepted Sosigines' observation and instituted the leap year in what we now call the **Julian calendar.**

Of course, the tropical year is actually slightly shorter than 365.25 days. The time of the vernal equinox now moved more slowly through the calendar, but occurred progressively earlier in the year. In 1582 Pope Gregory, following the advice of the astronomer Clavius, introduced the **Gregorian calendar.** He deleted 3 leap years every 400 years by making century years not divisible by 400 into ordinary years. The average calendar year, over a 400-year span, becomes 365.2425 days, which is still slightly too long. A modified Gregorian calendar in which the years exactly divisible by 4000 will be ordinary years will give an average length of 365.24225 days. Although it is over 2000 years before this modification will be required, it will result in a calendar in which the seasons will get out of step with the calendar by only one day every 19,600 years.

KEY TERMS

celestial sphere
astronomical horizon
diurnal circle
celestial poles
celestial equator
zenith
nadir
latitude
parallel of latitude
longitude
meridian of longitude
prime meridian
altitude
celestial meridian
declination
parallel of declination
hour circle
local hour angle
vernal equinox
right ascension
sidereal time
local apparent solar time
local mean solar time
standard time
daylight saving time
universal time
sidereal year
tropical year
obliquity
solstice
equinox
Julian calendar
Gregorian calendar

Review Questions

1. If a star were located exactly on the celestial equator, how would it appear to move in the sky?
2. If a star were located exactly at the North Celestial Pole, how would it appear to move in the sky?
3. Where would you have to be on the earth to see stars rise straight up from the horizon?
4. The North Star, Polaris, is within one degree of the North Celestial Pole. If Polaris appears about 40° above your horizon, where are you on the earth?

5. The following names refer to markings on a globe of the earth. For each, list the corresponding name on the celestial sphere.

   (1) equator

   (2) North Pole

   (3) latitude

   (4) longitude

6. Latitude 0° and declination 0° occur on the terrestrial and celestial equators. These circles, in turn, are naturally determined by the terrestrial poles and the celestial poles. But longitude 0° and 0 hours right ascension are arbitrarily defined and were selected for convenience. What defines longitude 0° and 0 hours right ascension?
7. How many degrees does the celestial sphere appear to rotate in one hour? In four minutes?
8. The sun's diameter in the sky is about one-half degree. How long does it take for the sun in its daily motion to appear to move its own diameter across the sky?
9. What is sidereal time?
10. Why has the keeping of accurate time traditionally been the task of astronomers?
11. What time is measured by a sundial?
12. Why is a solar day longer than a sidereal day?
13. Why don't we use local mean solar time in our everyday lives?
14. Why do standard time zones run generally north-south, not east-west?
15. The time zones for New York and California differ by three hours. When it is noon in New York, what time is it in California?
16. Why do different constellations appear at different seasons of the year?
17. What causes the seasons?
18. If you are north of the Arctic Circle in June (not exactly at the North Pole), in what direction would you look to see the midnight sun?
19. How are the sidereal and tropical years defined?
20. Why do we have leap years?

Discussion Questions

1. Visitors to the tropics sometimes comment on how quickly it gets dark after sunset there. Why does it get darker faster near the equator?
2. Which is easier to find by observing the stars, your latitude or your longitude? Why?
3. Suppose you note the time when the vernal equinox transits your meridian. Two hours later,

   (1) What is your local sidereal time?

   (2) What is the right ascension of a star on your meridian?

   (3) What is the hour angle of the vernal equinox?

   (4) What is the local apparent solar time?

4. About how many sidereal days are there in one year?
5. Why doesn't apparent solar time run at a constant rate?
6. What is the cause of the difference between the sidereal and tropical years?
7. If you studied the heavens at the various seasons now, what differences would you note if you could return in 10,000 years?
8. The Tropic of Cancer is 23½° north of the equator, and the Arctic Circle is 23½° south of the North Pole. What are the unique properties of these two parallels of latitude?

5

Earth's Nearest Neighbor

Of the countless objects in space, the moon is the earth's nearest neighbor, being only about 400,000 kilometers away. In fact, the earth and moon, linked together by mutual gravitation, are sometimes referred to as the double planet.

The moon has always played on the emotions of people. Who isn't familiar with the legendary werewolf and his transfiguration under the power of the full moon? And the term **lunacy,** which is used in reference to madness, stems from the Latin word for moon, **luna.**

Most scientists rule out the possibility that there are forces exerted by the moon which are direct causes of human behavior; but nonetheless, no one can argue that the moon has not inspired people in various ways. For example, it has been, at least in part, the inspiration for "The Moonlight Sonata," "Blue Moon," "Moon Shadow," "Claire de Lune," and a host of other musical compositions, some more memorable than others.

This is not to say that nothing on earth is directly affected by the moon, for its gravitational force plays the major role in raising our ocean tides. In a quite subtle way, these tides have had an interesting and direct effect on marine life. We have observed that fossils of certain coral organisms are arranged in distinct layers that represent annual growth. The layers consist of narrow bands that represent monthly growth. These narrow bands are made up of still narrower ridges, or bands, that represent daily growth.

By various methods of dating, particular coral fossils can be classified according to the geological periods. Counting the number of the narrowest bands in a layer representing annual growth gives a measure of the number of days in the year during the geological period in which that layer was formed.

Such measurements, which were first made in the 1960s, indicate that the number of days in a year is slowly but steadily decreasing. About 400 million years ago there were nearly 400 days in a year, whereas at present there are only about

365. Astronomers have known since Newton's time that the length of the day is increasing due to a slowing of the earth's rotation. And in fact, about 100 years ago, the rate of slowing of the earth's rotation was determined from locations of solar eclipses which had been recorded over the previous 2000 years. From this measurement, we can calculate that about 400 million years ago, when the earth was spinning faster, there should have been nearly 400 days in the year, in excellent agreement with the fossil corals.

Furthermore, it has been concluded that the slowing of the earth's rotation which leads to this increase in the length of the day is related to the frictional forces of the tides raised, for the most part, by the moon. So for millions of years, the tiny corals, in oceans now extinct as well as in those which presently exist, have had their schedules altered by the earth's nearest neighbor, the moon.

Above all of its other effects on us, both direct and indirect, lay the moon's invitation to visit another world. But people could only dream about visiting the moon until October 4, 1957, when the Soviet Union launched an artificial satellite, Sputnik I, into earth orbit and shocked the world into the realization that the quest for the moon had begun in earnest.

On Christmas Eve of 1968, only eleven years after Sputnik I, the world was invited to the moon by astronauts Frank Borman, William Anders, and James Lovell, who became the first men to escape the bonds of earth and become bound to another world. As they orbited the moon, they, along with the TV audience, viewed the lunar landscape made up of mountains, craters, and valleys. For the first time, people watched as the earth rose and set over the lunar horizon.

Six months later, on July 20, 1969, Neil Armstrong stepped from his spacecraft and became the first man to walk on another world. In walking on the moon, he took a walk for Copernicus, Kepler, Galileo, Newton, and men like them—men with great vision who laid the groundwork for the exploration of the universe.

The Moon's Phases and Appearance

The orbital motion of the moon relative to the earth, along with its illumination by the sun, gives rise to the moon's continually changing appearance in the sky. The amount of the moon's surface which is seen to be illuminated by the sun characterizes the phase of the moon. No matter what the phase of the moon, that is, no matter what fraction of its surface is seen bathed in sunlight, the same side of the moon always faces the earth.

The moon's period of rotation is equal to its period of revolution about the earth. The sidereal period of the moon—that is, the length of time it takes to make one complete circle about the sky relative to the stars—averages $27^d7^h43^m11^s.5$, or nearly twenty-seven and one-third days, and is called the **sidereal month.** The length of the sidereal month varies over a range of seven hours because of the perturbations upon the moon's motion about the earth, primarily caused by the sun. The moon also rotates once each twenty-seven and one-third days about an axis inclined at roughly six and one-half degrees to the plane of its orbit about the earth. Because these two motions are synchronized, roughly the same side of the moon is presented to an observer on the earth throughout the month.

This phenomenon is demonstrated in Figure 5–1, which schematically represents one revolution of the moon about the earth. Assume that the little flag represents a flag left on the moon by an astronaut. Also assume that on day zero, the astronaut placed the flag so it would point directly toward the earth. Now identify a distant star, off the diagram to the left, which just happens to be in the same direction as the earth, as viewed from the moon at that instant. The flag points toward both the earth and the star. Almost fourteen days later, the moon's orbital motion has carried it to the other side of its orbit, but the moon's rotation has likewise caused it to go through half a turn. The flag now points away from the star, in other words, toward the earth. On the twenty-first day, the moon has revolved three-quarters of the way around its orbit, but it has also turned three-quarters of the way on its axis; and again the flag points toward the earth. Finally, after twenty-seven and one-third days, the moon has made one full revolution and one full rotation. The flag and the earth are, therefore, once again aligned with the distant star.

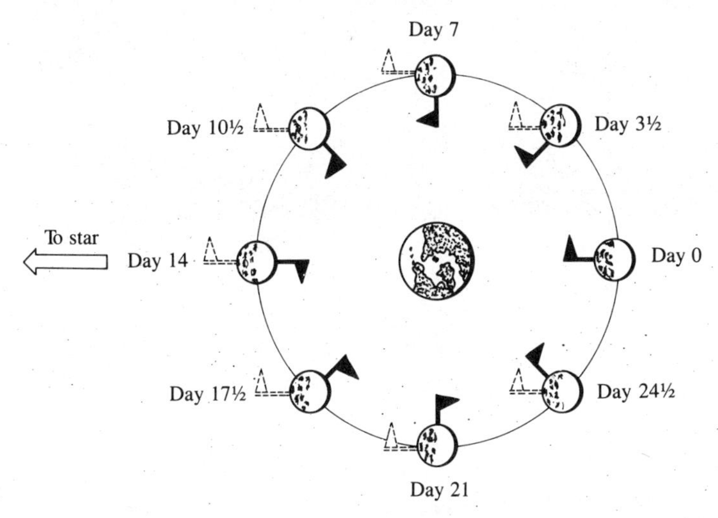

Figure 5–1 Comparison of the moon's periods of rotation and revolution.

Careful observations of the moon show that precisely the same face is not presented to the earth throughout the month. This variation occurs because the six-and-one-half-degree tilt of the moon's axis of rotation allows us to see alternately beyond its north and south poles. Also, since the moon's orbit is eccentric, it speeds up and slows down in its orbital revolution, but its rotation remains constant. The moon therefore appears to rotate slightly too much and then slightly too little for its corresponding orbital motion. We, in effect, see a bit around the moon's equator—first to one side, then the other. We can also see a tiny amount around the moon's edge because of the earth's size and rotation. At moonrise and moonset we are looking at the moon from two well-separated points in space and see a little to the east, then a little to the west of the moon. These effects result in an apparent wobbling of the moon as we see it in the sky, as shown in Figure 5–2. This wobbling is called the moon's *libration*. Because of libration, about 18 percent of the moon's surface alternately appears and disappears throughout the month. The same 41 percent of the lunar surface is always positioned toward us, and there remains 41 percent of the moon's surface that can never be seen from earth. This hidden part first became known by photographs sent back from spacecraft passing behind the moon.

Figure 5–2 Lunar libration.

Since the sun supplies the light by which we see the moon, the appearance of the moon depends upon its location in the sky relative to the sun. Consequently, the phases of the moon are related to the period of time which the moon takes to make one apparent revolution about the sky relative to the sun. This period must be longer than the sidereal period, as shown in Figure 5–3. Assume that when the earth is at position one on its orbit, the moon, the sun, and a distant reference star are all in a line. One sidereal month later the earth will be in a new position in its orbit, while the moon will once again be lined up with the distant star. But the moon would have to travel farther in its orbit about the earth to lie in the same direction as the sun.

The interval of time for the moon to line up with the sun twice in succession is called the **synodic month** and averages about $29^d12^h44^m2^s.9$, or about twenty-nine and one-half days. It varies in length by more than half a day both because of perturbations in the moon's orbit and because of the variations in the daily movement of the earth about the sun due to the eccentricity of the earth's orbit.

One-half of the moon is always illuminated by the sun, but the half of the moon seen from earth usually will not coincide with the illuminated half. The half visible from the earth will contain none, part, or all of the illuminated side, depending upon where the moon lies in its orbit relative to the sun. Thus, as shown in Figure 5–4, when the moon lies in the same direction as the sun, the side of the moon toward the earth is dark. This phase of the moon is said to be **new.**

As the moon moves eastward in the sky relative to the sun, part of its sunlit side will come into view, and the amount of the dark side facing the earth will diminish. For one-quarter of a synodic month, the moon will appear as a **crescent,** the sunlit portion growing larger as the moon moves farther eastward. Under favorable observing conditions, even the dark portion of the moon is faintly visible, shining by **earthshine,** or sunlight reflected from the earth. An astronaut standing on the night side of the moon at this time would see a large, bright, nearly full earth in the sky, shining almost fifty times brighter than a full moon seen from the earth. When the moon reaches a position ninety degrees away from the sun, it appears half dark and half bright. Its phase is then known as first **quarter phase.** The word quarter is used because the moon has made one-quarter of its movement about the sky relative to the sun. It is at quarter phase that the lunar craters are most prominent because of the long shadows they cast at that time.

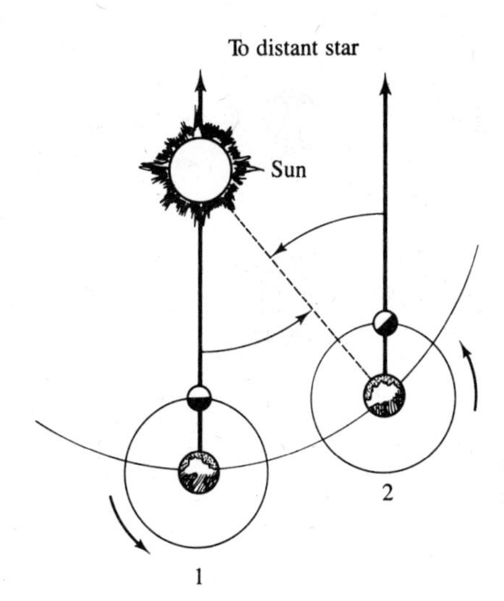

Figure 5–3 Comparison of the synodic and sidereal periods of the moon.

As the moon continues its monthly movement, we see more of its illuminated portion. Its disc appears more than half illuminated, and it enters what are called its **gibbous phases.** Finally, when it lies opposite to the sun, one-half of a synodic month has passed; and the moon is in the **full phase.** The entire disc is now bright.

During the next half-month, the phases reverse—gibbous to last quarter, to crescent, and, finally, to new again. During the first half of the month, when the illuminated portion appears to grow each day, the phases are called **waxing.** In the second half, when the bright part of the moon appears to continually diminish, the phases are called **waning.**

We can also describe the appearance of the moon in terms of the "age of the moon." The number of days since the last new moon occurred is called the *moon's age*. At first quarter, the moon is roughly seven and one-half days old, and about one-fourth of a synodic month has passed. At last quarter, the moon is about twenty-two days old, and three-quarters of a synodic month have passed. Specifying the age of the moon permits a quick determination of the fraction of its disc which will appear bright.

Knowing the phase of the moon, you can always determine where to look in the sky to see it at a given time. In Figure 5–5, the waxing crescent moon is singled out for study. The moon is shown at age three and three-fourths days, that is, halfway between new and first quarter. The moon is therefore about forty-five degrees to the east of the sun. It will appear to follow the sun across the sky by three hours throughout the day. When this particular crescent moon is highest in the sky, the sun will be about three hours past noon; or it will be three o'clock in the afternoon, assuming daylight saving time is not in effect. Approximately three hours after sunset, the moon will set.

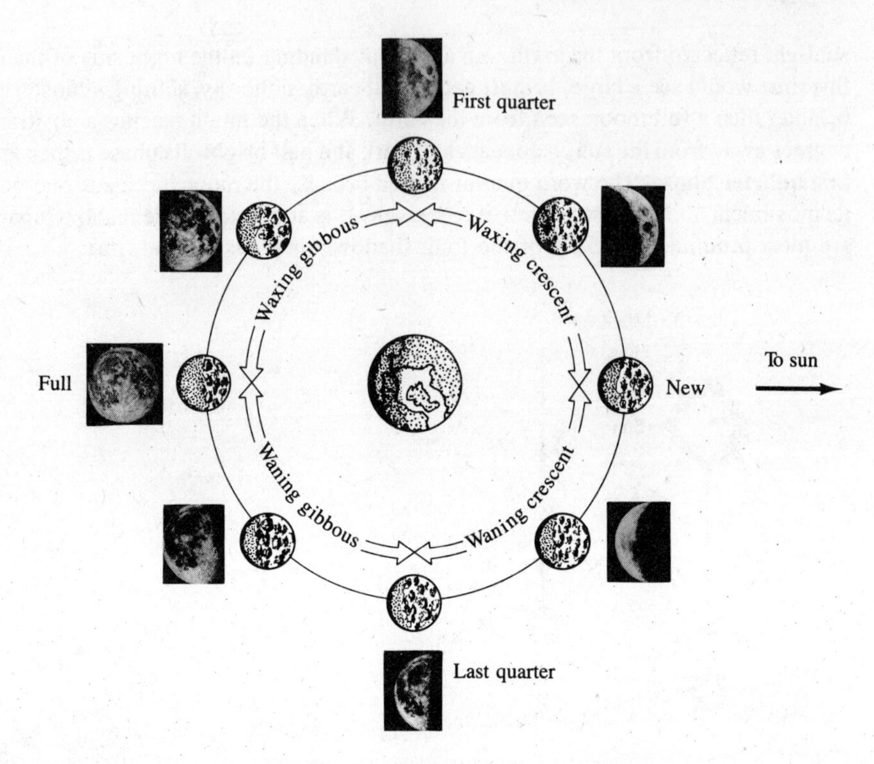

Figure 5–4 Phases of the moon.

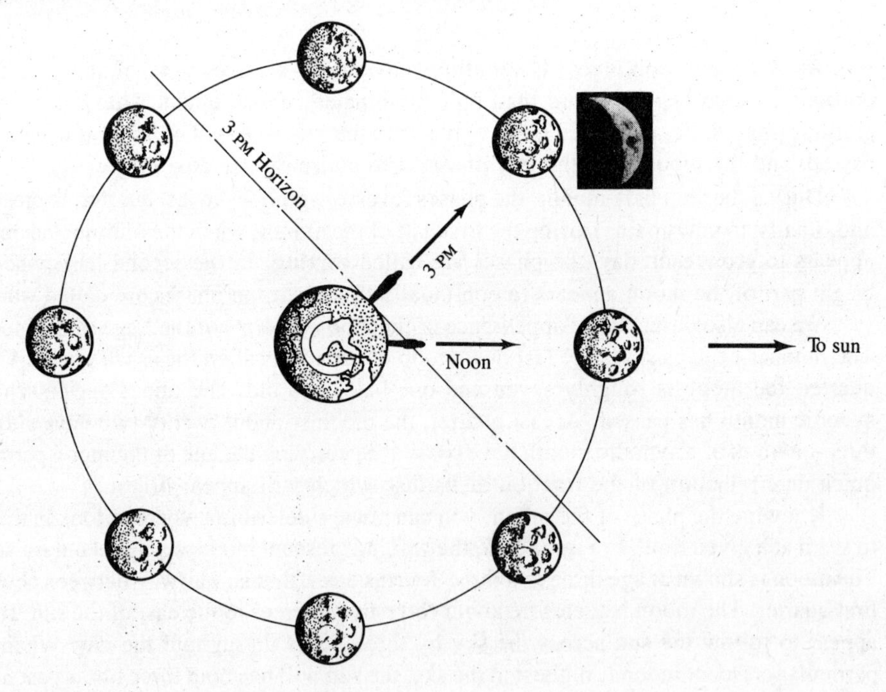

Figure 5–5 Locating the waxing crescent moon in the sky.

The moon transits the observer's meridian about forty-nine minutes later each succeeding day. In one-half of a synodic month it will have fallen behind the sun by a full twelve hours. Consequently, at full phase, it sets twelve hours, or one-half day, after sunset. Thus, the full moon must rise at sunset and transit at midnight. After its orbital motion has carried the moon to the opposite part of the sky relative to the sun, its continued motion will place it to the west of the sun. The moon will now appear to set before the sun. The waning crescent moon, corresponding to the waxing one in Figure 5–5, will therefore set around three hours before the sun and will have reached its highest point in the sky about three hours before noon, or at nine o'clock in the morning. Thus, the waxing crescent moon is best seen in the early evening, while the waning crescent moon is best seen in the early morning. When the moon reaches the new phase again, the synodic cycle starts anew; the moon once again appears to follow the sun for half a month until full moon, after which it spends half a month appearing to precede the sun across the sky.

Eclipses of the Sun and Moon

The moon's orbit about the earth is inclined at slightly more than five degrees to the earth's orbit about the sun. Because of the inclination of the planes of the two orbits, the moon usually will not pass directly in front of the sun at new moon or through the earth's shadow at full moon. Figure 5–6, exaggerated for clarity, shows the moon's shadow passing above the earth at new moon and the moon passing below the earth's shadow at full moon. Roughly six months after the situation shown in the diagram, the earth's orbital motion will have carried it and the moon to the other side of the sun. The plane of the moon's orbit will still be tipped in nearly the same direction with the same inclination. Thus, the shadow of the moon will now pass below the earth at new moon, and the moon will pass above the earth's shadow at full moon. Halfway between these two extreme positions, the shadow of the moon can point directly toward the earth at new phase, and the moon can pass through the earth's shadow at full phase. In the first case, where the moon blocks our view of the sun either partially or totally, we have an eclipse of the sun. In the second case, when the full moon is darkened by the earth's shadow, we have a lunar eclipse. Thus, roughly every six months, it is possible for eclipses to take place, when the sun lies close to the direction of the line of intersection of the moon's orbit with the earth's orbit.

A more convenient way to predict eclipses is shown in Figure 5–7. The moon's orbit is plotted on a star chart along with the ecliptic. Due to the inclination of its orbit, the

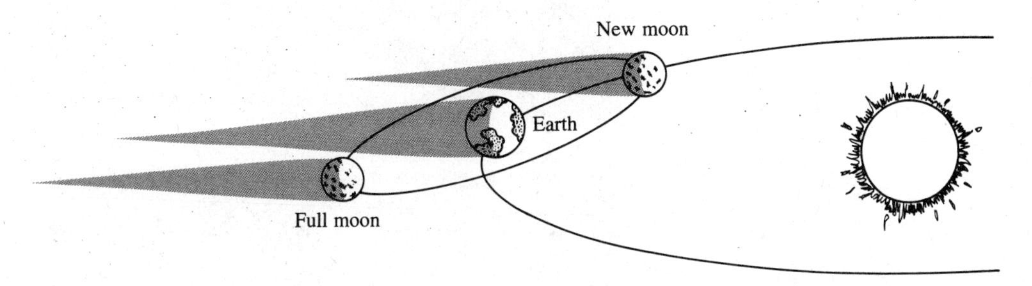

Figure 5–6 An exaggerated side view of the earth-moon system.

moon ranges above and below the ecliptic by five degrees and, of course, crosses the ecliptic twice a month. The points where the moon's path on the sky crosses the sun's apparent annual path are called the **nodes** of the moon's orbit. These two points lie along the line of intersection of the moon's orbit with the earth's orbit. They therefore define the positions in the sky near which the sun must be located for eclipses to occur.

The position of the sun on the ecliptic on April 14, 1972, is indicated in Figure 5–7. On that same date there was a new moon. But, as shown by the chart, the moon's position did not coincide with the sun's position, and there was no eclipse. Three months later, on July 10, the sun was very close to a lunar node. Therefore, the new moon on that date caused an eclipse of the sun. Roughly two weeks later, when the moon was full, the sun had not moved far enough away from the node for the moon to escape passing through the earth's shadow, and a lunar eclipse took place.

Since both the moon and the sun have apparent diameters close to one-half degree and the earth has a diameter of 12,700 kilometers, the sun does not have to be precisely at a lunar node for an eclipse, either solar or lunar, to occur. The most complete eclipses, of course, take place when a new or full moon happens when the sun is exactly at a node. For approximately a month, two weeks on either side of a node, the sun is close enough to the node for an eclipse to take place. One, two, or three eclipses may happen during this time, depending upon the particular circumstances. If there is more than one eclipse, they will occur at close to two-week intervals, alternating solar and lunar eclipses as the moon alternates between new and full phases. Slightly less than six months later, the sun will have moved to the vicinity of the other node, and more eclipses can happen.

The interval between the sun's proximity to one node and then to the other is less than six months because of the **regression of the nodes.** The nodes move slowly westward along the ecliptic, taking nearly nineteen years (actually 18.6 years) for one complete trip along the ecliptic. This rotation of the plane of the moon's orbit, causing its line of intersection with the earth's orbit to change direction continually, is another perturbation in the moon's motion, due mainly to the sun. The net result of the regression, in relation to eclipses, is that the sun's apparent eastward movement, combined with the westward

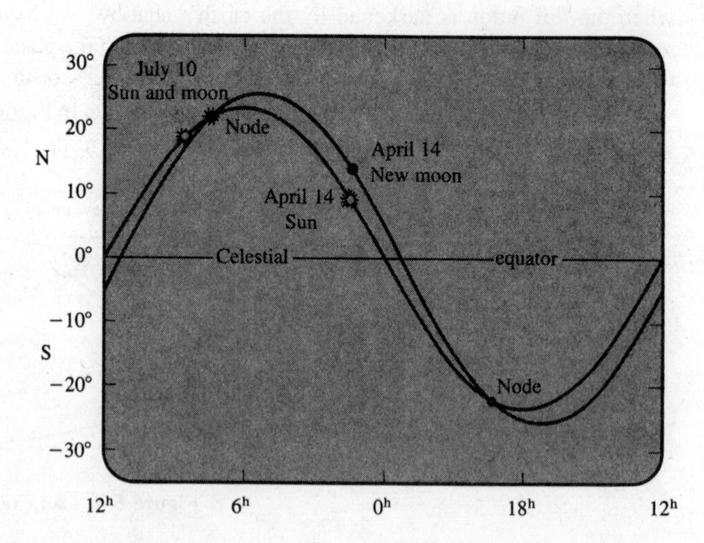

Figure 5–7 A diagram of the moon's orbit, showing the nodes.

movement of the node, allows the sun to travel from a node back to that node approximately three weeks earlier each year. Thus, in 1981, the intervals when eclipses may take place fall in January and July, while in 1984 they fall in November and May.

There are two distinct regions in the shadows cast by the earth and moon. The cones behind the earth and moon, pointing away from the sun and labeled **umbra** in Figure 5–8, are the shadow regions where all direct sunlight is excluded. The **penumbra,** on the other hand, is an inverted cone toward the sun, where sunlight is only partly excluded. Consider only the moon's shadow for the moment. If an observer happens to be located in the small area where the moon's umbra touches the earth, the disc of the sun is completely blocked from view, and he can see a **total solar eclipse.** An observer in the larger penumbral region would see the sun's disc partially covered by the moon, a **partial solar eclipse.**

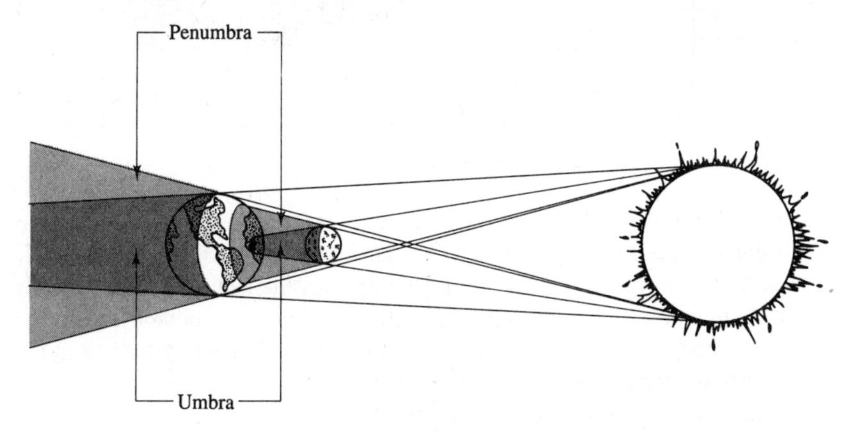

Figure 5–8 The umbra and penumbra for the shadows of the earth and the moon.

As the moon orbits eastward about the earth, its shadow also sweeps eastward through space. At new moon, with the sun near a node, the moon's shadow will sweep across the earth's surface faster than the surface of the earth is carried eastward by the earth's rotation. The shadow of the moon will thus define a path across the earth moving from west to east at speeds of 1700 kilometers per hour near the equator and at even greater speeds at high latitudes. Figure 5–9 shows the eclipse track for the July 10, 1972, solar eclipse. The narrow dark line is the path of the umbra across the earth. Anyone inside this narrow band, where the umbra touched down, could see a total solar eclipse. Because of the great speed of the shadow, this totality is short-lived. At maximum, a total eclipse can last for seven and one-half minutes. Of course, at the other extreme, it can be instantaneous. On July 10, 1972, totality was about 2.7 minutes. Under the most favorable circumstances, for an observer at the earth's equator, the width of the path of totality has a maximum value of only 269 kilometers. This path is surrounded by a band defined by the moon's penumbra which extends to around 3000 kilometers on either side of the path. In these regions, an observer can see a partial eclipse. Less of the sun is covered as you move away from the central regions. For observers fortunate enough to be in the path of totality, the penumbra sweeps over before the umbra arrives. They can see the sun go through progressively greater stages of partial eclipse. These *partial phases,* as they are called, last for one or two hours. After the fleeting moments of totality, there is another one- to two-hour period of decreasing partial phases until the sun is completely visible again.

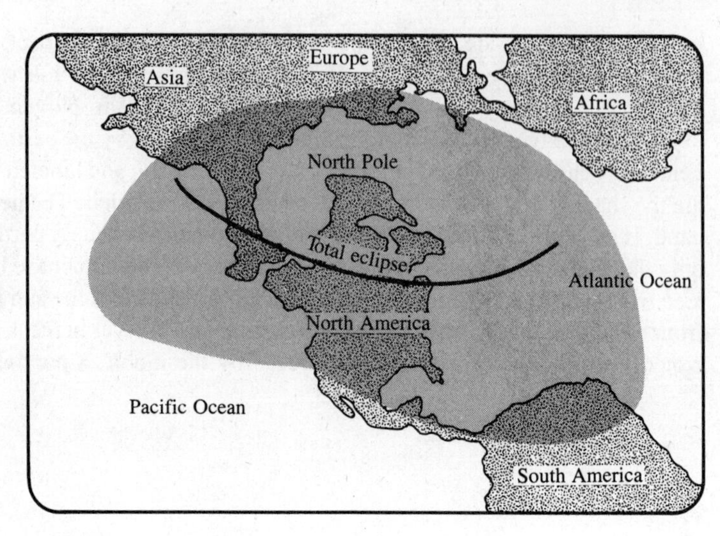

Figure 5–9 The solar eclipse path of July 10, 1972.

The appearance of the sun during an eclipse is shown for a shallow partial, a deep partial, and a total eclipse in Figure 5–10. The partial eclipses are not particularly interesting to see, but a total eclipse is entirely different. As the last of the sun's disc is covered, the sun's tenuous outer atmosphere, called the *corona,* suddenly becomes visible. The delicate pearly white of the corona can only be properly appreciated at the time of an eclipse. No photographs ever seem to do justice to its striking appearance to the eye. Many astronomers have invested great amounts of time and effort traveling to remote and often difficult parts of the world where eclipses chance to occur. The study of the sun during eclipses, as the moon appears to sweep across the lower atmosphere of the sun in the first and last few seconds of totality, has led to much of the fundamental knowledge needed to understand the sun's outer layers. Coronal studies still require observations of total eclipses. The sun is such an interesting object and we have so much yet to learn about it, as we will see later in this book, that astronomers will probably continue chasing eclipses for many years to come.

The moon's orbit about the earth and the earth's orbit about the sun are both eccentric. Therfore, the distance of the moon from the earth continually varies throughout the month; and, because of the varying distance from the sun, the length of the moon's umbra undergoes an annual variation. On the average, the lunar distance and the umbral shadow length are such that the umbra is not long enough to touch the surface of the earth, as shown in Figure 5–11. In other words, because of the moon's varying distance and the earth's varying distance from the sun, the apparent diameters of the moon and sun also vary. The average apparent diameter of the moon is slightly smaller than that of the sun. The moon fails to cover the sun completely, leaving a ring or *annulus* of light, more often than it completely covers it for a total eclipse. This kind of eclipse, shown in Plate 4B, is called an **annular eclipse.** Total solar eclipses are even more infrequent than our previous discussion may have implied. Not only must the sun, moon, and earth be properly aligned, but the particular locations of the moon and, to a lesser extent, the earth in their orbits also help determine whether there will be a total eclipse.

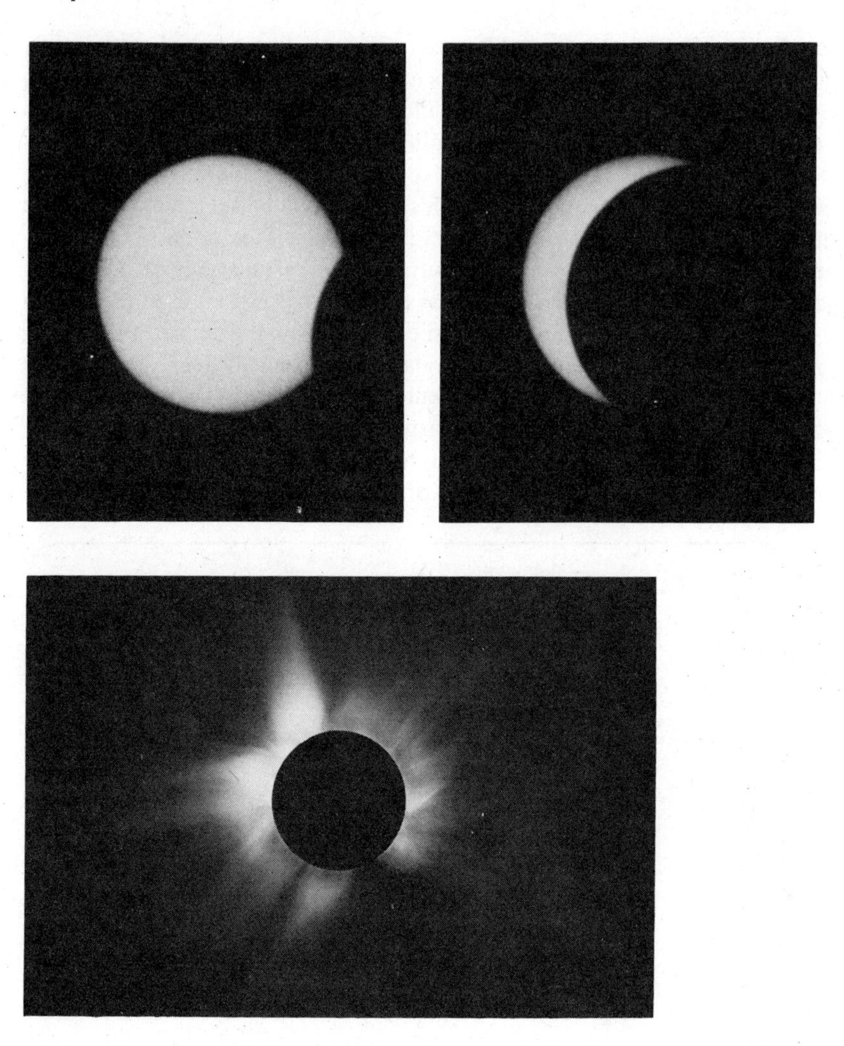

Figure 5–10 Partial and total solar eclipses.

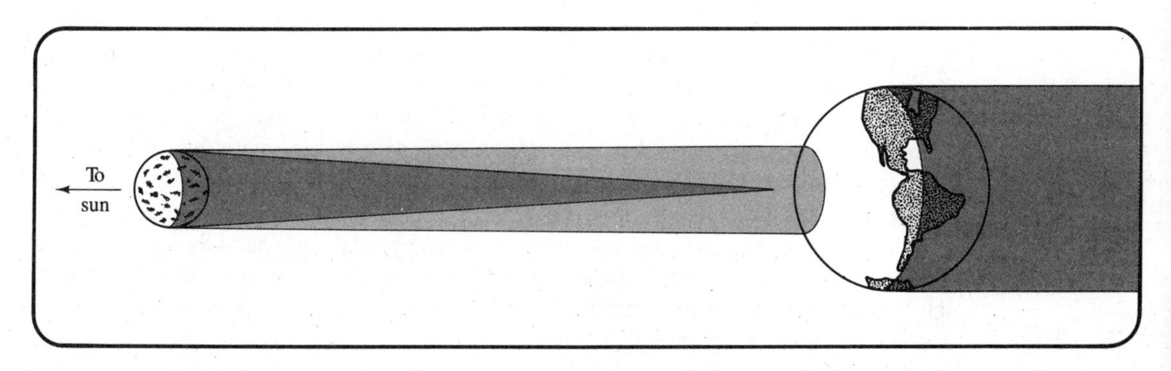

Figure 5–11 Circumstances for an annular solar eclipse.

Figure 5–12 demonstrates three possible cases of lunar eclipses, which occur when the moon enters the earth's shadow. If the moon is in the penumbra of the earth's shadow, there is a penumbral eclipse. In this case, some direct sunlight reaches all of the moon facing the earth, and the entire moon merely diminishes in brightness. Most penumbral eclipses pass by unnoticed and are therefore usually not counted as real lunar eclipses. When the moon is partially immersed in the earth's umbra, part of the moon is removed from direct sunlight, and we have a partial lunar eclipse. Finally, if the moon is completely within the umbra, we have a total lunar eclipse. During an umbral lunar eclipse, the part of the moon that should be completely dark often has a faint red glow to it. The moon then glows because the earth's atmosphere refracts some sunlight into the umbra of the earth. But in passing through the atmosphere, the blue light from the sun is scattered much more than red light and so is removed. The light finally entering the umbra has a definite reddish hue to it. When the condition of the earth's atmosphere is just right, the moon, even at midtotality, remains visible as a blood-red disc in the sky. The color in the umbra can be seen in the photograph of a nearly total eclipse of the moon shown in Plate 4A.

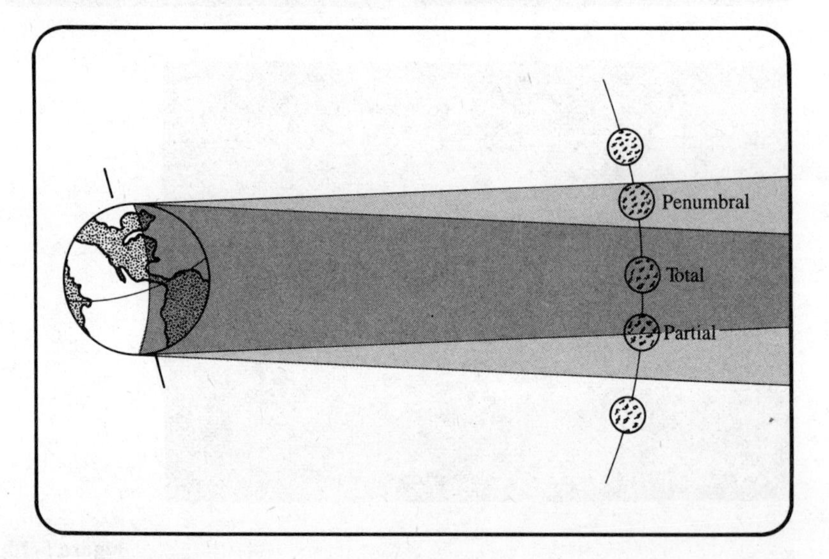

Figure 5–12 Circumstances for lunar eclipses.

Lunar eclipses are seen more frequently than solar eclipses. Anyone on the night side of the earth with good observing conditions can see a lunar eclipse as the full moon enters the earth's shadow. To see a solar eclipse, however, an observer must be close to the path of the moon's umbra as it makes its brief intercept or near-miss with the earth. On the average, from a fixed spot on the earth, a total solar eclipse can be seen only once every 400 years, but a total lunar eclipse can be seen every few years. Notwithstanding their limited visibility, solar eclipses—total, annular, and partial—are actually more frequent than umbral lunar eclipses. Figure 5–13 shows the extreme circumstances at which a solar eclipse and a lunar eclipse can occur. The range in position of the moon (on the left of the diagram) over which partial solar eclipses can occur obviously is larger than the range (on the right) where umbral lunar eclipses can occur. The range for solar eclipses is such that there must be at least one solar eclipse each time the sun draws near a lunar node, and there

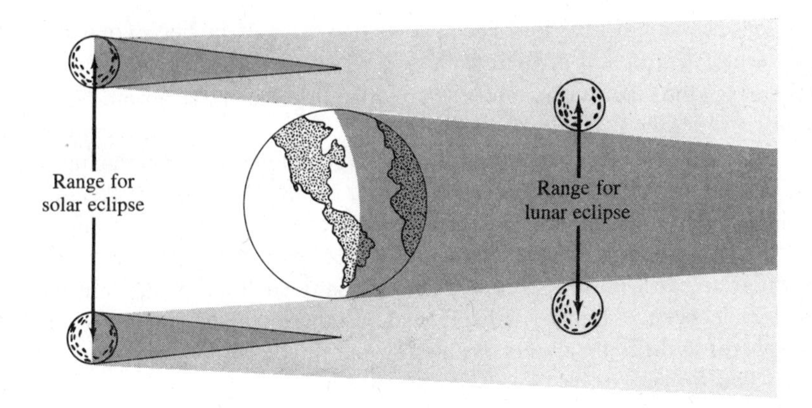

Figure 5–13 Extreme conditions for eclipses.

may even be two. The range for lunar eclipses is small enough that the sun can slip by the node between full moons without a lunar eclipse happening. Thus, at least two solar eclipses must occur every year, but no lunar eclipses need occur in any given year.

The Lunar Features

Most children have been intrigued by what, with a little imagination, appears to be the face of the old man in the moon. When Galileo first viewed the moon with his telescope, he saw in more detail than ever before the dark regions of the moon's surface that make up the face. Since these dark areas, easily seen in Figure 5–14, appear relatively smooth, Galileo thought they might be seas of water; he called them **maria** (the singular is *mare*),

Figure 5–14 The old man in the moon.

the Latin word for seas. Although it has long been known that there are no bodies of water on the moon, the term *maria* is still applied to these dark areas.

The brighter areas of the lunar surface are much rougher than the maria and are called the **lunar highlands.** As shown in Figure 5–15, the highlands are saturated with craters. Over 30,000 of these round depressions can be seen with a good telescope, ranging up to 240 kilometers in diameter. Vast numbers of much smaller craters have been discovered from spacecraft photographs of the moon. Most craters were formed by explosions caused by the impacting of celestial bodies upon the moon billions of years ago, in the early ages of the solar system. The earth must also have suffered many such collisions, but the craters so formed have been erased by erosion and weathering. The moon has no appreciable atmosphere; so the lunar craters, without severe weathering, remain more or less as they were when formed.

Figure 5–15 Lunar highlands.

Bodies randomly impacting on the moon would have formed craters over the entire surface. The fact that so few craters now appear in the maria leads to an important conclusion regarding these smooth areas: after most cratering events were over, large lava flows formed the maria, covering the former impact craters with molten rock. As the sea of lava cooled and solidified, it took on the appearance of a solid lake, as shown in Figure 5–16. The relatively few craters on the mare surfaces must have been formed after the lava flows had cooled. Some maria have more craters than others; the more heavily cratered a mare region, the earlier its lava flows must have solidified, and hence the more time there was for later impacts to cause craters.

Figure 5–16 A lunar mare with evidence of lava flow.

Many of the maria have fairly circular shapes, with mountain ranges on their borders. They look like oversized craters, or basins, formed by very massive objects striking the moon. Such a severe impact should have caused deep fractures in the underlying rocks. When radioactivity deep below the surface of the moon released enough heat to melt the lower rock, the lava could form and well up through the fractures to flood the basins.

Evidence for more than one lava flow in a mare is found in the so-called **rills** on the surfaces of some maria. The rills look like valleys cutting into the maria. Figure 5–17 is a photograph of one of the walls of Hadley Rill, taken during the Apollo moon program. Notice the layering indicated by the arrows. This layering is most easily interpreted as successive lava flows, meaning that more than one stage of molten rock formed this particular mare. Another example showing that lava spread in stages is shown in Figure 5–18, where one flow has partially covered another, older flow. We can see just how far the second flow spread before cooling and solidifying.

Prominent volcanic craters, similar to those found on the earth, are not observed on the lunar surface. Thus, it seems unlikely that the immense lava flows required for the maria came from volcanic activity similar to that of the earth. There are some features called *domes* which look like rounded hills, some even having what appears to be a tiny crater at the top. Solidified lava flows coming from these craters and running down the sides of the domes are not seen, however, so it is difficult to assign the source of the maria lava to them. There are also features called *wrinkled ridges,* shown in Figure 5–19, running across the relatively smooth maria surfaces; these might be places where molten rock welled up through cracks in the lunar crust.

Figure 5–17 Layering in Hadley Rill.

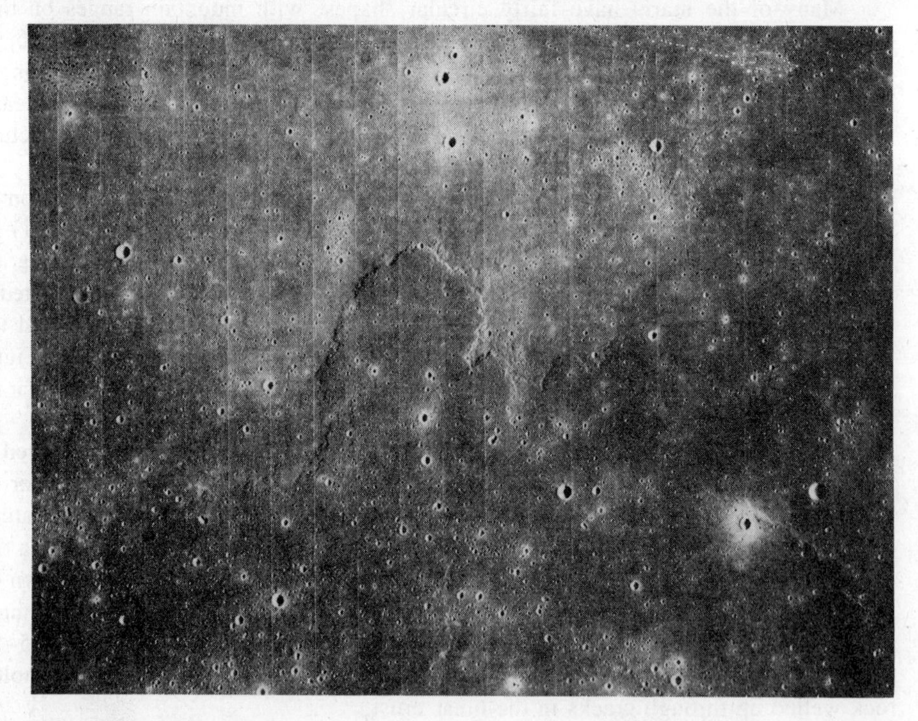

Figure 5–18 Overlapping lunar lava flows.

Plate 3A Laser light passing through a diffraction grating.

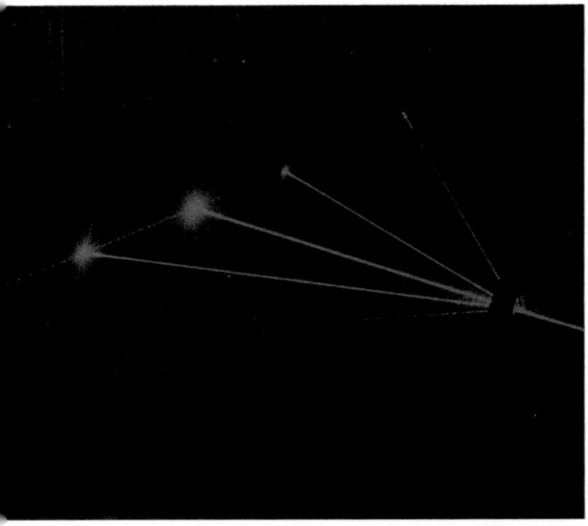

Plate 3B White light passing through a diffraction grating.

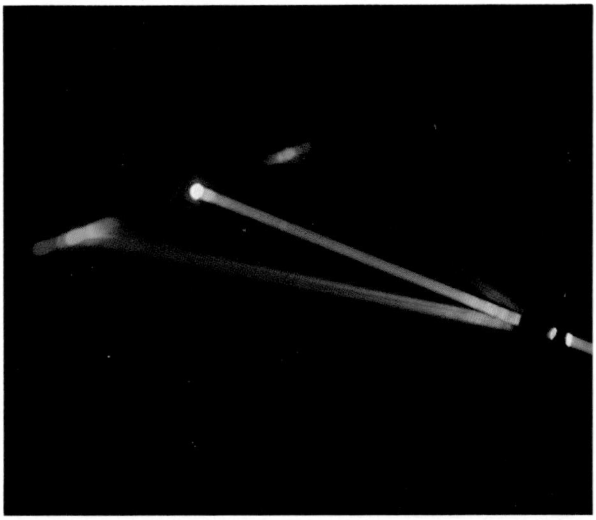

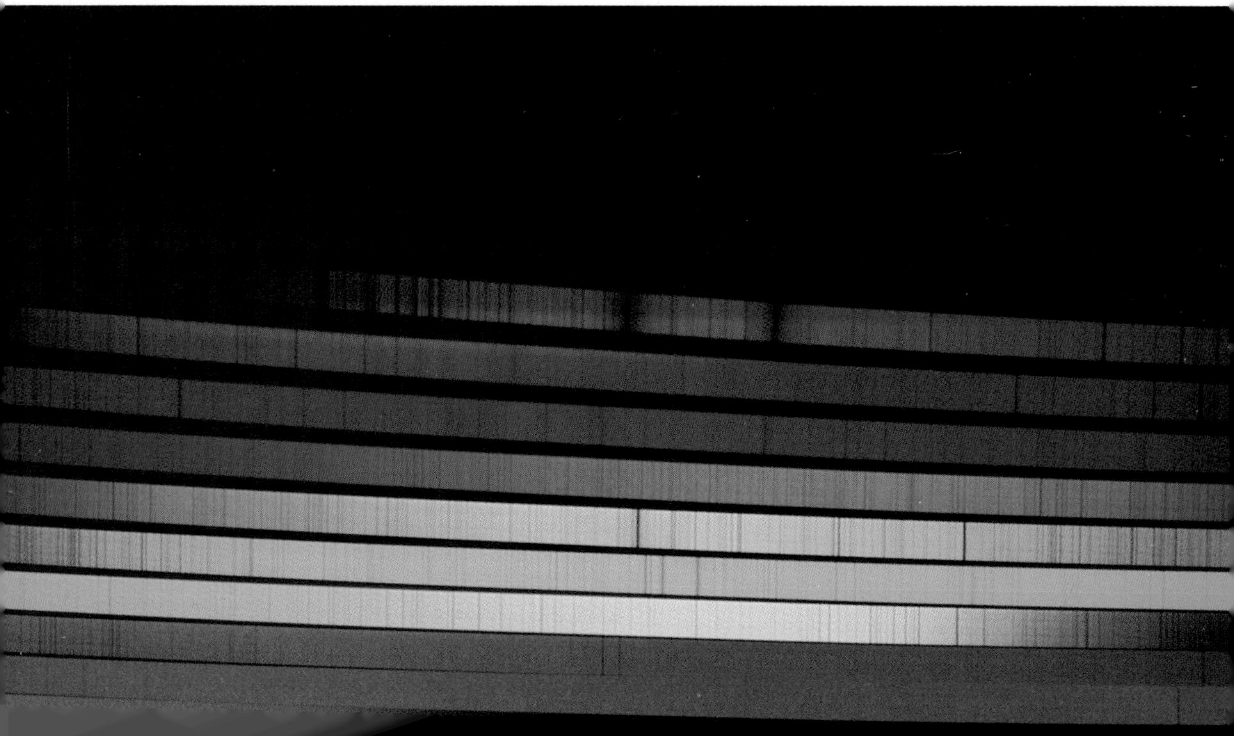

Plate 4A Partial eclipse of the moon.

Plate 4B Annular eclipse of the sun.

Lunar Earthside

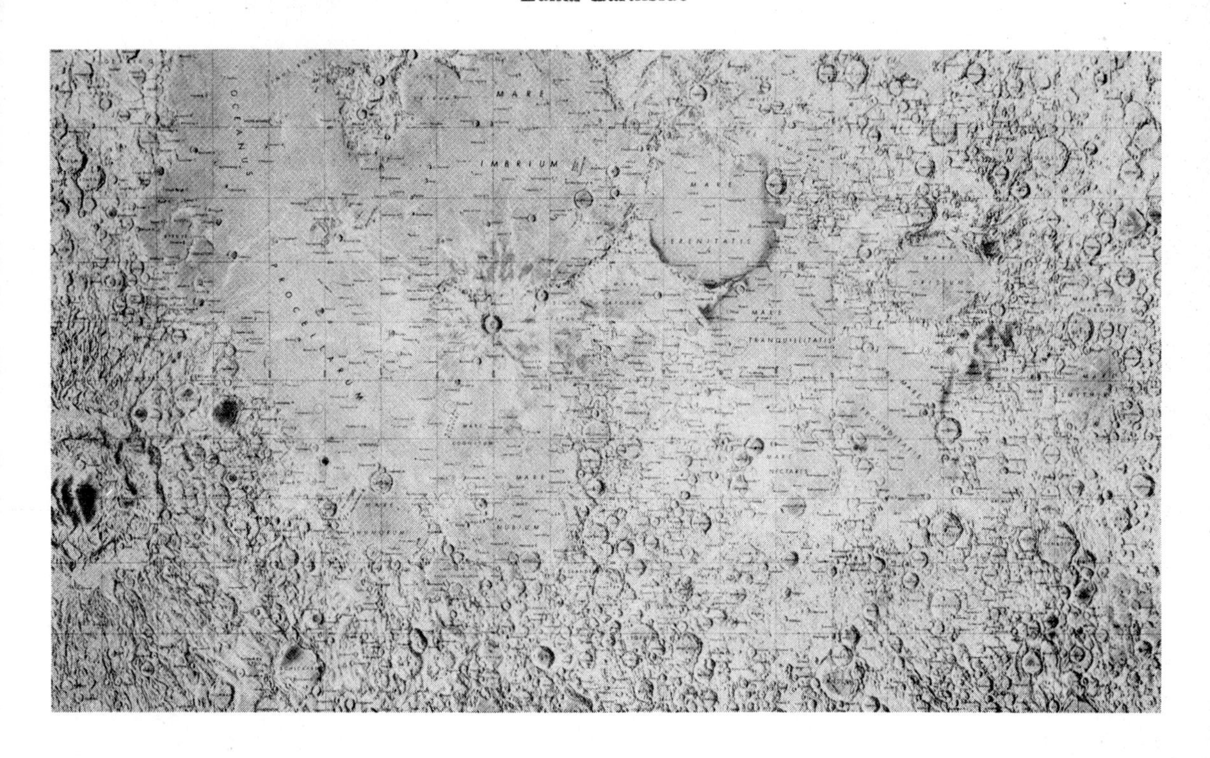

Lunar Farside

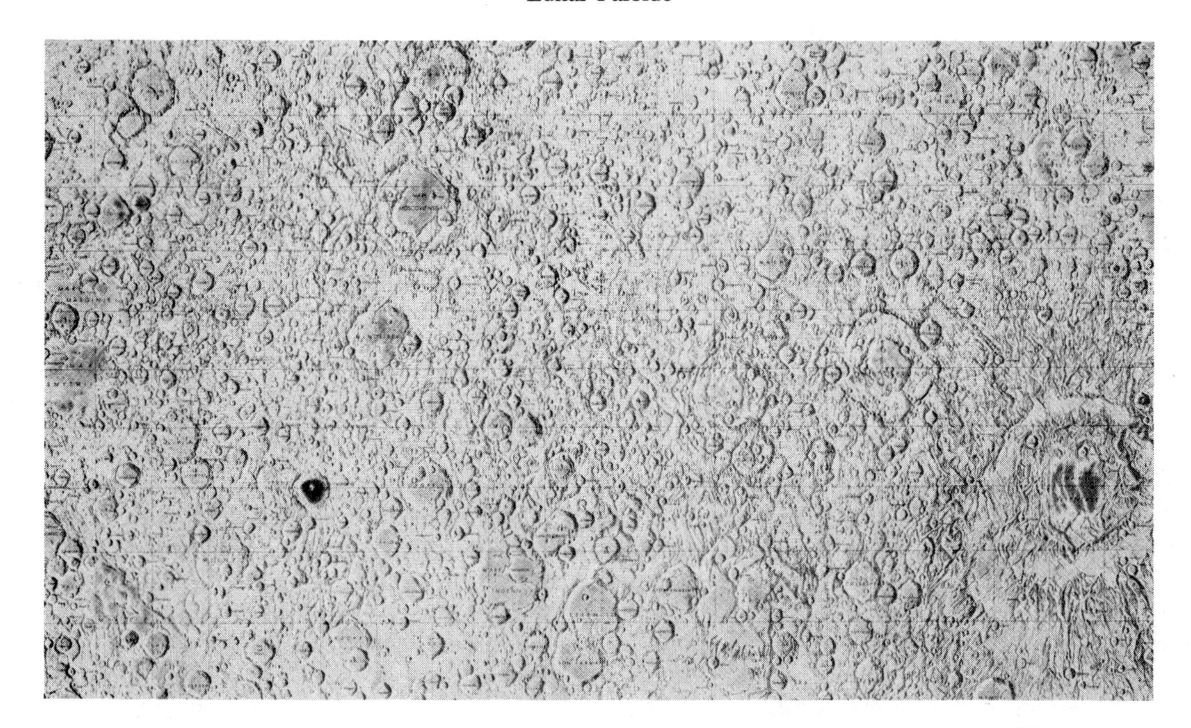

Figure 5–19 Wrinkled ridges on the moon.

Examination of lunar rocks and soils returned to the earth from six Apollo landings and three Soviet automated spacecraft have filled in two important gaps in our previous knowledge about the moon; we now know the composition of mare and highland rocks, and we can now date the actual time when each mare solidified. Compared to earth rocks, the lunar rocks have four important differences:

1. All returned lunar rocks were formed by the cooling of molten materials. No rocks found on the moon were similar to shale or limestone, which are formed on earth by deposits under water.
2. Earth rocks usually contain at least one or two percent water, but lunar rocks do not have even a trace of embedded water. Just why the moon formed from such dry material is still not understood.
3. Earth rocks have been exposed to free oxygen, while moon rocks have not. As a result, tiny crystals of pure iron occur in the moon's molten rocks (as shown in Figure 5–20), while any such crystals that may have once existed on the earth have long since rusted.
4. Compared to earth rocks, lunar rocks generally contain elements which have high-temperature boiling points, such as calcium, aluminum, and titanium. Lunar rocks are depleted in elements which have low boiling points, such as sodium and potassium. This implies that either the matter which formed the moon was considerably hotter than the matter which formed the earth or that the moon's surface was subjected to quite high temperatures after the moon was formed. In either case, sodium and potassium would be lacking, either by not condensing in the first place or by being vaporized later.

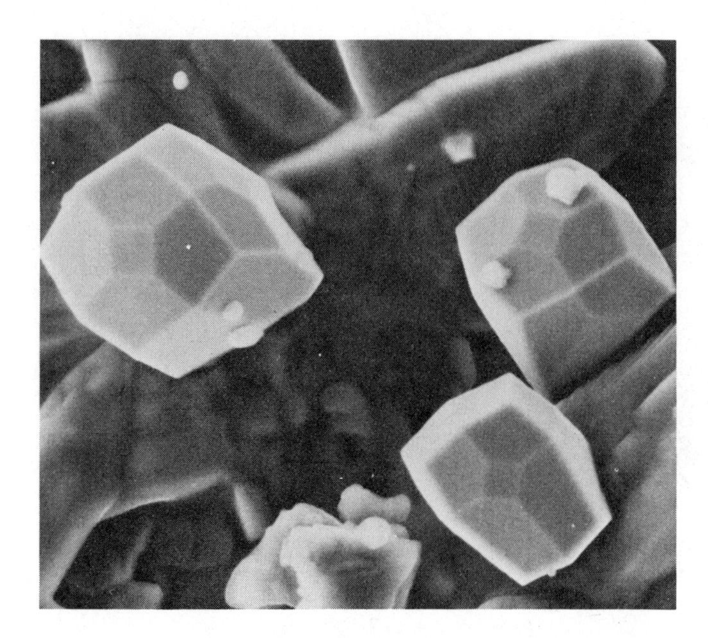

Figure 5–20 Microscopic iron crystals in a lunar rock sample.

If we compare mare and highland rocks, we find that the mare rocks have more iron and magnesium and less calcium and aluminum, giving them their darker color. The mare rocks are also denser. This increased density, incidentally, helped cause spacecraft orbiting the moon to experience an additional acceleration when passing over the maria, due to the extra mass in these rocks.

The ages of lunar rocks confirm the conclusion that the more heavily cratered the surface, the older it is. These ages measure the length of time since the rocks were last molten. Mare rocks vary between 3.1 and 3.8 billion years old, making the oldest of these rocks about the same age as the oldest earth rocks. Rocks from the highlands are mostly 4 billion years old, although a few fragments returned from Apollo 17 apparently are 4.6 billion years old, close to the generally accepted age of the solar system itself. Since so many highland rocks have ages of 4 billion years, it appears that cratering, and melting produced by the heat of the impacts, was frequent 4 billion years ago. And yet maria 3.8 billion years old have far fewer craters than the highlands, so the rate of impacts on the moon dropped rapidly around 4 billion years ago, to be followed by several hundred million years of lava flows which formed the maria in the great basins. We can reconstruct what the moon probably looked like in past epochs, as shown in Figure 5–21. The left painting shows the heavily cratered moon 3.9 billion years ago, with several large basins. In the right picture, by 3.1 billion years ago the basins have been filled in by lava flows. The bottom painting shows the present appearance of the moon, with some relatively recent craters formed on the maria.

Seismographs left on the moon during the Apollo program were used to measure the thickness of the lunar crust. Spent rockets and lunar modules were deliberately crashed onto the moon and the travel times of the resulting disturbances through the moon were measured. The results, analyzed by the same methods applied to seismic studies of the earth, indicate that the moon has a relatively thick crust for its size. The crust is about 60 kilometers deep on the side of the moon facing the earth, and over 100 kilometers thick on

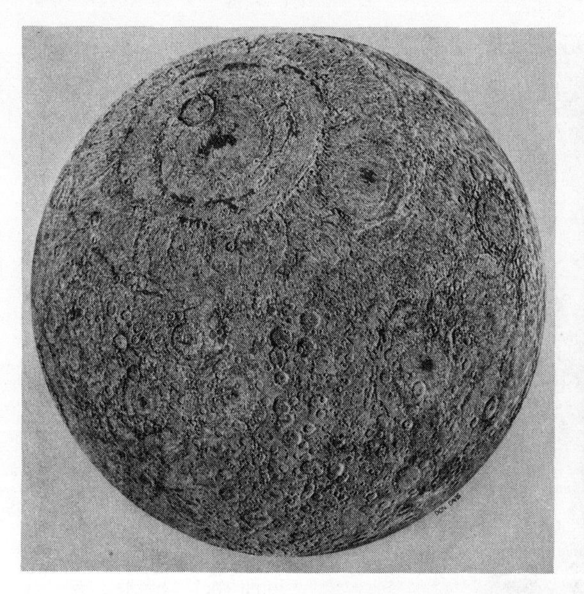

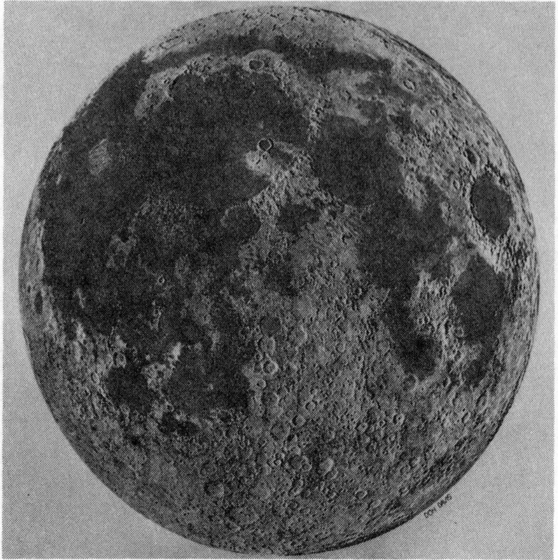

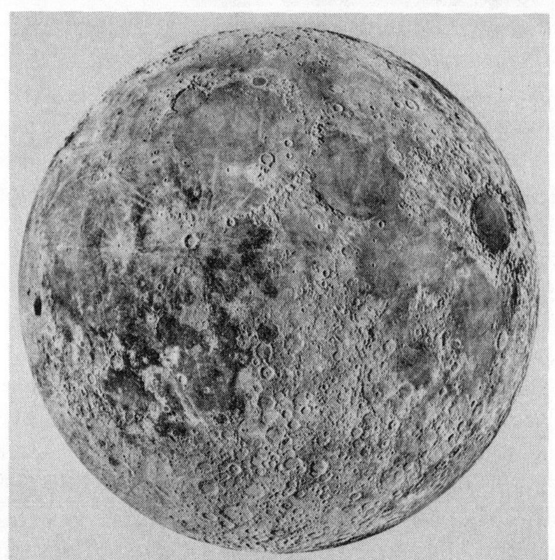

Figure 5–21 A reconstruction of the lunar appearance 3.9 and 3.1 billion years ago contrasted with its present appearance.

the far side. The maria occur almost entirely on the near side of the moon, which somehow may result from the crust being relatively thin there. But we still await an understanding of why the maria formed more frequently where the crust is thinner and why the crust is so different on the near and far sides.

Below the crust is a denser mantle of rock, extending downward more than 800 kilometers. About 3000 moonquakes are detected each year, most coming from the bottom regions of the mantle, 600 to 800 kilometers below the surface. But, since an average moonquake is very weak (only one millionth as strong as an average one on earth), our view of the moon as a nearly dead body is still quite accurate. Moonquakes often repeat when the moon passes through the same point in its orbit, suggesting that tidal strains caused by the earth's gravitational pull actually trigger the moonquakes.

The nature of the moon below the mantle remains almost totally unknown. There is, however, one fact about the core regions which can be deduced by noting that the average density of the moon is only slightly higher than that of the rocks at its surface. The surface rocks have a density quite like that of the surface rocks on the earth, around three grams per cubic centimeter. But the average density of the earth is five and one-half grams per cubic centimeter, thus indicating that it must have a very dense core to raise the average density so high. The moon, obviously, does not have a large, dense core, although the possibility remains that a small iron-nickel core may exist at the center of the moon. But in any case, compared to the earth, the interior of the moon has much less iron.

Theories of Lunar Formation

We do not yet have the full story of the present state of the moon, but that does not mean that we cannot think about its origin. The Apollo program has given us information by which to test the theories already proposed for the formation of the moon. One theory, called the *fission theory,* proposes that in the early stages of the earth, as the denser materials settled inward to form the earth's core, the earth had to begin spinning more rapidly. Eventually the earth reached a rotation rate that caused a part of it to be thrown off. This piece cooled and became the moon. Unfortunately, the difference in composition of the lunar and earth rocks (such as the lack of water in lunar rocks), coupled with the evidence that moon material experienced a higher temperature than the matter which formed the earth, argues against the idea that the moon was once part of the earth. Also, the earth would have to have been rotating much too fast to have slowed to its present rotation period. The fission theory is probably the least accepted explanation of lunar origin.

Another theory, which can be called the *accretion theory,* proposes that the earth formed from a large cloud of material. As this cloud condensed, some material was left over in a large ring, which in turn accreted to form the moon. The differing chemical composition of lunar and earth rocks again prevents us from accepting this model of the moon's origin.

The so-called *capture theory* presents a nice way to circumvent the composition problems. Let the moon form in some other part of the solar system where there could be chemical differences. Then, if its orbit brought it close enough to the earth, the earth and sun both pulling together could cause it to go into orbit around the earth. But the orbit along which the moon would have to move for such a capture to occur is a very unusual one, and thus the odds that the moon could have been captured through a close pass by the earth are extremely small.

All three of these theories face one common problem—to account properly for the changing size of the lunar orbit. The earth's ocean tides are caused mostly by the gravitational attraction of the moon, which is twice as effective as the sun in raising tides. The two high tides therefore tend to align with the direction of the moon. The earth rotates beneath the moon and hence beneath the tides. The friction between the rapidly spinning earth and the tides acts as a brake which slows down the earth's rotation. Accurate measurements show that the length of the day is increasing 0.0016 seconds each century. The tides are slightly advanced by this drag, and they in turn pull unequally on the moon, causing it to move into a larger orbit.

At the present time, the radius of the moon's orbit is increasing by about three centimeters per year—an insignificant effect for one year, but not for millions of years. The moon must have been about four percent closer 380 million years ago than it is at present, and the earth's rotation rate was greater, the day being only twenty-two hours long according to a present-day clock. It would appear that even further in the past, say two billion years ago, the moon's orbit must have been so small that the moon would have raised enormous tides. Also, the earth would have been spinning very rapidly. The energy caused by tidal friction would have heated the waters of the ocean, maybe even boiling the oceans away. Certainly great disruptions in both the lunar and terrestrial surfaces would have occurred. However, no strong evidence of this kind of great disruption has yet been found. So, no matter how the moon formed, we must still explain where it was a mere two billion years ago.

KEY TERMS

sidereal month
synodic month
earthshine
new moon
crescent phase
quarter phase
gibbous phase
full moon
waxing
waning
node
regression of the nodes
umbra
penumbra
total eclipse
partial eclipse
annual eclipse
maria
lunar highlands
lunar rills

Review Questions

1. *True or false:* The moon rotates on its axis.
2. As seen from the moon, would the earth go through phases?
3. What would be the phase of the moon if it is on your meridian at 5:00 P.M.?
4. At what time will the moon be highest in the sky when its phase is full?
5. Can a moon in narrow crescent phase be seen from temperate latitudes at midnight?
6. Suppose one evening as the sky gets dark, you notice the moon low in the east, looking like a crescent. How could you decide if the moon is in crescent phase or if it is being partially eclipsed?
7. What is a node of the moon's orbit?
8. Why do eclipses occur on the average three weeks earlier on successive years?

9. What do the words *umbra* and *penumbra* mean?
10. *True or false:* The two nodes of the moon's orbit are exactly opposite to each other in the sky.
11. Solar and lunar eclipses sometimes occur about two weeks apart. What causes this two-week interval?
12. Groups of eclipses occur about six months apart. What causes this six-month interval?
13. When the moon is exactly in line between the earth and sun, the resulting eclipses can be either annular or total. What primarily determines which kind of eclipse it is?
14. Why doesn't the moon's umbra contain red light like the earth's umbra does?
15. Why can more lunar eclipses than solar eclipses be seen from any one place on earth?
16. Why are parts of the moon heavily cratered and other parts relatively free of craters?
17. What are lunar rills?
18. What evidence is there that the lava flows on the moon occurred in several stages?
19. The earth appears to have a dense core of iron and nickel. How do we know the moon doesn't also have an appreciable dense core?
20. What have analyses of lunar rocks told us about differences in past temperatures on the earth and moon?
21. Which should have had more impact craters form on its surface at one time or another, the earth or the moon?
22. What are the three theories of the origin of the moon?

Discussion Questions

1. Why is the synodic month longer than the sidereal month?
2. If the moon is on your meridian at 5:00 P.M. on one day, about what time would it be on your meridian the following day?
3. Why doesn't new moon always occur on the first day of each calendar month?
4. Apollo astronauts landing near the center of the side of the moon toward earth timed their voyage to arrive soon after first quarter phase. Why?
5. What requirements are needed to get the longest possible total eclipse of the sun?
6. One of the longest solar eclipses of the century occurred on June 30, 1973, in North Africa. On December 24, 1973, the moon was again exactly between the sun and earth. What kind of eclipse occurred then?
7. How can we deduce that most of the impacts making craters on the moon occurred during the first billion years of the moon's existence?
8. If the heat to melt a mare came from a large impact on the lunar surface, you would expect a single flow of lava after the impact. Does this agree with the evidence of the nature of the lava flow?
9. How have results of the Apollo program affected the arguments over the origin of the moon?
10. What roles have tides played in the history of the earth-moon system?

6

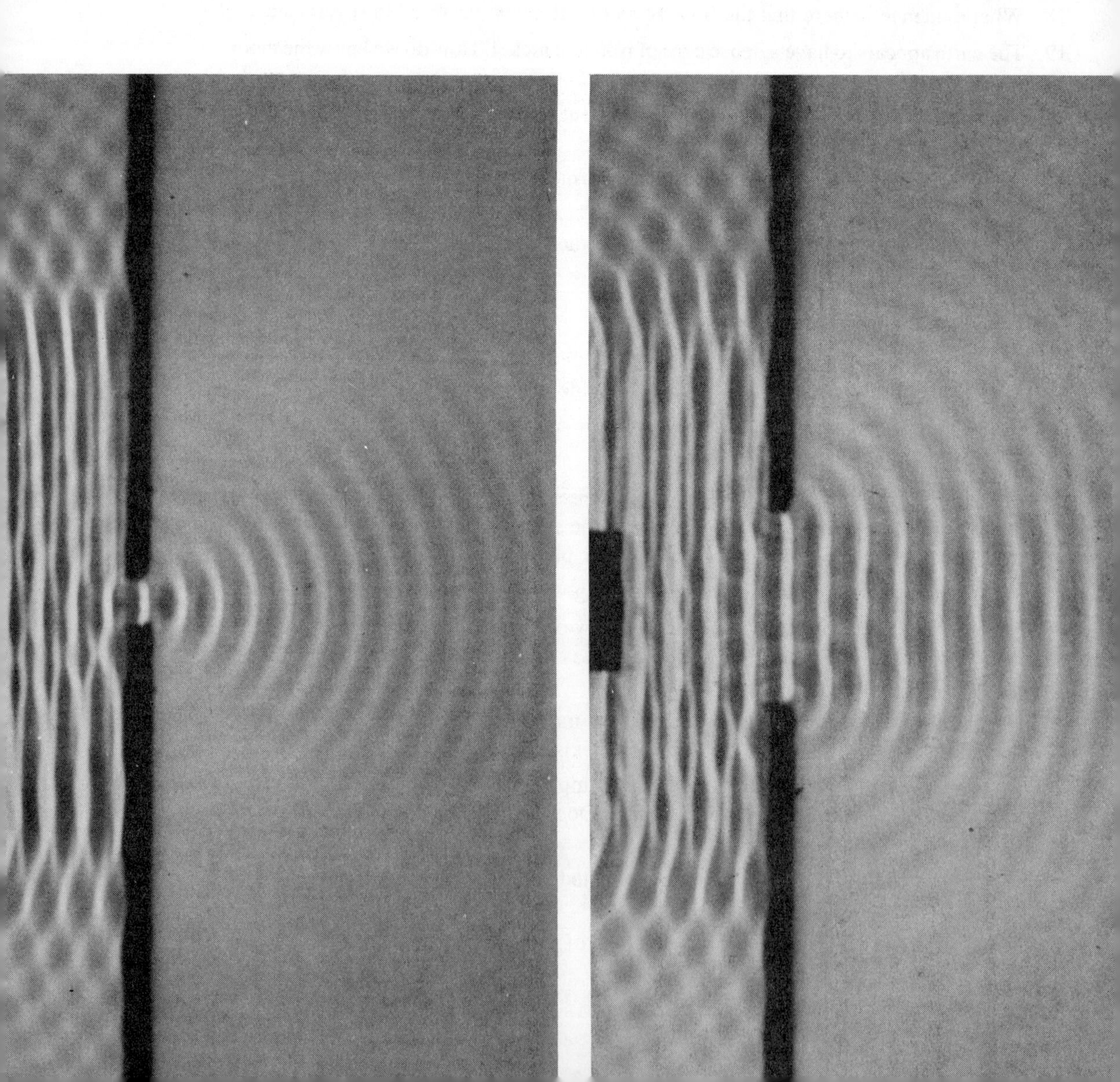

The Nature of Light

Our first perception of the universe was through the sense of sight. People looked up and saw the sun, moon, planets, and stars. Some of these objects, the sun and stars, are luminous—they emit light. Others, the moon and planets, are nonluminous. They would not be visible if they were not in the presence of a luminous object, the sun.

Today it may seem simple or even obvious to think of a luminous body as being the source of energy which, upon reaching the eye either directly or by reflection from another body, causes the sensation of sight. However, there were some scholars of ancient times who believed that the eye itself is the source of energy which helps to illuminate observed objects.

According to Plato, a stream of divine fire emanates from the eye. This stream combines with a similar stream from the sun; when the combined streams reflect from the various objects in the universe or on the earth, they enter the eye and cause sight. Plato's theory leads to some interesting effects after the sun sets. In his own words, "When the external and kindred fire [that is, the sun] passes away in the night, then the stream of vision is cut off . . . the eye no longer sees, and we go to sleep, for when the eyelids are closed, which the gods invented to preserve the sight, they keep in the eternal fire."

In fairness to Plato, he did have some idea that bodies other than the sun have the ability to cause the sensation of sight. He must have seen a lamp or open fire at night some time in his life. Where he went wrong was in his notion that the eye itself emits energy which in part gives form to light.

Matter can emit light even if there are no eyes to see it. In fact, some radiation emitted by matter, such as radio waves, cannot even be detected by the human eye. The invisible radiations have the same general character as visible radiation but differ in the response that they excite in our eyes.

But what is the process by which matter emits light? How are matter and radiation connected? Christian Huygens, a seventeenth century Dutch scientist, con-

Diffraction of waves through a small and a large opening

cluded that light is somehow associated with the motions of tiny particles which comprise material objects. He wrote, "One sees that here upon earth [light] is chiefly engendered by fire and flame which contain without doubt, bodies that are in rapid motion . . . one sees that when light is collected as by concave mirrors, it has the property of burning as a fire does, that is to say, it disunites the particles of bodies."

Huygen's intuition that light results from the agitation of particles in the emitting material had him on the right track. However, it was difficult to understand just how the agitation of particles was related to light until the work of later researchers, particularly Michael Faraday and James Clerk Maxwell, in nineteenth century Britain demonstrated that light is related to the electrical properties of matter. The emission of light was shown to result from the acceleration of electrically charged particles in which the particles lose energy. The absorption of light, on the other hand, takes place with the acceleration of electrically charged particles in which the particles gain energy. Light itself was supposed to radiate from an emitting particle in the form of a continuous wave, similar to the propagation of waves on the surface of water.

By the turn of the century, however, it became obvious that this theory of light and its interaction with matter was not totally correct. The color spectrum of light emitted by heated objects in laboratory experiments differed strongly from the color spectrum predicted by the theory. The theory itself was simple enough, proposing that the light emitted by an object was formed by the continuous accelerations and changes in energy of agitated charged particles oscillating within the matter. Scientists were in a quandary over the disagreement between theory and observation.

In 1901 Max Planck, a German scientist, put forth a revolutionary new idea regarding the interaction of matter and light. He suggested that oscillating charged particles do not continuously lose or gain energy, but rather they lose or gain energy in discrete jumps. Thus, in the emission process, rather than a continuous light wave being generated, the radiation is emitted in discrete amounts of energy, or photons. Each photon corresponds to a discrete change in the energy of an oscillating charge as it jumps from one energy value to a lower one.

Planck's new concept of the interaction of light and matter reintroduced an old notion of Newton that light has a particle nature. The realization that the energy of particles which emit and absorb light changes in discrete jumps leads to the concept that the basic building blocks of matter, atoms, have structures that involve discrete sets of energy levels.

These concepts have proven particularly important in astronomy, because it is through their application to the radiation received from outer space that we have developed our knowledge of the matter of the universe, in all of its forms.

Except for the occasional meteorite which manages to penetrate the earth's atmosphere, the samples of moon rock returned by the astronauts, and the analysis of surface material on other planets made by space probes, our knowledge of the universe is based upon radiation which we receive from the objects in it. The study of astronomy, therefore, requires us to have a knowledge of the character of light to interpret properly the radiations which supply our basic information.

The Wave Theory of Light

One way of describing light, which explains many of the observed phenomena associated with this radiation, is to assign it a wavelike character. The wave theory of light can be described by an analogy—surface waves on water. When a stone is dropped into the quiet water of a pond, waves travel outward from its point of entry in ever-increasing circles. The waves, although diminished in strength, eventually strike the shore; and the original disturbance, which may have occurred many meters away, influences the shoreline. In a similar way, a light source can be pictured as a source of light waves which travel across the intervening space and, upon entering an observer's eye, cause a sensation indicating a distance source of energy.

One property of light which can best be explained by the wave theory is the phenomenon of diffraction. **Diffraction** occurs whenever a wave strikes an obstacle which partly blocks its path. A wave disturbance is always found where the shadow of the obstacle would be. The wave appears to bend around the obstacle and, more properly, is said to be diffracted. Figure 6–1 shows a comparison of a water wave and a light wave. Notice on the top how the water waves, after passing through the entrance to the harbor, spread with diminished strength into the sheltered portions of the harbor. On the bottom, a laser beam, consisting of light of only one wavelength, was focused on a narrow slit in an opaque card—the analogue of the harbor entrance. Instead of only finding light on the screen directly behind the slit, the light has been spread by diffraction into a much larger region. This diffraction phenomenon in light is taken as evidence for the wavelike behavior of light.

If two or more slits are used to impede the progress of a wave, we obtain an interesting result which can be used to measure a property of the waves. In Figure 6–2, we see two photographs of a ripple tank. Waves are moving upward across the surface of the water. On striking the obstacle with two narrow openings, the waves move into the sheltered part of the water. But the waves spreading from the two slits **interfere** with each other and show a pattern of varying strength.

When the two interfering waves add in phase, that is, oscillate in unison, the water oscillates with twice the height that it would with only one wave present. In other regions, the two waves combine exactly out of phase. That is, when one wave tries to make the surface move upward, the other wave tries to make it move downward by the same amount; and the water surface remains stationary. These regions which are exactly in and out of phase occur along lines spreading outward from the barrier. A few of the directions in which the waves are intensified are shown on the photographs. An object on the water would bob most vigorously along these directions where the maximum disturbances of the water's surface occur. Notice that the lines of maximum disturbance are spread by larger angles in the upper photograph than in the lower one. Notice also that the

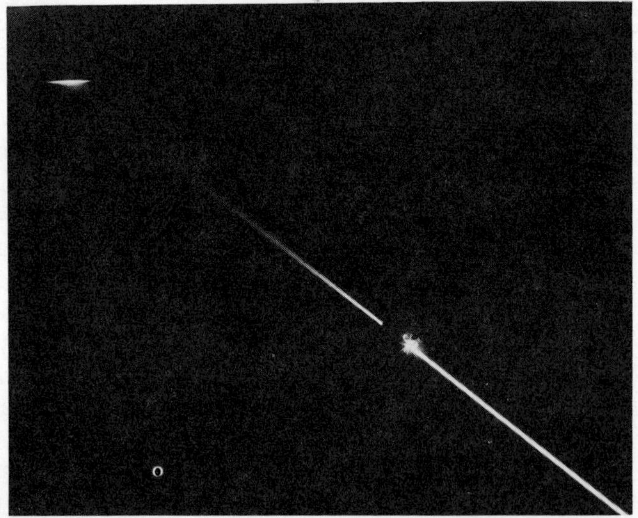

Figure 6–1 Diffraction of water waves by a harbor entrance and light waves by a slit.

wavelength of the original waves, that is, the distance between successive wave crests, is greater in the upper photograph. The greater the wavelength, the more the regions of maximum disturbance behind the slits are spread. If we know the spacing of the slits, measuring the angle of spreading permits the wavelength of the wave to be calculated. The combination of the diffraction effect and the subsequent interference effect from multiple slits can thus be used to determine wavelength.

The same process occurs with light. Figure 6–3 shows what happens when a laser beam is directed at two slits. The light beyond the slits is spread out into a pattern of regions of varying intensity, just as the water waves were in the ripple tank. A central maximum directly behind the two slits is flanked on either side by regions of enhanced

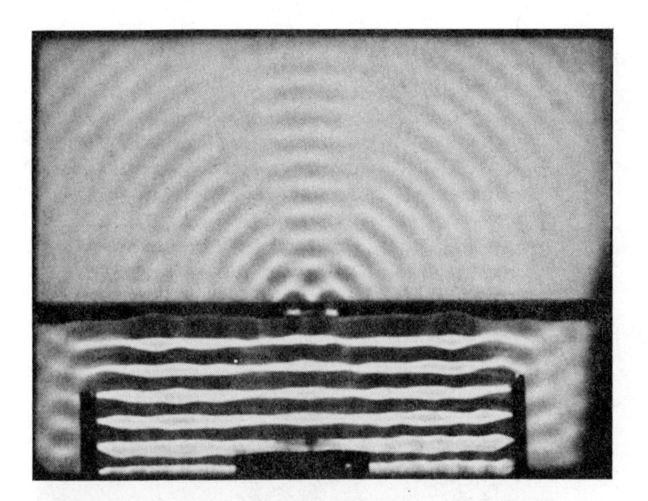

Figure 6–2 Diffraction and interference of water waves by double openings.

intensity which diminish in brightness as their distance from the central peak increases. If more slits are used, the pattern becomes sharper; and, as the distance betweeen the slits is made smaller, the regions of the bright intensity are spread farther apart. A device having a large number of slits, usually hundreds to thousands per centimeter, is called a *diffraction grating*. The pattern formed when a laser beam shines on a diffraction grating is shown in Plate 3A.

If white light from an ordinary light bulb is directed toward a diffraction grating, we obtain another very interesting result. Directly behind the grating, a central maximum of white light is formed, with two bands of colored light on either side of it, as shown in Plate 3B. The bands of light are violet closest to the central maximum and red farthest from it. In between, all of the other colors of the rainbow can be seen. This experiment demonstrates two features of light: first, that white light is made up of a combination of light of all colors; and second, that red light has a longer wavelength than violet light. The latter must be true since the wave theory predicts that the waves of greatest wavelength will be deviated by the greatest amount by a diffraction grating. The wavelength of light in any part of the colored band can be calculated if the spacing of the slits is known and the angle by which the light was deviated is measured. Light of a particular color always turns out to

Figure 6–3 Diffraction and interference of light waves by a double slit.

have the same wavelength, no matter what grating spacing is used. Therefore, in the wave model of light, wavelength determines the color of light. The wavelength of light of any color visible to the human eye is very short. The red light of the laser beam used in Plate 3A has a wavelength that would require almost 16,000 successive waves to add up to one centimeter. These wavelengths are so short that they are frequently noted in **angstrom units** rather than centimeters. It takes 100 million angstroms to make one centimeter.

Our eyes are sensitive to a restricted range of wavelengths, appropriately called *visible light*. There are also radiations of other wavelengths, not visible to the human eye but having the same fundamental nature as visible light. Figure 6–4 shows these different regions of wavelength. **Ultraviolet** radiation, which is sometimes called *black light* because you cannot see it, is shorter in wavelength than violet light. Even shorter are **X-rays** and shorter yet are **gamma rays.** Beyond the red end of visible light comes the **infrared,** or the radiation that we sense as heat from a heat lamp. The microwave and **radio** regions occur at even longer wavelengths. These radiations are all part of the same phenomenon, but merely have different wavelengths.

Because all of these radiations travel with the same speed in the absence of matter, the color or kind of radiation can also be designated by giving the frequency of the radiation. Figure 6–5 schematically compares a short- with a long-wavelength radiation. Light travels at nearly 300,000 kilometers per second (actually 299,792.5 kilometers per second). Therefore, there must be more wave crests in the distance traveled by light in one second for short-wavelength radiation than for long-wavelength radiation. The number of waves in this distance, or, if you prefer, the number of waves passing a stationary observer in one second, is called the **frequency** of the radiation. Since frequency is the number of waves passing a given point in one second and the wavelength is the length of one wave, multiplying frequency by wavelength equals the distance the light will travel in one second, that is, its speed. By convention, infrared and shorter radiation are usually designated by wavelength. Microwave and radio radiation are usually designated by frequency.

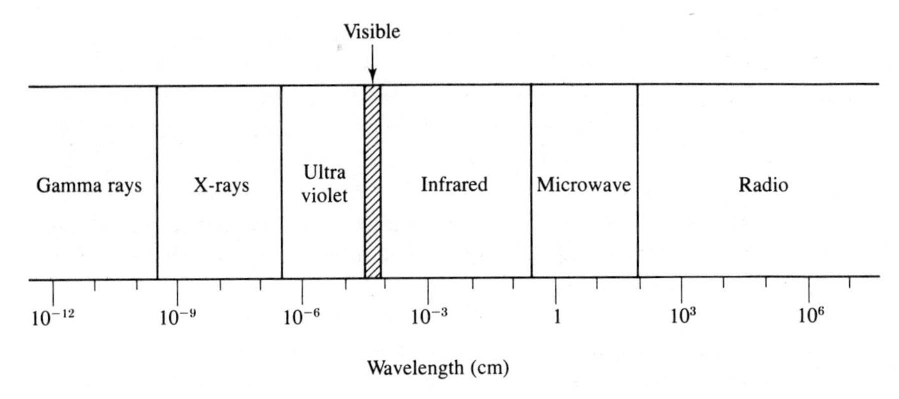

Figure 6–4 Wavelengths of the different radiation regions.

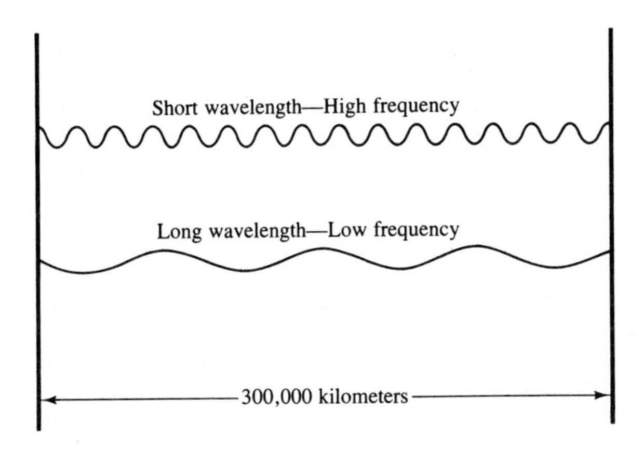

Figure 6–5 The relation between frequency and wavelength.

The Particle Theory of Light

An alternate theory of light, known as the particle theory, assumes that light consists of a stream of particles, or bundles of energy, called **photons.** The energy of each photon is proportional to the frequency of the light with which it is associated. Therefore, photons of blue light are much more energetic than photons of red light. In this theory, a light source throws off huge numbers of photons moving with the speed of light in all directions. When some of these photons enter our eyes, they cause a sensation which signals the existence of the light source.

The particle theory can be used to describe a phenomenon which the wave theory cannot fully explain. This phenomenon, called the **photoelectric effect,** occurs when high frequency radiation, say ultraviolet light, strikes a metal surface; and **electrons,** or tiny particles of electricity, are emitted from the metal. The more intense the light, the more electrons are emitted. When the light strikes the metal surface, some of the radiant energy is absorbed and transferred to the electrons in the metal. If an electron gains enough energy, it can escape from the metal. There is, however, one feature of the photoelectric effect which is impossible to explain using the wave theory. If low frequency radiation, say red light, shines on a piece of metal, no electrons are emitted—even though the brightness of the red light may exceed by far that of ultraviolet light which does cause electrons to be emitted. Einstein was awarded the Nobel Prize for his explanation of this feature of the photoelectric effect using the particle theory. He proposed that an electron can be ejected only when it receives enough energy from a single photon. If the photon is energetic enough (as in the case of ultraviolet light), the electron picks up enough energy to escape the metal. But with red light, the photon energy is too low to eject an electron. Since the photon-to-electron interaction is a one-to-one interaction, no matter how many photons are present, no electrons are emitted.

We are thus faced with a problem: Is light wavelike or particlelike in character? To explain diffraction, it must be wavelike; but to explain the photoelectric effect, it must be particlelike. The accepted answer today is that light behaves both ways; it has a dual nature. More complex mathematical descriptions have been devised which combine the wave and particle characters of light into one system, but we will use that simple description—wave or particle—which is most easily applied to the problem at hand; because, in the last analysis, they have been shown to be equivalent. Thus, the color of light can be given in terms of wavelength or the energy of a photon. The brightness of light of a particular wavelength will be measured by the height, or amplitude, of the light wave, or, alternatively, by the number of photons streaming past per second.

Spectra

Light, or radiant energy, is created by matter. The color of the light, or more specifically, the brightness of the light at each wavelength, will depend upon the physical conditions of the matter which generates the light. In astronomy we invert the process; that is, by analyzing the light we receive from an object, we deduce the physical condition of the object. The analysis of the light begins by spreading the light into its component colors. Most modern devices use diffraction gratings to disperse or spread the light. The spread of colors resulting from this dispersion is called a *spectrum*. Figure 6–6 shows the three kinds of spectra which we can observe. In the first spectrum, the colors gradually and continuously change from blue through red across the spectrum. Such a spectrum is called a *continuous spectrum*. In the middle spectrum, energy (or if you wish, photons) is found only at certain wavelengths. This is called an *emission* or *bright-line spectrum*. Finally, the last one looks like a continuous spectrum but has regions showing a deficiency of photons; that is, it shows sudden drops in brightness over a restricted wavelength region. This kind of spectrum is known as a *dark-line* or *absorption spectrum*. Our task, for the moment at least, is to explain how matter can generate light with these three rather different kinds of spectra.

The generation of light occurs at the atomic level of matter. An atom is the smallest amount of matter which can retain its usual chemical properties. Atoms are so small that their structure cannot be viewed using even the best microscopes. To visualize what happens in the emission and absorption of light by an atom, we imagine a model of the atom. The simplest model of an atom pictures it as a miniature solar system, where electrons orbit around the atomic nucleus, similar to the way planets orbit the sun. The nucleus contains almost all of the atom's mass. At this point, even in our simple model, the similarity to the solar system disappears, because it is not gravity which binds the system together but rather electrical force between the positively charged nucleus and the

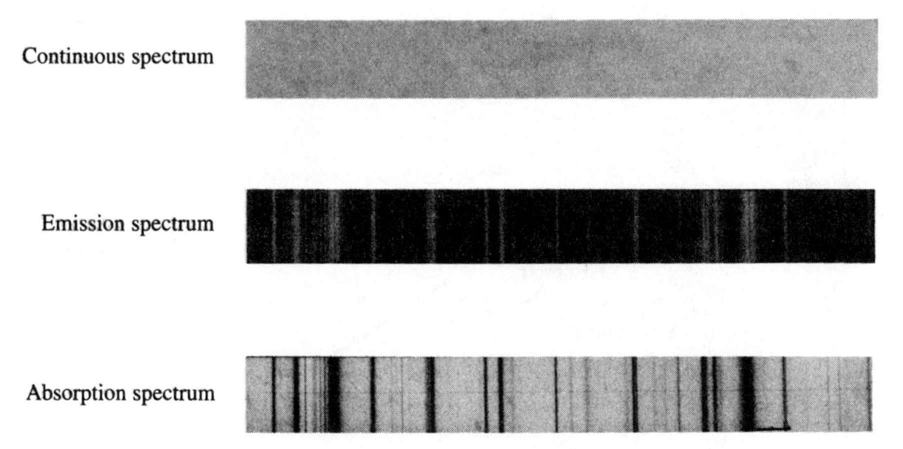

Figure 6–6 The three types of spectra: continuous, emission, and absorption.

negatively charged electrons. Another difference is that the electrons can only exist on certain prescribed orbits. In the solar system, at least in principle, we could move a planet to any size orbit by using rockets to adjust its velocity.

To understand what causes an electron to change from one permitted orbit to another permitted orbit, we turn to the analogy of launching a space satellite from the earth. If a small rocket and hence a small amount of energy is used, the satellite goes into a small orbit about the earth. If a larger and more energetic rocket is used, the satellite will enter a larger orbit. We therefore say that the larger orbit is a high-energy orbit and the small orbit is a low-energy orbit. Similarly, the smaller the permitted orbit in an atom, the lower is its corresponding energy. When an electron moves from a low- to a high-energy orbit, energy must be supplied to cause the electron to move outward against the electrical attraction of the nucleus. When the electron moves from an outer to an inner orbit, energy must be released from the atom.

To apply our model, consider the simplest atom, hydrogen, which consists of a nucleus with one positive charge and one orbiting negatively charged electron. Figure 6–7 is a schematic drawing of our model applied to the hydrogen atom showing only the first three permitted orbits in the atoms. The actual dimensions of the system, of course, are very much smaller than shown—of the order of one hundredth of a millionth of a centimeter, or 10^{-8} centimeters. If a photon collides with an atom, it can give its energy to the atom and be destroyed only if it has *just the correct energy* to cause the electron to move from one inner permitted orbit to an outer permitted orbit. Thus, a photon corresponding to a wavelength of 1216 angstroms can cause the electron to undergo a transition from the first to the second orbit, and a photon of 1026 angstroms can cause an electron to go from the first to the third orbit. A photon corresponding to a wavelength between these two values will pass right by the atom, because the photon does not have the correct energy to cause a transition between permitted orbits. For a hydrogen atom with its electron in the first orbit, a photon of wavelength 912 angstroms or less can cause the electron to achieve escape velocity and be ejected from the atom altogether. When an atom loses one or more of its normal number of electrons, it is called an ionized atom, or an **ion.**

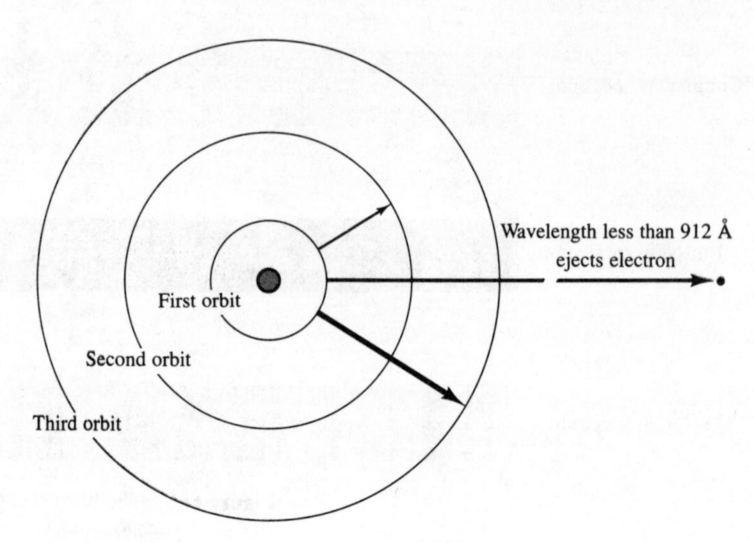

Figure 6–7 Permitted electron orbits of the hydrogen atom.

Of course, the opposite process can also take place. If an electron spontaneously moves from the second orbit to the first orbit of a hydrogen atom, energy is given up in the form of a photon, and that photon has a wavelength of 1216 angstroms. If an electron recombines with a hydrogen ion and lands on the innermost orbit, a photon of wavelength 912 angstroms or less will be released, depending upon how fast the electron was moving just before it recombined with the ion.

The atoms of the other chemical elements are all more complicated than hydrogen. They all have more than one electron; and while their atomic models are consequently more complicated than that of hydrogen, still only certain orbital arrangements for the electrons are permitted. Each arrangement has a particular energy associated with it. Therefore, a change from one orbital arrangement to another corresponds to a particular change in energy and, hence, to a photon of a particular wavelength.

The absorption spectrum can now be explained. If a beam of white light, which has a continuous spectrum, is passed through an appropriate sample of gas and the transmitted light is spread into a spectrum, we see an absorption spectrum. Most of the photons entering the gas pass right through, but many of those having just the right energies, and hence wavelengths, to lift the electrons of the gas atoms to higher energy orbits will be absorbed. A photon which is absorbed is destroyed when it gives up its energy to change the arrangement of the electrons in the atom which captures it. The beam of light which leaves the sample of gas therefore has lost photons at selected wavelengths, and absorption lines appear in the spectrum.

An atom which has just absorbed a photon is said to be *excited* to a higher energy. Such an atom has a strong tendency to return spontaneously to a lower energy orbital arrangement. When this return occurs, a photon is created, because energy must be given up in the process. A photon emitted that way can be given off in any direction. Therefore, if we examine the sample of gas from the side, we will see a faint spectrum. This spectrum, however, will be an emission spectrum. Each emission line will have exactly the same wavelength as a corresponding absorption line we observed when we looked directly at the light through the gas. This property holds true because the energy required to excite an atom from one permitted electron arrangement to another is exactly the same amount of energy released when the atom returns to its original arrangement.

A gas can also be made to give off an emission spectrum without shining a light on it. The temperature of a gas measures the speed of the random motions of the particles of the gas. As the temperature of a gas increases, the motion of its particles becomes more vigorous. When the gas is hot enough, some of the atoms will be moving fast enough to collide with enough force to cause a change in the arrangement of their electrons. The atoms are thus said to be excited by collision. Again, the excited atoms can de-excite by spontaneously returning to a lower energy arrangement, giving off a photon. The gas consequently glows with its characteristic emission spectrum.

The difference between absorption and emission lines can be demonstrated by looking at a gas flame in which pieces of ordinary table salt have been placed. Figure 6–8 shows what happens. In the upper diagram, an observer examining the salted flame, which gives off yellow light, will see a pair of bright yellow emission lines. These lines are caused by the sodium atoms in the salt which was vaporized in the flame. If, as indicated in the lower diagram, we now place a bright enough light behind the flame and view it again, we see the continuous spectrum of the bulb; but it is now crossed by two dark lines where we saw the yellow emission lines before. The sodium is still emitting the lines just as it did earlier because of its high temperature, but it is also absorbing light from the continuous spectrum of the bright light. The lines, although as bright as before, now appear dark in contrast to the light in the spectrum of the light bulb at nearby wavelengths.

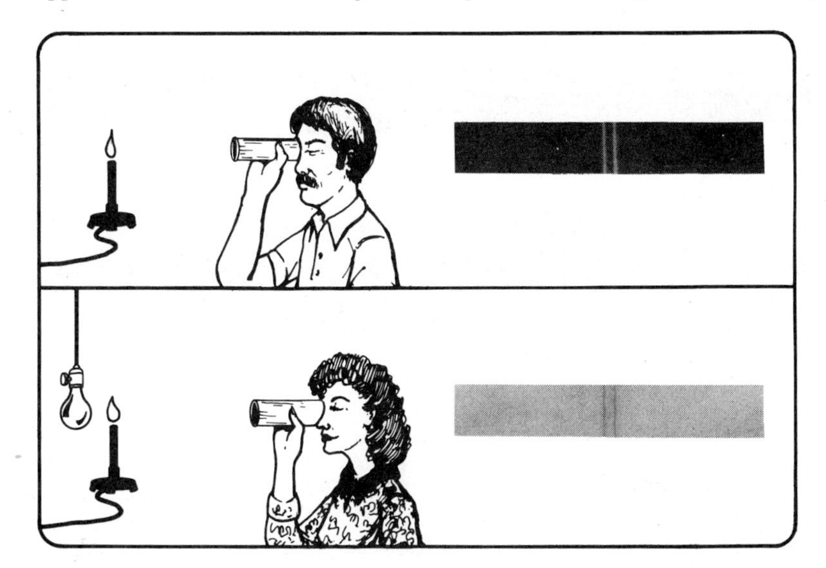

Figure 6–8 Emission and absorption by sodium atoms.

If we were to increase the amount of salt in the flame, the number of sodium atoms which could absorb or emit photons would be increased. Thus, the number of both absorbed and emitted photons would increase. This finding leads to the important principle that the more strongly something absorbs light, the more strongly it can emit light.

The differences in the chemical elements result from the number of positive charges on the nucleus of each element. Each chemical element, therefore, also has a unique number of electrons. The permitted orbits of the electrons in an atom depend upon both the total charge on the nucleus, which binds the electrons to the atom, and the repulsive forces between its negatively charged electrons. Each element consequently has a distinct and separate set of electron arrangements and associated energies. As shown in Figure 6–9, this uniqueness gives rise to distinctive emission and hence absorption spectra for each of the elements. The spectra of the chemical elements, as well as those of their ions, have been studied in the laboratory, and the wavelengths of their prominent spectral lines have been listed in tables. An ion of an atom has an entirely different spectrum from that of the neutral atom because the electron orbits readjust after one or more of the usual number of electrons are removed in the ionization process. Comparing the wavelengths of the spectral lines found in a star's spectrum with the values in the wavelength tables will identify the chemical elements in the star. A photograph of the spectrum of the sun, with lines from several elements, is shown in Plate 3C.

Figure 6–9 Emission spectra of four chemical elements.

The Doppler Effect

Spectral lines also can tell something about the movement of the object from which the spectrum was obtained. Figure 6–10 compares a stellar spectrum which shows absorption lines of iron with an emission spectrum of iron. The emission spectrum came from an iron arc at the telescope using the same spectrograph used to obtain the stellar spectrum. Notice that the stellar absorption lines are displaced slightly from the positions of the corresponding emission lines. The emission spectrum is called a *comparison spectrum* because by comparing it with the stellar spectrum we can determine the wavelengths of the stellar lines. All of the stellar lines in this example are shifted to shorter wavelengths by about 0.01 percent in wavelength from their expected positions as listed in the wavelength tables. The shift in the lines is a result of motion between the source and the observer. This effect was first described by Christian Doppler in the 1800s and is usually called the Doppler effect, or **Doppler shift.**

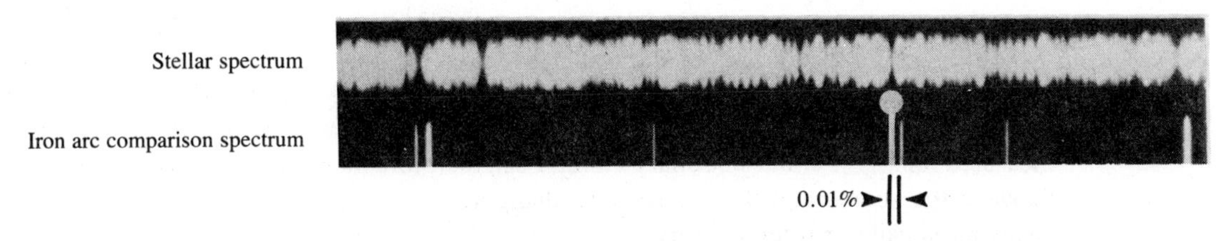

Figure 6–10 Doppler shift in a stellar spectrum.

The Doppler effect occurs in any wave phenomenon—sound, water waves, or light. For example, if you stand by a railroad crossing as a train passes by with its horn sounding, you hear a change in the pitch of the horn as the train first approaches and then recedes from the crossing. While the train is approaching, the pitch of the horn is higher; while it is receding, the pitch is lower.

We can understand the Doppler effect by considering what would happen with surface waves on water generated by a vibrating object in contact with the water. If the object is held over a fixed point and the frequency of its vibration is constant, a series of equally spaced concentric waves will expand outward from the vibrator. The spacing between adjacent waves along an outward radius from the vibrating object will correspond to what we have called wavelength. Now, if the vibrator is moved toward the right at a constant speed while its vibration rate remains constant, we would see a pattern similar to that in Figure 6–11. Each wave crest is identified with the position of the vibrator when that crest was formed. Since the vibrator is moving uniformly to the right, the center of each succeeding wave is displaced to the right by the same amount. For an observer on the surface of the water to the right (in the direction of motion of the vibrator), the waves would seem to arrive with a greater frequency (higher pitch) than they would if the vibrator were stationary. The observer would measure the wavelength to be shorter than expected from a knowledge of the frequency of the vibration. If the observation of the waves were made from the left (on the opposite side of the moving vibrator), just the opposite effect would occur—lower frequency and longer wavelength. At right angles to the line of motion, the frequency and wavelength measured will be that expected from the

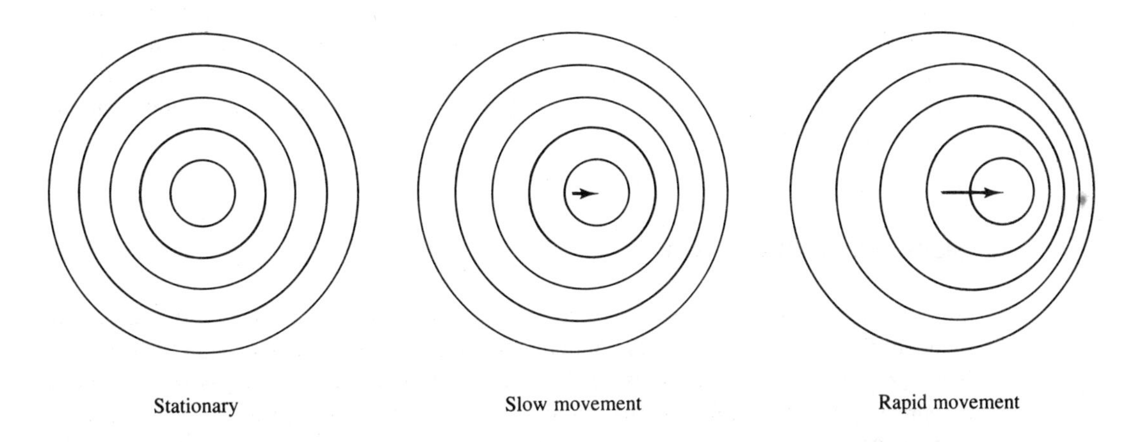

Figure 6–11 The Doppler effect.

frequency of the vibrator, as long as the vibrator moves slowly compared to the velocity of the waves across the surface of the water.

If the vibrator is moved faster across the water, the effect on the measured wavelength will increase. Comparing the measured wavelength with the wavelength measured for a similar but stationary vibrator can therefore measure how fast the distance between the source and the observer is changing. The rate at which this distance is changing is called **radial velocity.**

This same effect occurs with light waves. The shift we saw in stellar lines in Figure 6–10 indicates that there was relative motion between the star and the telescope used to collect the light from the star. In this case, since the observed wavelengths are decreased, the lines are said to be shifted to the blue. The distance between the star and the telescope must be decreasing. If the distance were increasing, the wavelengths would lengthen; that is, the lines would be shifted to the red.

When the velocity of the source is small compared to the velocity of light, the fractional change in wavelength is equal to the radial velocity expressed as a fraction of the velocity of light. We can write an expression which states this simply:

$$\frac{\Delta\lambda}{\lambda} = \frac{v}{c}$$

where $\Delta\lambda$ = the observed change in wavelength
λ = the laboratory wavelength
v = the radial velocity of the source
c = the velocity of light

In Figure 6–10, since the stellar lines are shifted in wavelength by 0.01 percent to the blue, the star and the observer are approaching by 0.01 percent of the velocity of light, or 30 kilometers per second. Had the lines been shifted to the red by the same amount, the source and the observer could be separating by 30 kilometers per second. Always remember the Doppler shift does not tell how fast an object is moving, but merely how fast an object appears to be approaching or receding. It is not possible to tell from the Doppler shift alone whether the star or the observer is moving or if the shift is due to the combined motions of both. If the relative speed of the star and the observer is small compared to the speed of light and the movement is perpendicular to the line of sight to the star, no Doppler shift will be seen. A zero Doppler shift does not imply a lack of motion, but only that the distance between the source and the observer is not changing when the observation is made. For example, if the star and the observer are moving in parallel directions with the same speed, no Doppler shift will occur.

Black Body Radiation

We have been able to account for both emission and absorption lines with our simple model of the atom. Can we use that model to explain a continuous spectrum? The answer is yes, and one way is as follows. Imagine a collection of atoms in a gas which we can compress into a decreasing volume. At first, when the atoms occupy a large volume, each atom, except at the instant of collision with another atom, is essentially isolated. Usually, then, its electron arrangement is exactly the one we have already described. When atoms are squeezed close enough together by increasing the pressure, the collisions between atoms become very frequent. Most of the electron orbits are altered from the case of an isolated atom. The well-defined spectral lines seen with a gas at low pressure are consequently replaced by broad, fuzzy lines. Figure 6–12 shows these two cases, as well as what happens at even greater pressure. Ultimately, when the atoms are close enough together, the influence of neighboring atoms upon each other becomes so severe that the spectral lines completely broaden and disappear. Thus, in a solid, a liquid, or a gas under very high pressure, the spectra lines are completely smeared out, and we get a continuous spectrum.

Figure 6–12 Change in an emission spectrum with pressure.

For example, an ordinary light bulb emits a continuous spectrum even though the filament is made of tungsten. If we wished to see the emission spectrum characteristic of tungsten, we would have to vaporize the normally solid filament of the bulb. When we examine moonlight with a spectrograph, we see the spectral features of the atmosphere of the sun which supplies the light, not those expected from the elements of the lunar crust. To make a spectral analysis of the chemical composition of moon rocks, they must be brought back to an earth laboratory where they can be vaporized, so that the constituent atoms can absorb or emit at their characteristic atomic wavelengths. Although we cannot determine the precise chemical composition of a continuous emitter from its spectrum, we can determine something about its temperature.

We begin by noting that the ability of a body to emit radiation is proportional to its ability to absorb light, as shown in Figure 6–13. The left-hand photograph shows a piece of black carbon rod and a piece of white-coated metal in an oven at room temperature. The carbon appears black because it absorbs nearly all of the visible light which falls on it, while the white-coated metal reflects almost all of the light striking it. The carbon rod is a good absorber; the white-coated metal is a poor absorber. Now look at the right-hand photograph. The oven was used to heat both bodies to the same temperature, 820°C. The

Figure 6–13 Emission of energy by a good absorber and a good reflector.

piece of black carbon, the good absorber, now emits more light than the good reflector—the best absorber is also the best emitter.

We call a perfect absorber—one that would absorb all radiation falling upon it—a *black body.* This concept is theoretical because in practice nothing has yet been found to have this property completely. Being a perfect absorber, a black body must also be a perfect emitter. It is possible to mathematically describe how the theoretical black body would radiate as its temperature is changed. Max Planck, who first derived the correct mathematical expression for the radiation, could do so only after he had assumed that light is emitted in photons and that the energy of a photon is equal to a constant, now called Planck's constant, times the frequency of the wave associated with the photon.

Planck's law of **black body radiation** can be used to calculate the energy radiated per unit area at each wavelength by a black body at any given temperature. The temperature used in Planck's law is expressed on the absolute temperature scale and is given in degrees Kelvin (K). The **Kelvin temperature** of a body is measured from absolute zero, the temperature at which a black body radiator would cease all radiation. Since absolute zero is 273 degrees below the freezing point of water, a temperature given in the Celsius system can be converted to an absolute temperature by adding 273 to the Celsius temperature.

The plots of the energy distribution for a black body— in other words, the description of its continuous spectrum—at two temperatures are shown in Figure 6–14. These curves show that as the temperature of the black body is increased, it radiates more energy at every wavelength. In addition, the wavelength at which the body radiates the maximum energy shifts to shorter wavelengths as the temperature increases. Planck's law predicts that the only time a black body would not radiate any energy is when it has a temperature of absolute zero. Even at room temperature (about 293K), the body would radiate, but at long wavelengths beyond the visible region. At a temperature of 1093K, the temperature of the carbon rod in Figure 6–13, the black body would radiate some energy in the visible region. Figure 6–14 shows that the great majority of its radiant energy would still be in the infrared and beyond. At a temperature of 4000K, the energy at each wavelength would increase, but more so at the shorter wavelengths. At this higher temperature a much greater fraction of the body's radiant energy would now be in the visible part of the spectrum.

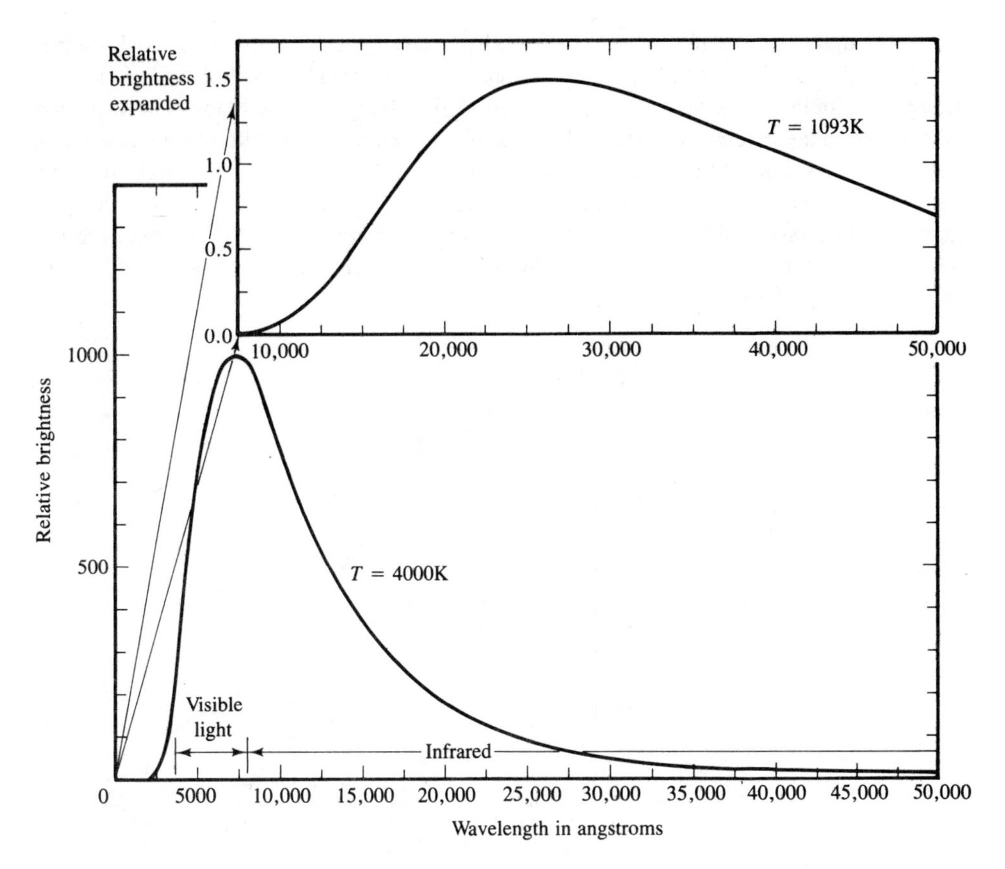

Figure 6–14 Planck's law.

The total amount of energy radiated from each square centimeter of a black body can be found by adding up the contributions which are predicted at each wavelength by Planck's law. When this process is carried out, the result agrees with Stefan's law, which states that:

$$E = \sigma T^4$$

where E = the energy radiated per unit area over all wavelengths

σ = a constant known as Stefan's constant

T = the temperature of the black body expressed in the absolute temperature scale

The total radiant energy, being dependent on the fourth power of the temperature, increases very rapidly as a black body is heated. For example, if the temperature is doubled, the body will radiate sixteen times as much energy; if the temperature is increased 3 times, the radiation rate will increase 81 times.

Stefan's law and Planck's law can both be used to measure the temperature of a radiating body, which is most fortunate from an astronomical viewpoint because the stars are too far away to sample directly. We can only study the radiation from the star. Furthermore, even if we could reach the stars, most stars are so hot that a thermometer placed in one would immediately be vaporized.

Although stars are not perfect black bodies, they are close enough to black bodies that Planck's law gives good estimates of their surface temperatures. In Figure 6–15, the energy distributions of several stars have been plotted along with the Planck curves which best fit the stellar spectra. The sun, which is represented by the middle plot, corresponds best to a black body temperature of about 5800K. The sun and other stars near its temperature have a yellowish color. A star with a temperature of 4000K has most of its radiant energy in the red and infrared and consequently appears red to the eye. On the other hand, a star with a temperature of 8000K emits much more blue light than the sun and appears white in color.

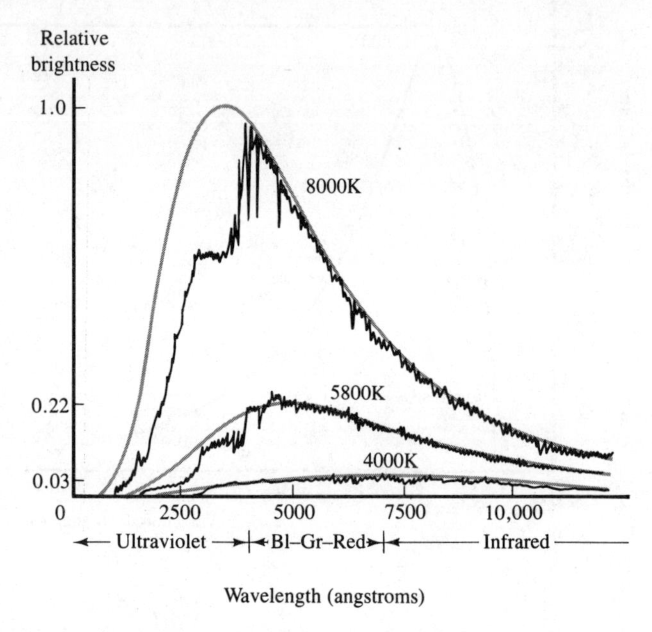

Figure 6–15 The comparison of Planck's law with the radiation of stars at three temperatures.

These last observations lead to another and quicker way of determining the temperatures of the stars: observe their colors. The red stars are relatively cool, yellow stars have intermediate temperatures, and the hottest stars appear bluish-white to the eye. There is one difficulty in applying this method to the determination of stellar temperatures, however. If there is any material between us and the star being observed, the star's apparent color can be changed, making it appear redder. For instance, when you look at the sun at sunset, it appears very red compared to when you see it high in the sky. The sun does not cool off at sunset as its color might indicate; it is simply that, at that time, the sunlight which has passed through more layers of the earth's atmosphere has had a considerable part of its short wavelength radiation removed. In later chapters, we will refine our method of stellar temperature measurements from color measurements and will also find out how to correct them for interstellar material which might lie in our line of sight.

Finally, remember that the model of the atom that we have used to explain the radiation from matter is an oversimplified one. While in principle it accounts for the basic processes, more refined models are required to interpret some of the more complicated features seen in spectra.

KEY TERMS

diffraction
interference
wavelength
angstrom unit
ultraviolet light
X-rays
gamma rays
infrared light
radio waves
frequency
electron
photon
photoelectric effect
ion
Doppler shift
radial velocity
black body radiation
absolute (Kelvin) temperature

Review Questions

1. Why does astronomy depend on analyzing light more than other sciences do?
2. What is diffraction?
3. How can interference aid in analyzing light?
4. When red light shines on a group of slits, the pattern of bright regions is spread farther apart than the pattern produced when yellow light shines on the same slits. What does this effect demonstrate?
5. How does the wave model of light explain color?
6. What colors are the longest and the shortest wavelengths of visible light?
7. What color is the visible light with the highest frequency?
8. Which has a higher frequency, X-rays or radio waves?
9. What is a photon?
10. What is the photoelectric effect? Why is it significant in the study of light?
11. How does Einstein's explanation of the photoelectric effect account for the fact that light with a long wavelength does not cause electrons to be emitted from a metal surface?
12. List three ways in which a photon of red light differs from a photon of blue light.
13. If one photon has twice the wavelength of a second photon, how do their frequencies and energies compare?
14. Describe the three basic types of spectra.
15. Why can't an atom absorb a photon of any wavelength?
16. Why do the spectral lines from hydrogen (one positive charge in the nucleus and one electron) have different wavelengths than those of helium (two charges in the nucleus and two orbiting electrons)?
17. What is the Doppler effect?
18. Is it possible for two observers to measure a different Doppler shift for the same object (an automobile, for example)?

19. If lines in a star's spectrum show a Doppler shift to shorter wavelengths, what does that tell us about the star?
20. If two stars each are approaching the earth at the same speed, but one star is twice as far away, how will their Doppler shifts compare?
21. What is radial velocity?
22. Why does the sun glow with its own light, while the visible light from planets is just reflected sunlight?
23. Is a bluish-white star a relatively hot or relatively cool star?
24. What two effects occur as a black body gets hotter?
25. If a black body is heated to three times its previous absolute temperature, how much does the amount of its radiation change?

    (1) 1/81 as much as before
    (2) 1/3 as much
    (3) It stays the same
    (4) 3 times more than before
    (5) 81 times more than before

Discussion Questions

1. Give some examples of diffraction and interference of sound waves.
2. In our model atom and its interaction with light, what process is similar to the photoelectric effect?
3. Why are the wavelengths of spectral lines from a given atom the same regardless of whether they are emission or absorption lines?
4. Why does a hot gas glow, while a cool gas doesn't?
5. Suppose you carefully measure the wavelengths of lines in the spectrum of Mars. What Doppler shifts would you expect a month before opposition of Mars, at time of opposition, and a month after opposition of Mars?
6. Compare two situations: (1) when an observer is stationary and a star moves toward him at 100 kilometers per second, and (2) when the star is stationary and the observer moves toward it at 100 kilometers per second. Will the observer see the same Doppler shift in both cases?
7. What condition is sufficient for a group of atoms to give off only a continuous spectrum? In what three states can matter be to satisfy this condition?
8. If you see a red star, how could you decide if the star is cool or if the star is actually quite hot and the red color results from a Doppler shift due to the star's motion away from us?

7

The Tools of Astronomy

The development of modern astronomy has been strongly influenced by the development of instruments for the detection and analysis of light. Particularly, we think of the telescope and its development almost entirely in terms of use by astronomers. Although the telescope was not invented by an astronomer, within a year after the first one was constructed by a Dutch optician, Galileo had built his own and used it to study the heavens.

Galileo actually made several telescopes. Each one was small enough to be held by hand. Later astronomers, eager to discover more objects in the universe and to see those already known more clearly, needed larger telescopes. Holding a large system of lenses or mirrors rigidly in place so that it can be pointed toward a small point of light in the sky is not easy. The early telescope builders devised some ingenious methods to handle this problem—some more successful than others.

One attempt to build a large telescope was made in seventeenth century Danzig by Johannes Hevelius, a beer brewer by day and an astronomer by night. Hevelius constructed a long, thin telescope, the length of which was about half the length of a football field. The telescope was fixed to a tall, vertical mast and was raised or lowered by adjusting many ropes. Pointing the telescope required the coordinated efforts of a mob of men.

Even though Hevelius wrote of the ease with which the telescope could be used, it is doubtful that it could have been held steady. Even a gentle breeze would have caused the telescope to vibrate wildly. Edmund Halley, an English scientist and close friend of Isaac Newton, probably best described Hevelius' telescope with the single word "useless."

William Herschel, who discovered the planet Uranus, made several large telescopes, the largest of which was over twelve meters long. This giant reflecting telescope had a main mirror with a diameter of 1.2 meters aperture. No larger telescope was constructed in the seventeenth or eighteenth centuries.

Herschel's giant telescope was complicated and ungainly. The mirror was sup-

The Hale telescope at Palomar Mountain

ported by a huge open wooden structure. While the telescope could be pointed in any direction by a system of ropes and pulleys, it could not continuously follow the diurnal motion of the stars. Oliver Wendell Holmes captured the emotional impact of this fantastic contraption when he said, "It was a mighty bewilderment of slanted masts, spars and ladders and ropes, from the midst of which a vast tube, looking as if it might be a piece of ordnance such as the revolted angels battered the walls of heaven with, according to Milton, lifted its mighty muzzle defiantly to the sky."

During the twentieth century, the art of building large telescopes has become very sophisticated. Much of our present knowledge of the universe has been gained by using these large telescopes, many of which are in the United States. For the development of these instruments and for what we have discovered because of them, we owe a huge debt to one man, George Ellery Hale.

In 1892 Hale began to solicit funds for what would be the world's largest operational telescope, a 40-inch (one-meter) aperture refractor, and for what he hoped would become the world's finest observatory. Hale was an astronomer at the University of Chicago at the time. He eventually obtained a financial commitment from Charles Yerkes, the Chicago streetcar tycoon.

Yerkes was not a man of impeccable character. He had been arrested on a charge of embezzlement in Philadelphia and was often suspected of questionable financial dealings. Legend has it that Yerkes was actually broke when approached by Hale. However, he may have assumed that he could convince prospective creditors of his solvency by freely making a commitment of several hundred thousand dollars to the University of Chicago for the sole purpose of gaining new knowledge of the universe. What a fine way to give the impression that he was a man of such wealth that he could afford to support that noblest endeavor of the human race—the search for knowledge—with complete disregard for financial reward!

In addition, Yerkes was an ambitious man who wanted attention. A world's fair was being held in Chicago, and the telescope was to be exhibited there before being moved to its permanent site. Yerkes was delighted by the thought of having "his" telescope featured at a world's fair. In his glee, some of his motives for funding its construction showed through. One newspaper reporter concluded, "Mr. Yerkes is trying his prettiest to ride into Chicago's sacred temples [of society] on a telescope."

Whatever the reaction of high society, in scientific circles the 40-inch telescope was considered a masterpiece. The lenses were figured by Alvin G. Clark, an optical expert renowned as the world's greatest lens maker. As a site for the telescope, Hale chose a bluff on the north shore of Wisconsin's Lake Geneva, about 120 kilometers north of Chicago. This remote site was chosen to be away from pollution and bright city lights.

During the construction of the observatory, Yerkes allegedly tried to renege on his financial commitment to Hale several times. It was not that he did not want it to be named Yerkes Observatory; he just did not want to pay for it. But Hale held Yerkes fast to the agreement, and finally, with Yerkes' money, the observatory was dedicated in 1897.

By the time the finishing touches were being applied to the building, Hale was not overly fond of Yerkes, to say the least. The workers on the project also had little love for the man because of the many times he had threatened to take away their jobs by pulling out his financial support. No one knows whose idea it was, but as another legend has it, caricatures of Yerkes appeared in the decorative molding on the pillars at the observatory's main entrances. Each caricature showed a bee perched on Yerkes' nose, symbolizing how he had been "stung" for the funds.

Still another legend says that Yerkes caught wind of the meaning of the bees before the observatory's dedication. Many wealthy and influential people, including John D. Rockefeller, founder of the University of Chicago, were to be present at the grand event. Fearful that the story of the bees would be leaked and he would be made to look the fool, Yerkes had workers chisel every single bee from every single nose on the observatory pillars. Maybe the legend about the bee on the nose is true, and maybe it isn't, but any visitor to the Yerkes Observatory can easily see chisel marks on the noses on the masonry pillars.

The story of George Ellery Hale does not end with the building of the Yerkes Observatory. He had barely settled into a research routine with the Yerkes telescope when he revived an old dream for a still larger instrument. He studied different locations and decided on Mount Wilson, rising 5900 feet (1800 meters) above the Los Angeles basin, as an ideal site for an observatory. Through his efforts, several telescopes were placed there, the largest being a 100-inch (2.5-meter) diameter reflecting telescope. This telescope helped us find the key to the structure of the universe.

Hale then began to solicit financial support for an even larger telescope, the great 200-inch (5.0-meter) reflector which he wanted to put on Palomar Mountain. However, the Second World War intervened and the 200-inch telescope was not put into operation until ten years after Hale's death. Today, fittingly, the Mount Wilson and the Palomar Mountain observatories are combined under the name Hale Observatories.

After Galileo proved the advantages of using a telescope to look at the sky, the telescope became an essential tool of astronomy. It is so closely associated with astronomy that the mention of the word *astronomer* usually brings to mind someone keeping an all-night vigil, peering through the end of a large telescope. While this image may have been true at one time, the invention of photography, the development of spectroscopy, and the use of photoelectric cells have changed the methods by which astronomers use telescopes. Today, observations using radio waves, infrared and ultraviolet light, and X-rays have led to instruments that often have little outward resemblance to optical telescopes. All of these instruments, no matter how they are constructed, have one fundamental use—to collect the weak radiation received from different objects in the sky and allow it to be measured.

Image Formation

Optical telescopes are imaging devices. They form a small picture, an image, of an object being studied. In this sense, they are very similar to cameras used to take photographs. In fact, astronomical telescopes are frequently used as large cameras to take photographs of celestial objects.

The diagram of a simple box camera in Figure 7–1 illustrates how a lens forms an image. The light reflected or given off by an object, an arrow in this case, enters the camera through its transparent lens. The glass lens refracts or bends the light passing through it. If the front and back surfaces of the lens are sections of a sphere, all of the light from one point of the object converges to a corresponding point (or nearly so) on the image. Thus, in Figure 7–1, the light rays from the tip of the arrow come together to form the tip of the arrow in the image. All other points on the object are similarly imaged, thus building a complete image of the object. An image formed in this way is always inverted. A telescope which forms an image using a lens is called a **refractor,** or **refracting telescope,** because the lens functions by refraction of light.

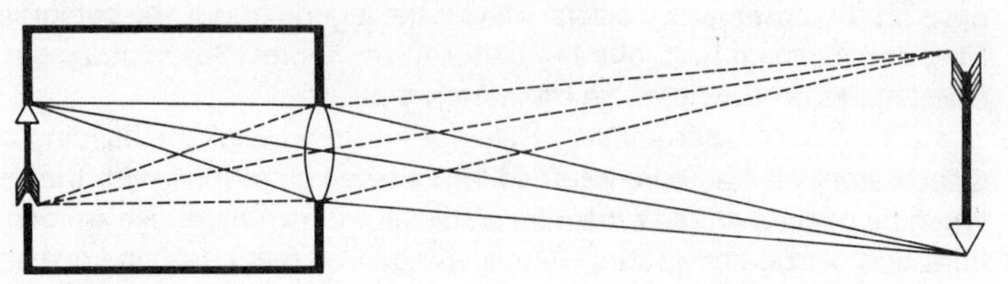

Figure 7–1 Image formation in a simple camera.

The place behind the lens where the sharpest image of the object is formed is called the *focus* of the lens. As the distance to the object increases, the focus moves closer to the lens. Finally, as the object is at a very large distance, the image is focused at a distance behind the lens called the **focal length** of the lens. Since all objects observed with astronomical telescopes are at great distances, their images always are found at one focal length from the lens.

Notice in Figure 7–1 that the angular size of the object and its image are the same when measured at the center of the lens. Thus, the linear size of the image for any distant object depends directly on the focal length of the lens. Consequently, in Figure 7–2, the

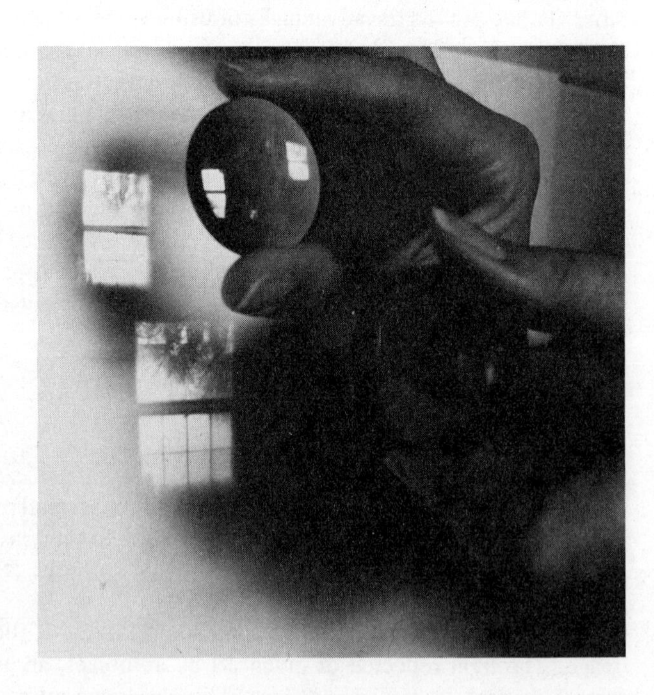

Figure 7–2 The size of an image depends upon the focal length of the lens.

images of the same window formed by a short and a long focal length lens are small and large, respectively. For example, if one lens has a focal length three times that of another, it will form an image which is three times larger than that formed by the short focal length lens.

Simple lenses, like those we have been talking about, have a major fault. Since glass refracts light of short wavelengths more strongly than light of long wavelengths, blue light from the object will be focused closer to the lens than red light. This confusion can be corrected by making a compound lens, in which two lenses made of different kinds of glass are combined. One lens converges light while the other diverges light, but less strongly. The net result is still a converging lens capable of forming an image. By choosing appropriate glasses and grinding the lenses properly, the blue and red rays can be made to come to almost the same focus. Such a lens is said to be *color-corrected.* All large astronomical lenses are color-corrected.

As shown in Figure 7–3, a concave mirror can also be used to form an image. A mirror has a major advantage over a lens because light of any color is reflected by a mirror exactly the same as light of any other color. Therefore, a mirror is completely color-corrected without any special treatment. A mirror also has some other advantages over a lens. First, there is only one surface to grind; while in a color-corrected lens, at least four surfaces must be formed. The glass in a lens must not have any internal imperfections such as bubbles; but imperfections in the glass in a mirror are not important, because the light does not have to pass through the glass. Also, a lens can only be supported by its edges, while a mirror can be supported across its back surface. All telescopes larger than about forty inches (one meter) in diameter use mirrors, because much larger lenses would sag under their own weight. Consequently, the largest telescopes are always **reflectors,** that is, **reflecting telescopes.**

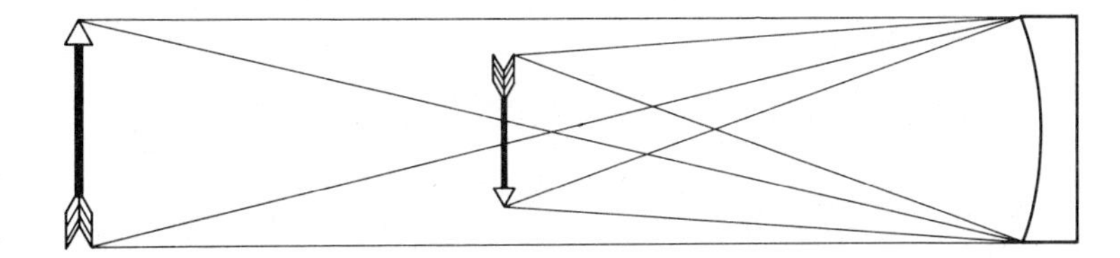

Figure 7–3 Image formation by a mirror.

Unfortunately, reflecting telescopes also have a fault—they form excellent images only near the center of the field of view. The largest reflector in use today cannot give a good focus across the entire moon, which is only one-half degree in diameter. Refractors have much larger fields of view, but they have color defects. A third kind of telescope, called a Schmidt telescope after the man who invented it, combines the best of both the reflector and the refractor. It consists of a very thin lens, called a *correcting lens,* and a mirror. The correcting lens deviates the incoming light just enough to correct parts of the image away from the center of the field of view. Because it is very thin and does not deviate the incoming light very much, the correcting lens causes only negligible color effects. The major deviation of the light required to form the image is accomplished by the mirror. The Schmidt telescope thus has the large field of view of a refractor with the excellent color-handling properties of a reflector.

Visual Telescopes

While most large telescopes are seldom used for direct visual observation (except to make sure they are pointed at the right object), many smaller telescopes are built for direct viewing. The large ones at observatories which have open nights for the public are, of course, pressed into visual use on those occasions.

To understand how a telescope is used with the eye, refer to Figure 7–4. This diagram shows a refracting telescope, although most of what we will say about visual uses of a telescope will also hold for reflecting telescopes. The main lens (or in a reflector, the main mirror) is called the **objective of the telescope.** It forms an image of the object at its focus. To help the eye see the image clearly, a second lens, called an **eyepiece,** is placed behind the focus of the objective. The human eye cannot focus on objects at very short distances. The eyepiece allows you to view the image formed by the objective at very short range without straining your eyes. The eyepiece is placed at a distance behind the image equal to its own focal length. As Figure 7–4 shows, the light from each point in the image departs from the eyepiece in parallel bundles of rays. The normal eye is most relaxed when it forms an image from parallel light, so the eye is able to see comfortably the image formed by the objective, which now appears to have a very large angular size. Therefore, it appears to be enlarged. The combination of the objective and the eyepiece consequently causes the object which you are viewing to appear larger than it would without the telescope.

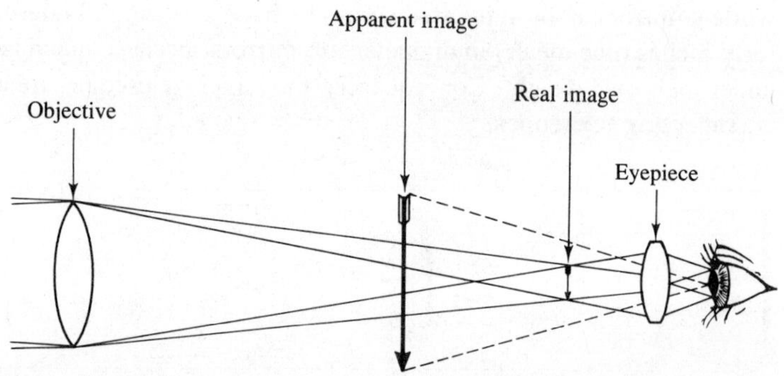

Figure 7–4 Angular magnification by a visual telescope.

The **magnifying power** of a visual telescope indicates the amount by which the object appears to be enlarged. There is an easy way to calculate the magnification. It is given by the expression

$$\text{Magnifying power} = \frac{F_o}{F_e}$$

where F_o = the focal length of the objective
F_e = the focal length of the eyepiece

Without proving the equation, it is easy to see how it works. The longer the focal length of the objective, the larger the image it forms. Also, the shorter the focal length of the eyepiece, the closer the eye is to the image and the larger it appears. The combined effects determine the magnification. Most visual telescopes are supplied with interchangeable eyepieces, thus allowing an observer to select the desired magnifying power.

Note that the eye still sees the image upside down. In telescopes made for bird watching and in binoculars, more optical parts are placed between the objective and the eyepiece to turn the image right side up. Astronomical telescopes are always inverting, however, because the inconvenience of upside-down images is ignored to avoid losing any light in more optical parts.

At first glance, it seems that we could obtain any magnification, without limit, by using progressively shorter eyepieces with a given objective. Thus, if we were to use an objective with a 100-centimeter focal length and an eyepiece with a 10-centimeter focal length, the magnification of the telescope would be 10. With a 1-centimeter eyepiece, the magnification would be 100; a 0.5-centimeter eyepiece would give a magnification of 200. Unfortunately, two other effects limit the useful magnification: one dependent upon the telescope and one upon the air through which we view the object. First, consider the limitation of the telescope itself.

Several views of Saturn are shown in Figure 7–5. The naked-eye view, which is power one, shows no detail at all. At 50 power we begin to see the planet's rings. The 200-power view shows the rings and Saturn itself clearly. The 1000-power view, while it is larger, does not give five times as much detail because we have also magnified the inherent fuzziness of the image.

The fuzziness of the image results mainly from the diffraction of light. As we saw in Chapter 6, whenever light passes through an opening, diffraction causes the light waves to spread out. This spreading consequently causes a blurring of the image. Remember also that the smaller the opening, the more pronounced is the diffraction effect. Therefore,

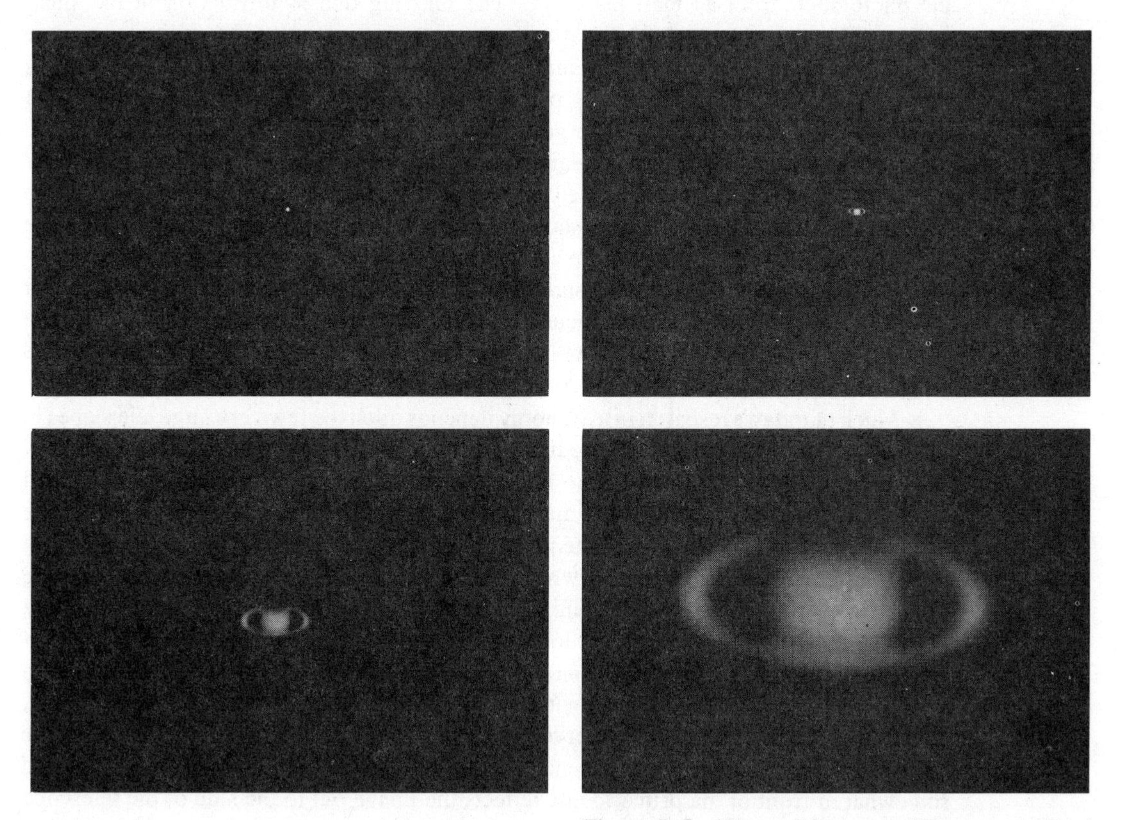

Figure 7–5 Views of Saturn at different magnifications.

Figure 7–6 Resolution of a double star.

telescopes with small diameter objectives have more inherent blurring than telescopes with large diameter objectives. Once the magnification of the telescope has been increased to the point where the fuzziness is noticeable, increasing the magnification even more does not help you see more detail.

Two photographs of a double star—that is, two stars very close together in the sky—are shown in Figure 7–6. Note that the two stars are clearly separated in the photograph taken with a large telescope, while all we see is one large fuzzy image in the photograph from a small telescope. The blurring due to diffraction in the small instrument is so bad that the two images appear to merge into one. The ability to separate two adjacent points of light, and hence the ability to see detail, is called the **resolving power** of a telescope. The resolving power is limited by the wave nature of light. Large diameter objectives are required to get the best results.

Besides giving better detail, a large diameter objective also allows us to see fainter objects. As larger objectives are used, the number of stars which can be seen in a given part of the sky increases. More stars are seen with a larger objective because the light-collecting area of the objective can concentrate enough light to the images to make them bright enough to see or measure. The **light-gathering power** of a lens or mirror increases with the square of its diameter. Therefore, doubling the diameter of an objective would allow you to see stars only one-quarter as bright. Increasing the diameter 100 times would allow you to see stars 10,000 times fainter.

Much of today's research in astronomy depends upon the few very large telescopes in the world because only they have the light-gathering and resolving power needed to study faint celestial objects in detail. The large instruments are all reflectors, so we should explain how you gain access to the images they form. A concave mirror always forms the image of an object on the side of the mirror on which the object is located. Figure 7–7a shows the **prime focus** of a mirror, where the focus would naturally occur and where the best images are formed. On medium to large telescopes, a photographic plate can be placed at the prime focus without blocking out too much incoming light. In the largest telescopes, with diameters more than ten feet (three meters), the observer can sit inside the telescope at the prime focus, without a large light loss. For smaller instruments, however, even the back of the observer's head would block the light and extinguish the image. Newton proposed a way around this, as shown in 7–7b. A small flat mirror, placed somewhat in front of the prime focus, reflects the image out to the side of the telescope where it can easily be viewed. This arrangement is appropriately called a **Newtonian**

focus. Another common arrangement, shown in Figure 7–7c, particularly useful when heavy devices like spectrographs are attached to the telescope, is called a **Cassegrain focus.** It is achieved by placing a convex mirror near the prime focus. This mirror lessens the convergence of the light and reflects it through a hole cut in the objective where the light ultimately comes to a focus.

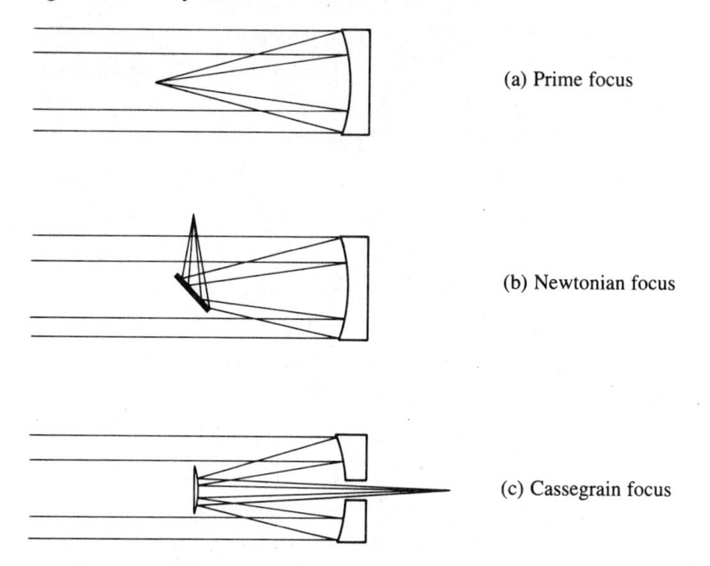

Figure 7–7 Focus arrangements in a reflector.

Notable Optical Telescopes

The largest refractor in the world is the one-meter, or more commonly the 40-inch, telescope at Yerkes Observatory (see Figure 7–8). The objective is a doublet, with a total weight of over 200 kilograms. Notice the mounting of the telescope. The two axes upon which the tube is mounted permit the tube to be pointed everywhere in the sky. One of the axes, called the *polar axis,* is paralled to the earth's axis of rotation. Consequently, a clock can be used to drive the telescope accurately about its polar axis, opposite to the earth's rotation. This rotation keeps an object in the telescope's field of view for an extended time.

The most famous reflector in the world is the 5-meter, or 200-inch, telescope at Palomar Mountain (Figure 7–9). It has a different type of mounting than most other telescopes because of the enormous weight it must support. It is still a polar mounting, however, and the telescope can follow a star with its accurate clock drive. The Palomar telescope was the first in which the observer could sit in a cage at the prime focus.

The largest Schmidt telescope, 1.22 meters or 48 inches, is shown in Figure 7–10. It is also located at Palomar and was built to be used along with the large reflector. Because the large reflector has such a small field of view, it would take around 100 years to examine in detail the entire sky visible from Palomar. The Schmidt, with an excellent field of view, completed a survey of the sky in three years. This survey is used to locate interesting regions to study with the big reflector.

Figure 7–8 The 40-inch (102-centimeter) Yerkes refractor.

Figure 7–9 The 200-inch (5-meter) Hale reflector.

Figure 7–10 The 48-inch (122-centimeter) Schmidt telescope at Palomar Mountain.

The structure shown in Figure 7–11 houses the McMath solar telescope at Kitt Peak National Observatory. It obviously is constructed much differently than the telescopes we have just discussed. The objective of the telescope is stationary; it is placed in a deep hole extending below the slanting section of the structure. At the top of the high end of the building is a large flat mirror which rotates to reflect sunlight down to the objective. The

Figure 7–11 The McMath solar telescope at Kitt Peak National Observatory.

objective, which has a focal length about the length of a football field, forms an image of the sun nearly one meter wide in an observing room near ground level. The huge, bright image of the sun can be studied in minute detail from the stationary observing room.

Uses of a Telescope

A telescope and the measuring and detecting devices attached to it are used to make one or more of three basic measurements. The most obvious is to measure the position of objects in the sky, or, if you wish, to determine the direction from which the light is received. Another use is to measure the brightness of the celestial objects, or, in other words, to measure how much energy is received. And finally, the telescope and its attachments are used to measure the quality of the light received. By that we mean how the light is distributed according to wavelength.

Although major telescopes are rarely used with the eye, they are frequently used as large celestial cameras. There are two major advantages. First, we get a permanent record, for accurate measurement and comparison with other photographs. Second, by using a very long exposure, we can get photographic records of objects which are much too faint to be seen with the eye. Most astronomical photography is done with emulsions on glass plates because they are rigid and give the most stable negatives. These plates can be used to measure the position of one star relative to the others on the plate. Measuring the blackness and size of the images tells about the stars' relative brightnesses. If we place a piece of colored glass, a *filter,* in front of the plate, we get information about the relative amount of energy received in different colors.

Comparing two photographs of the same celestial region taken at different times lets us measure the motions of the stars relative to each other and identify stars that change in brightness. Photographs also allow us to see details of structure in many extended celestial objects much too faint to be seen with the eye.

The most precise measurements of the light received from an object are made using the photoelectric effect. Figure 7–12 diagrams a photoelectric photometer, that is, a light-measuring device. The light from a single star is isolated by an opaque shield with a small hole in it placed at the focus of the telescope. This light is then focused by an auxiliary lens on a photoelectric cell. The cell produces an electrical current which is precisely related to the light which falls on its sensitive surface. By placing a color filter before the cell, we can also measure the amount of light received from the star in different parts of the spectrum.

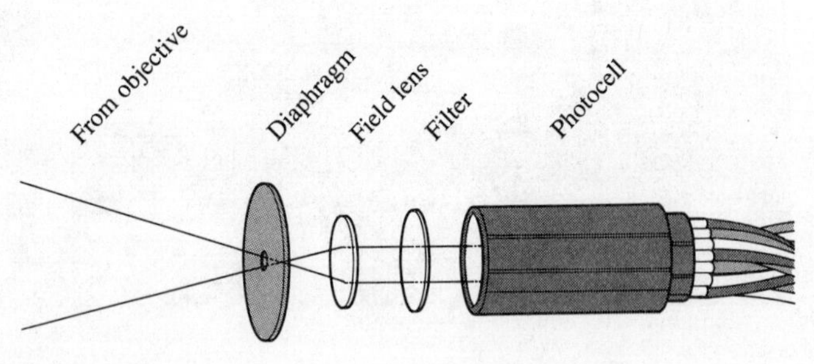

Figure 7–12 Components of a photoelectric photometer.

We can study the distribution of light by wavelength with a spectrograph. A schematic diagram of a spectrograph is shown in Figure 7–13. Light enters the spectrograph through a narrow slit. This light is then reflected in a parallel beam by a concave mirror toward a diffraction grating. The light reflected by the grating is dispersed, with light of different wavelengths leaving in different directions. A second mirror focuses this dispersed light onto a photographic plate. The plate records an image of the spectrograph slit in each color in the light of the star. This spectrum can then be studied to find the star's temperature, its radial velocity, the chemical elements present, and many other facts. In some spectrographs, sensitive electronic devices are substituted for the photographic plates, giving systems which can detect the spectra of fainter objects and attain greater precision.

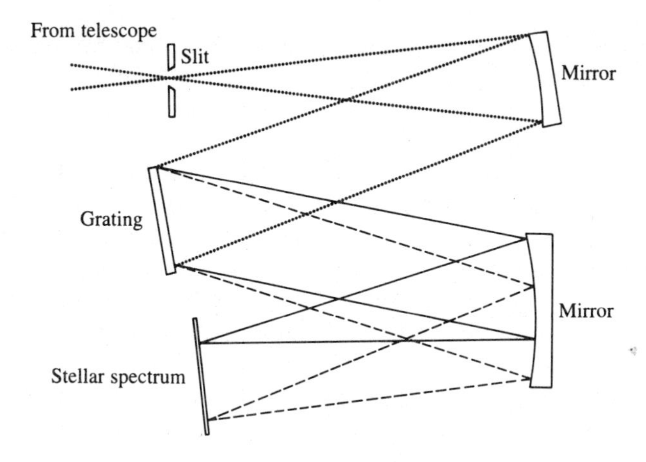

Figure 7–13 Diagram of a grating spectrograph.

The Influence of the Earth's Atmosphere

It is rather pleasant to be surrounded by air to breathe, but it is an astronomical nuisance. The earth's atmosphere, which is in constant movement and varies in temperature, causes small and continual changes in the direction of starlight passing through it. This causes the twinkling of stars and, more disturbing, a continual dancing of the stellar image. More often than not, the resolution which can be attained with a moderate to large telescope is limited by the atmosphere before diffraction becomes important. Observatory sites are therefore chosen because of the usual condition of the atmosphere above them. High mountain locations are chosen to get above the lower and denser parts of the atmosphere.

The faintest stars which can be photographed from the earth are also limited by the atmosphere because of scattered artificial light and a natural glow from the night sky. When we make a long exposure using a telescope, the light of the night sky builds up a background fog which eventually swamps the faintest images. Thus our own atmosphere sets the limit of how deeply we can see into space.

Besides impairing images, the earth's atmosphere has another even more devastating effect; it strongly absorbs radiation in the X-ray and short ultraviolet region, as well as parts of the infrared, microwave, and radio regions. Figure 7–14 shows a plot of the

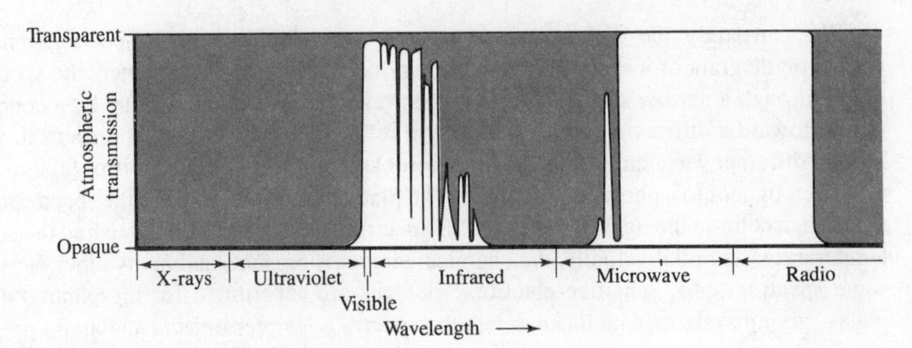

Figure 7–14 Transmission windows in the earth's atmosphere.

atmospheric windows, the regions of the spectrum in which the earth's atmosphere is transparent. Ground-based observations, even from high mountains, will always be unable to detect radiation in some parts of the spectrum, the very parts where some of the most exciting discoveries may yet be made.

Fortunately, we are now able to defeat the interference of the atmosphere. By suspending a telescope from a high-altitude balloon floating at thirty kilometers, with less than one percent of the earth's atmosphere above it, most of the atmospheric effects are eliminated. Observations from rockets and artificial satellites, like the Orbiting Astronomical Observatory, are completely free of atmospheric interference. Not only can any part of the spectrum be reached, but the telescopes can perform at their theoretical resolving power. Ultimately, when large telescopes are put into space, we should be able to detect fainter stars than ever before. Already space observations have led to the discovery of matter in the universe which exists under more extreme conditions than could be imagined only a few decades ago.

Radio Astronomy

In 1931 Karl Jansky, a Bell Telephone engineer, discovered radio noise from outer space. Not much effort was spent on this radiation until the 1950s, when electronic techniques were sufficiently developed to introduce radio astronomy in earnest. Radio observations have had a strong impact upon astronomy because they are particularly suited for finding matter under extreme conditions. Most stars which we can see radiate very nearly as a black body would radiate. A large part of the radiation is in the visible spectrum; a relatively small amount is in the radio portion of the spectrum. Many objects radiate much differently, however. They put out large amounts of radio energy because at least some of the matter in them is in a much different condition from that in the stars. The interpretation of these objects has profoundly affected astronomy as well as physics.

Radio telescopes come in many shapes and sizes. Some look like nothing more than a vast field of radio antennas. The radio telescope shown in Figure 7–15 resembles the reflecting telescopes used in optical work. The large concave mirror, often called a *dish,* is made of metal plates or wire mesh. The dish reflects radio waves, concentrating them on a sensitive receiver at the focus of the dish. The receiver can be turned to a particular frequency much as you tune your home radio. Instead of receiving the local news, the

Figure 7–15 The 42-meter radio telescope at the National Radio Astronomy Observatory.

radio telescope receiver measures the radio energy received from the part of the sky toward which it is pointed and at the frequency to which it is tuned. By tuning the receiver across a number of frequencies, a radio spectrum of a source can be measured.

Radio telescopes have to be much larger than optical telescopes, not only to gather more of the feeble radio energy falling upon them but to give them even modest direction-measuring capabilities. The resolution of any telescope depends upon the wavelength of the waves which it collects. Diffraction effects increase with the wavelength of the radiation. Since radio waves are hundreds of thousands of times longer than light waves, the resolving power of even large radio telescopes is less than that of a modest optical telescope. The largest radio dish in use today is the 300-meter dish at Arecibo in Puerto Rico. The dish is so large that it could not be made free-standing and still be self-supporting. Consequently, it was made by excavating a natural basin in the ground to give it the proper shape and lining it with a suitable metal reflecting surface. Even though it is 300 meters in diameter, its resolving power is only equivalent to a visual telescope a few millimeters in diameter.

To improve resolution, and still keep construction costs within reason, radio astronomers use interferometers. A radio **interferometer** consists of two or more connected radio telescopes. When it is done properly, the resolution of the system corresponds to the largest separation of its components, not to the size of the individual parts. Modern technology has even made it possible to use two telescopes, not physically connected, as an interferometer. The signal received at each telescope is recorded on magnetic tape along with a signal from an extremely accurate atomic clock. If the clocks at both telescopes have been carefully synchronized, the timing signals on the two tapes can be used to play the two tapes together just as if the telescopes had in fact been connected.

Spacings of 8000 to 10,000 kilometers have been used between telescopes in the United States and telescopes in Australia, Europe, and the Soviet Union. By using an interferometer approaching the size of the earth itself, radio astronomers have been able to get a resolving power exceeding that of even the largest optical telescopes.

Among the many interesting things discovered by radio astonomers are the complicated molecules like ethyl alcohol and formaldehyde found between the stars. They were found by identifying spectral lines emitted in the microwave region. The molecules could never have been found from optical observations because they exist in regions so cold that they cannot emit at visible wavelengths. It has also been possible to measure the surface temperatures of the planets by measuring their radio radiation. Visible light is not of much use in finding planetary temperatures because all we see is reflected sunlight, which (although slightly modified by the planet) is more characteristic of the sun than the planet itself.

Radar Astronomy

A radio telescope is completely passive; that is, it merely collects radiation just as an optical telescope merely collects light. It is possible to make a dish active by using a radio transmitter as well as a receiver at its focus. We then have a **radar** system. The technique of sending a radio signal and then receiving the reflected echo from something in the path of the transmitted signal is called *radar*.

For example, in using a radar system to study Mercury, a pulse of radio energy with a power of one million watts is sent toward Mercury. The dish, acting like a searchlight, focuses the radio energy toward the planet. The radio signal moves outward from the earth with the speed of light, spreading and getting weaker as it approaches Mercury. Nearly five minutes later, if Mercury is closest to the earth, the pulse strikes Mercury. Another five minutes pass and the reflected echo, with a power of only 10^{-21} watts, returns to the dish to be detected by the radar's receiver. By timing precisely how long it takes for the signal to make its round trip, dividing this interval by two, and multiplying the result by the speed of light, we can find the distance between the earth and Mercury. While this technique works well for nearby objects, the nearest star is so far away that a radar system billions upon billions of times more powerful than any available would be needed to detect it. Radar astronomy is therefore useful only in our solar system.

A detailed study of the radar echo from a planet can supply information about the surface of the planet. First, the echo is spread in time, because the portion of the echo from the part of the planet closest to the earth is received before the echoes of the more distant parts of the planet. There is also a spread in the frequency introduced into the echo whenever the planet rotates, because each portion of the planet introduces a different Doppler shift. By analyzing the time delays and Doppler shifts in the echo, we can construct a radar map of the reflection properties of a planet. Radar maps of Venus have been constructed in this way. They provided the first evidence of details of its surface, which is always hidden from view by its thick atmosphere.

Radar measurements have been used to find a surprising number of facts about the nearer planets. Besides the obvious measurement of planetary distances, radar has given us the planets' rotation speeds, the sizes, and surface roughness.

Space Astronomy

Besides using space vehicles to carry instruments above the earth's atmosphere for direct observations, spacecraft can be sent to the planets for close-up measurements. Eventually they may return samples of planetary material as they did for the moon.

The space probes have themselves been used to measure the masses of planets with greater precision than was possible when we were earth-bound. Whenever a spacecraft nears a planet, its motion is influenced by the planet's gravitational attraction. By measuring the exact acceleration induced, we can use Newton's law of gravitation to calculate the mass of the planet. The continuously changing velocity of the probe is found by measuring the frequency of a known radio signal sent from the space probe. The difference between the received frequency and the known frequency gives the Doppler shift and, hence, the velocity of the probe. The accuracy of these measurements is astounding. When *Pioneer II* made its closest approach to Jupiter, it reached a maximum velocity of 48,000 meters per second. The error in the measured velocity was only 0.3 millimeters per second, that is, only six parts in a billion. With this degree of precision, the accelerations can be calculated so accurately that we can determine the planetary masses with far greater certainty than ever before.

Whenever a probe passes behind a planet (as seen from the earth), the probe's signals have to pass through the planet's atmosphere to reach the earth. The gases of an atmosphere slow down and weaken radio signals passing through them. The amount by which the signals are retarded depends upon the density of the gases. Measuring this changing delay as the spacecraft sweeps behind the planet lets us calculate how much gas there is in the planet's atmosphere between us and the spacecraft at each moment. From these data, a mathematical model of the planet's atmosphere can be constructed, giving its temperatures and pressures in terms of height above the planetary surface.

The use of space vehicles has opened a whole new way of making astronomical measurements. These measurements, along with new and refined measurements from the earth, will surely make astronomy exciting and challenging for years to come.

KEY TERMS

refracting telescope (refractor)

reflecting telescope (reflector)

focal length

telescope objective

eyepiece

magnifying power

resolving power

light-gathering power

prime focus

Newtonian focus

Cassegrain focus

interferometer

radio telescope

radar

Review Questions

1. What is the difference between refracting and reflecting telescopes?
2. What is the focal length of a lens?
3. When we compare the images formed by two lenses of different focal lengths, what difference is there?
4. *True or false:* All telescopes that use visible light are designed so the astronomer can look through them.
5. How does the magnifying power of a telescope depend on the focal lengths of the objective and eyepiece?
6. How can the magnifying power of a telescope be easily and quickly changed?
7. Very large telescopes are expensive to build and operate. What is the major reason for building them?
8. Where is the eyepiece located for the Newtonian focus? For the Cassegrain focus?
9. If you have excellent quality mirrors or lenses in a telescope, what determines its resolving power?
10. What can you do to increase a telescope's resolving power without changing the size of the objective?
11. How does the resolving power of the 200-inch telescope using visible light compare with that of a radio telescope of the same diameter?
12. How is the photoelectric effect used in observing the stars?
13. Is a radio telescope a reflecting or a refracting telescope?
14. How are radio telescopes useful in studying the planets?
15. What properties of a radar echo can be used to construct a radar map of a planet?
16. *True or false:* The most distant object so far detected with radar is Proxima Centauri, the nearest star to the sun.
17. What is the most accurate way to measure the mass of a nearby planet?
18. How can you use radio signals sent to the earth from a space probe to find the amount of atmosphere around another planet?

Discussion Questions

1. Name some disadvantages of each of the two basic types of telescopes, refracting and reflecting.
2. What factors limit the detail you can see when you use a telescope?
3. Suppose you see in the sky a comet appearing as a very faint luminous patch to the naked eye. To see it best through a small telescope, should you use a small or large magnifying power?
4. Why aren't all observations of the heavens made through interferometers?
5. How do we use photographs of the sky?

6. Name some advantages of making astronomical observations from an artificial satellite.
7. How are radio telescopes useful in studying clouds of gas between the stars?
8. What can a radar experiment measure about another planet?
9. What is the difference between a radio telescope and a radar telescope? Which one would be used to measure the temperature of another planet and why? Which would be used to measure the rotation period of another planet and why?

8

The Planets

Three times before 1971, spacecraft from Earth flew by Mars and sent back television pictures. In each case, the spacecraft flew by just as one side of Mars rotated into view. The side of the Martian surface which was observed is cratered much like the moon's surface. No extraordinary features were seen.

No one could be sure, however, that Mars' surface would be similar on the other side, which had not yet been closely observed. So in 1971 Mariner 9 was sent into orbit about Mars to observe all sides of it. For the first few weeks, as luck would have it, virtually nothing could be seen because of a tremendous dust storm. When the storm subsided, scientists were shocked to see a huge crater peaking through the settling dust. Eventually they found the crater to be the top of a huge volcanic mountain, dwarfing any found on Earth. Subsequent pictures showed a tremendous canyon, far larger than the Grand Canyon, wrapping nearly one-fourth of the way around Mars. No one could have predicted this great canyon or the giant volcano on the basis of the appearance of the planet's other side.

Our technology is now so sophisticated that we can get a close-up view of the planets. It is hard to believe that only two centuries ago, even the existence of three planets was totally unknown. The last planet to be discovered, Pluto, was not found until 1930.

Until 1781, the known planets were Mercury, Venus, Earth, Mars, Jupiter, and Saturn. In that year William Herschel, a great English astronomer, discovered a seventh planet. Barely bright enough to be seen with the unaided eye, this planet had been observed before Herschel's time, but no one had noticed its slow progression relative to the stars.

Once Herschel had discovered the planet, it had to be named. Herschel wanted to call it Georgium Sidus in honor of England's king at that time, George III (the same George III who managed to lose the American colonies). Others wanted to name it Herschel. Imagine—planet Herschel! Fortunately, good sense prevailed. In keeping with the tradition of using the names of Roman gods, the newly found planet was named after Uranus, father of Saturn and grandfather of Jupiter.

Mercury, as observed by Mariner 10

After Herschel discovered that Uranus is a planet, it was continually observed. Its orbit was calculated. Old star charts were examined to map its previous path. Apparently something was affecting the planet because its motion was not consistent with its theoretical orbit, as calculated according to the law of gravitation. As a matter of fact, some scholars speculated that the gravitation law did not apply to an object so far away as Uranus. Others thought that the law held for Uranus as for the other planets, but that there was an unseen planet whose weak gravitational attraction on Uranus caused the observed deviations.

In 1843 John Couch Adams, a recent graduate of Cambridge University, set out to calculate the position of the supposed planet. In less than two years he had the answer. He wrote to the Astronomer-Royal at the Greenwich Observatory, Sir George Airy, giving him the coordinates of the unseen planet.

Sir George was not willing to interrupt the programs already underway at Greenwich to search for a planet that might not exist. Besides, Adams was a novice. Could he really handle the mathematical intricacies of such a calculation? To find out, Sir George decided to test Adams. He asked him for explanations of some of the perturbations of the orbit of Uranus that he himself had noticed. Adams did not bother to reply, and the matter was temporarily closed.

Meanwhile, a brilliant French mathematician named Urbain Jean Joseph Leverrier, unaware of Adams' work, attacked the same problem. Leverrier published a paper in June, 1846, announcing the position of the unseen planet. Within a month Airy received a copy of the paper. To his astonishment, he saw that Leverrier's prediction was within one degree of the position proposed by Adams almost one year earlier. Still, Sir George proceeded cautiously. Instead of ordering an immediate search for the new planet, he sent the same test to Leverrier that he had sent to Adams. Leverrier promptly and satisfactorily answered Airy's challenge.

At last Sir George decided that the time was ripe for a search. And yet, he was still unwilling to interrupt the programs already underway in Greenwich. Instead, he wrote to Professor Challis at Cambridge, suggesting that Challis search for the new planet with the large telescope there. Challis agreed and proceeded to accumulate some observational data that he intended to examine at his leisure. In the process of collecting these data, Challis had unknowingly seen the new planet twice, on August 4 and August 12, 1846.

Meanwhile, Leverrier was having his own ideas on how the search should be made. He wrote to Dr. Galle, the director of the Berlin Observatory, telling him in detail where to look for the new planet. Galle received Leverrier's letter on September 23, 1846. That very night, he pointed his telescope in the direction specified by Leverrier and saw—and knew that he was indeed seeing—the planet. On that night the planet was located in the constellation Aquarius. It seemed fitting that it should be named Neptune, after the Roman god of the seas.

In spite of Sir George's agonizing foot dragging, which probably cost Adams immediate recognition, the honors for Neptune's discovery go to both Adams and Leverrier. This discovery is one of the great achievements of the theory of gravitation. Adams and Leverrier, armed only with this theory and mathematical expertise, predicted the precise location of a far-distant world which no one had ever seen.

The last planet to be discovered, Pluto, was found in a rather strange way. Percival Lowell, at his observatory in Flagstaff, Arizona, thought there were perturbations in Uranus' orbit which could not be attributed to any known planet, including Neptune. Assuming that there was yet a ninth planet, he worked out its position on the basis of its alleged effect on Uranus. Lowell did not live to see it, but in 1930 an assistant at Lowell Observatory, Clyde Tombaugh, found Pluto within six degrees of the position Lowell predicted.

Later analysis of Lowell's calculations showed that they were in fact not applicable to Pluto and that its discovery was purely accidental. But it was no accident that the Lowell Observatory had systematically searched for a new planet. That search was carried out at Percival Lowell's insistence. So even though the discovery of Pluto ranks as one of the most astounding coincidences of science, Lowell is fittingly honored as its discoverer.

We think of the earth as a big place until we begin to consider it as a part of the solar system, which consists of the sun and all its companions. In order of size, Earth stands midway in the list of the nine planets. Mercury, Venus, Mars, and Pluto are smaller, while Jupiter, Saturn, Uranus, and Neptune are much larger. It also stands midway in the list of planets arranged by mass. All the planets, in turn, are dwarfed by the sun, which has a diameter of 109 Earth diameters and a mass of 333,000 Earth masses. The sun contains almost 99.9 percent of the mass of the solar system, and Jupiter alone contains nearly three-quarters of the remaining 0.1 percent.

Even more striking than these comparisons is the vast scale of the solar system itself. In reality, it is quite empty. If we reduce the solar system to the size of the island of Manhattan, as shown in Figure 8–1, it is easier to comprehend the relative proportions of the system. The sun would be a luminous sphere two meters in diameter in the middle of the island. Earth would be half the size of a golf ball, three city blocks away. And Neptune, on an orbit just crossing the extremes of Manhattan, would be size of a baseball.

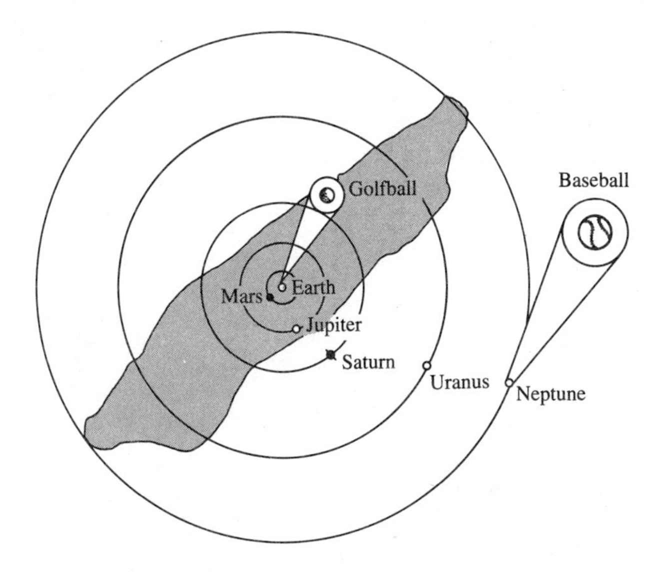

Figure 8–1 The relative scale of the solar system.

When we consider the great distances of the planets compared to their sizes, it is obvious why our knowledge of the physical nature of the planets has grown slowly.

Since the invention of the telescope, generations of astronomers have fought the weather and the general unsteadiness of our atmosphere to get short-lived detailed glimpses of the planets. Drawings, and later photographs, of the planets have been obtained, and maps of surface details constructed. But even for Mars, the planet usually showing the most detail, we had found relatively little exciting or new by these methods in the past several decades. Fortunately, the space program and radio and radar astronomy have changed that. The space probes have taken our instruments to the planets for close-up examination. Radio and radar techniques have let us discover facts about the planets that were forever hidden when we could only observe them at long range in visible radiation.

Mercury

Mercury is the closest planet to the sun and, at its brightest, the seventh brightest object in the sky. The very quality that makes Mercury so bright, its proximity to the sun, also makes it difficult to observe. The maximum elongation from the sun which Mercury can have is about 28°, so it always lies close to the sun in the sky. When Mercury can be seen in the late evening or early morning sky, it is close to the horizon. Its image in a telescope is usually poor because of the large path through the earth's atmosphere necessary to view it. Serious observers, therefore, usually observe Mercury in the daytime when it is high in the sky. The steadiness of the image improves, but the clarity of the faint markings decreases because of the competing daylight.

Persistent and careful observers have been able to see enough detail on Mercury to draw maps. In a telescope, Mercury looks somewhat like the moon. Mercury's markings led visual observers to conclude that it rotates once on its axis for every revolution about the sun. Its rotation was assumed to be synchronized with its revolution by tidal friction induced by the sun.

In the 1960s, radar signals were reflected and received from Mercury with a completely astounding result—Mercury rotates with a period of 59 days, not 88 days as everyone had thought. The theory of Mercury's motion was quickly reviewed; and sure enough, for a body on a highly elliptical orbit like Mercury's, a rotational period

Table 8–1
Mercury: Vital statistics

Size: 4880 km = 0.38 Earth
Mass: 0.055 Earth
Density: 5.4 g/cm^3
Rotational period: $58^{d}.6$
Orbit
  Semi-major axis: 0.387 AU = 57.9 million km
  Eccentricity: 0.206
  Inclination: $7^{\circ}.0$
  Sidereal period: $87^{d}.97$

two-thirds as long as its period of revolution is compatible with the sun's tidal influences on Mercury. Mercury has two permanent tidal bulges. When the planet is at perihelion, a bulge always faces the sun. The two bulges alternate on successive passages.

But how were the visual observers in error so consistently for nearly 100 years? It was one of those unhappy circumstances of nature. When the configurations of the earth, sun, and Mercury are best for telescopic observations of that planet, nearly the same face is presented toward the sun. After Mercury's true rotation period became known, we could show that the visual observations were, in fact, consistent with the shorter period.

We obtained the first detailed information about Mercury's surface in 1974 when *Mariner 10* passed by and returned television pictures of the planet. Figure 8–2 shows a *Mariner 10* picture. The surface features are amazingly similar to those on the moon. There are heavily cratered highlands and comparatively uncratered basins where previous craters have apparently been filled in with lava flows. A portion of the largest such basin discovered on Mercury appears in Figure 8–2 as a semicircle on the left-hand side. In size and shape, this basin closely resembles the largest mare basin clearly seen on the moon. The impact that produced this feature must have generated very strong disturbances in the planet, of a type similar to seismic waves generated on Earth by earthquakes. As these powerful waves spread through and around the planet, they would have come together on the opposite side of Mercury. This region, shown in Figure 8–3, contains a very irregular broken surface, probably resulting from the effects of the converging waves.

While most of Mercury's surface features are like the moon's, one distinct difference is the pattern of ridges and cracks on Mercury. Many examples show up inside the large basin in Figure 8–2. The cause of these patterns is most likely a compression of the surface

Figure 8–2 Large basin on Mercury.

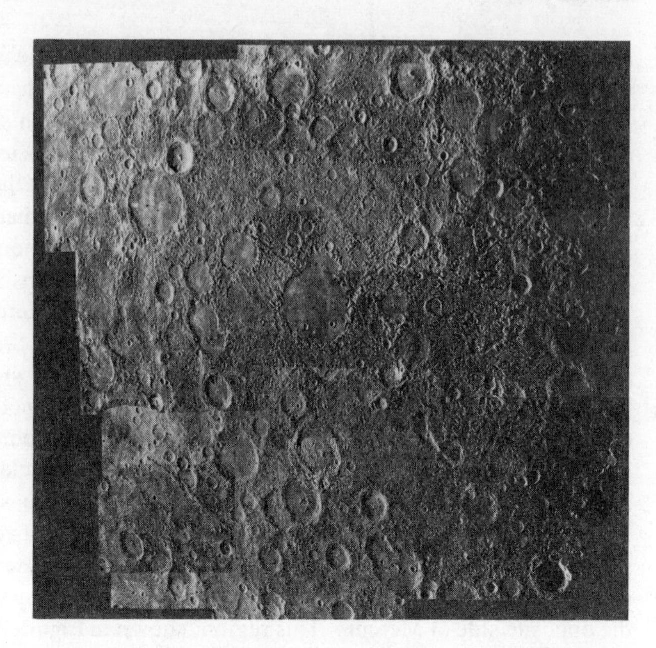

Figure 8–3 Disrupted region on Mercury.

of the planet, implying that the planet's radius has shrunk slightly (one or two kilometers) as Mercury cooled off after an initially hot beginning. The view of one such ridge, near the right-hand side of Figure 8–4, shows a cliff about three kilometers high, crossing completely through two older craters. The history of Mercury's surface appears to be more complex than the moon's.

A second and fundamental difference between Mercury and the moon is the average density. Mercury, with a density of 5.4 grams per cubic centimeter, is nearly twice as dense as the moon. To create this high density in a small planet requires a large dense core.

Figure 8–4 A cliff, three kilometers high, crossing craters on Mercury.

Consequently, Mercury is thought to have an iron-rich core with a diameter 80 percent of the diameter of the planet itself, thus making Mercury the most iron-rich of all the planets. As such a core cooled and solidified, it would have shrunk about the amount needed to explain the wrinkling of the surface.

We have long known that Mercury has little or no atmosphere. *Mariner 10* confirmed this when it found only a slight trace of helium there, with a pressure one trillion times less than the earth's atmosphere. It would have been surprising if much more of an atmosphere had been discovered. Because of its nearness to the sun, Mercury has a high surface temperature, near 700K, on its sunlit side. This is hot enough to melt lead or tin. Gas particles at the planet's surface would have relatively high velocities due to this temperature. And the surface gravity is so low that the velocity of escape from Mercury is only one-third that for Earth. The combination of the high temperature and the small velocity of escape means that Mercury could not retain an atmosphere for any appreciable time. In fact, the helium atoms around Mercury did not come from the planet but rather originated in the sun. The hot outer atmosphere of the sun steadily streams outward, forming a tenuous stream of hot ionized gas known as the **solar wind** that spreads throughout the solar system. A few of the helium ions in the solar wind manage to capture electrons, forming helium atoms. Some of the atoms are trapped by Mercury's gravity for perhaps a year before escaping back into space. Such a thin atmosphere means that Mercury, like the moon, has been undisturbed by erosion since the beginning of the solar system.

One of the most interesting observations of the Mariner mission to Mercury was the behavior of the solar wind near the planet. Much as a boat forms a bow wave as it cuts through water, a bow wave is produced as the ionized gas of the solar wind streams by Mercury. But this bow wave forms far above the surface of Mercury, as shown in Figure 8–5. This means that Mercury has a magnetic field; otherwise the bow wave would approach much closer to the planet's surface. The ionized particles of the solar wind are deflected by the magnetic field well before they reach the planet's surface. Because

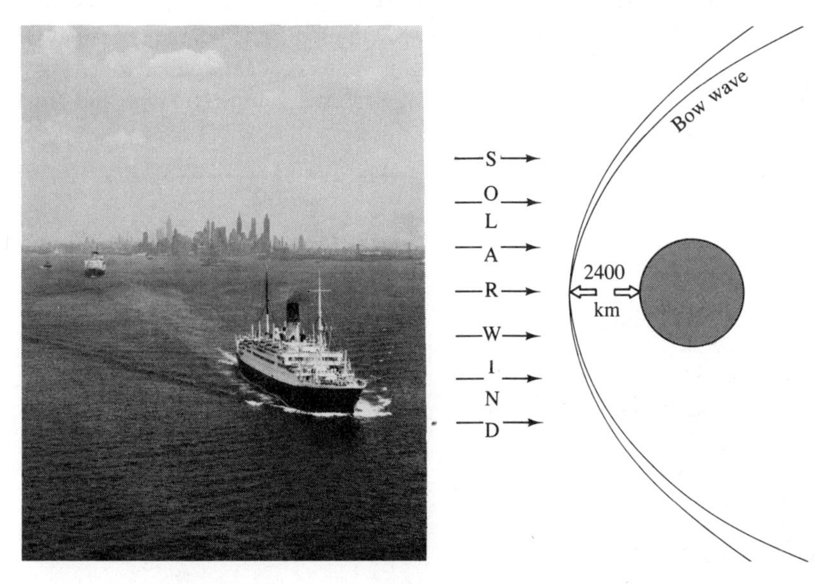

Figure 8–5 Photograph of the bow wave from a boat and diagram of the bow wave of Mercury in the solar wind.

magnetic fields are usually thought to be connected with large, rapidly rotating planets, finding even an extremely weak one about Mercury was completely unexpected.

Venus

Venus comes closer to Earth than any other planet, approaching as close as 40 million kilometers. At greatest elongation, Venus lies 47° away from the sun. Its maximum brightness, when it outshines everything in the sky except the sun and the moon, occurs at an elongation of about 39°. Venus is then in the crescent phase and can easily be seen by eye during the daytime, if you know where to look. When Venus is closest to Earth, it is unfortunately in the new phase. It cannot be easily seen except on infrequent occasions when it crosses directly in front of the sun and appears as a black dot silhouetted against the sun. (This next happens in the year 2004.) The great range in Venus' brightness is a combination of the effects of its phase and its varying distance from Earth during one synodic period. Like Mercury, Venus has no moons.

Venus is easy to observe telescopically during most of its synodic period. But except for its continually changing phases, it does not present a very exciting view because of the dense, cloudy atmosphere that completely hides its surface in visible light. To chart the surface, a radar telescope must be used. Radiation at radio wavelengths can penetrate the clouds, permitting the construction of a map showing how various parts of the surface reflect the waves. Even before the first map was completed, one of the first results of radar studies of Venus was the observation that the returned echo has a very small range of Doppler shifts. This demonstrates that Venus rotates slowly, once in 243 earth days. But, surprisingly, the rotation is retrograde, east to west, opposite to the direction of rotation of the sun and most planets. The spacecraft measurements showing that Venus does not have a significant magnetic field are easily understood in view of this slow rotation.

The radar maps show signs of past activity on the surface of Venus. Figure 8–6 shows a prominent dark streak, running nearly north-south, which may be a large rift valley. If this interpretation is correct, then internal motions in Venus probably pulled the surface apart there. Other areas of Venus show mountains up to three kilometers high and large

Table 8–2
Venus: Vital statistics

Size: 12,112 km = 0.95 Earth
Mass: 0.82 Earth
Density: 5.3 g/cm^3
Rotational period: $242^{\mathrm{d}}.9$ (retrograde)
Orbit
  Semi-major axis: 0.723 AU = 108.2 million km
  Eccentricity: 0.007
  Inclination: $3^{\circ}.394$
  Sidereal period: $224^{\mathrm{d}}.7$

smooth regions that may be similar to lunar maria. Unfortunately, however, the slow retrograde rotation of Venus causes it to turn almost exactly four times between successive inferior conjunctions. The radar mappings are done at these times, when Venus is closest to Earth, so only one side has been mapped so far.

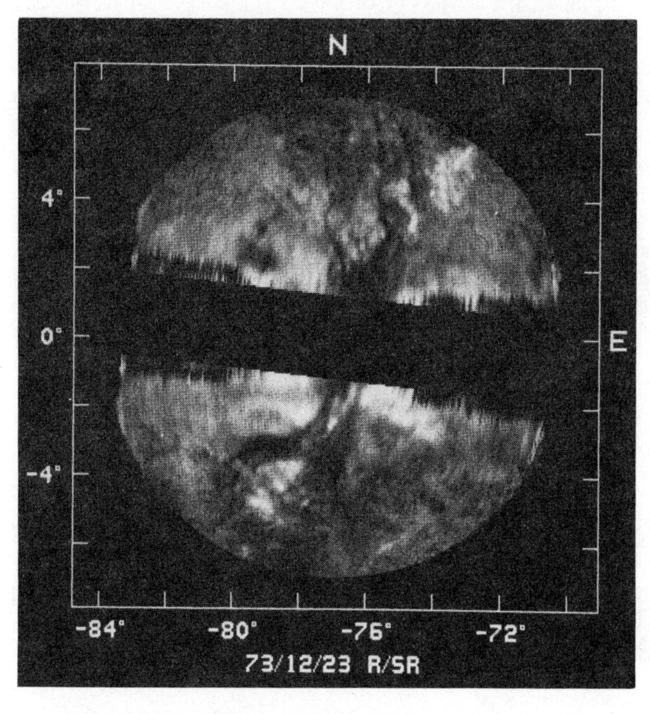

Figure 8–6 Radar map of Venus, showing a large rift.

The best ground-based photographs taken in violet light, like the one in Figure 8–7, show faint dark markings which change with time. *Mariner 10* sent back the pictures in Figure 8–8 and Plate 5A when it passed Venus on its way to Mercury. These pictures,

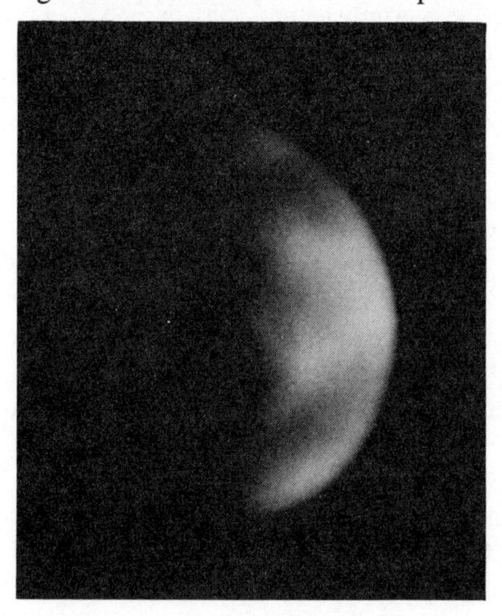

Figure 8–7 Venus photographed from Earth.

however, were taken in ultraviolet light and have been enhanced by computer to bring out detail. They show a rather complex circulation pattern. The three photographs in Figure 8–8, taken at seven-hour intervals, show how rapidly the clouds move around the planet. The same dark cloud patch, about 1000 kilometers across, is identified by an arrow in each picture. Like the surface of the planet, the clouds also move from east to west, but at a much faster rate, taking only four earth days to rotate once. North-south cloud belts are easily seen crossing east-west belts, particularly in the third picture of Figure 8–8, thus showing there are several layers of cloud in the atmosphere. The cloud belts at different heights in the atmosphere are driven by different winds. Also, the atmosphere at greater latitudes from the equator moves around the planet faster than it does near the equator (just the opposite of what happens on the sun and Jupiter).

Figure 8–8 Venus photographed from *Mariner 10,* showing cloud movement.

The temperature of the clouds, high in the upper layers of the atmosphere, can be found from their brightness at infrared wavelengths. It is only about 235K, or −38°C. The clouds are not made up of water droplets or ice crystals like those on earth, however, but apparently consist of droplets of almost pure sulfuric acid! The different gases in the atmosphere were measured by several Soviet space probes as they descended through the atmosphere on parachutes. Carbon dioxide, CO_2, dominates the composition, accounting for approximately 97 percent of the volume of the atmosphere. Most of the remaining gas is nitrogen, with traces of oxygen, carbon monoxide, and hydrochloric and hydrofluoric acids. Table 8–3 compares this composition with the atmospheres of the next three planets. But the most surprising aspect of Venus' atmosphere is how much of it there is. In size and mass, Venus is very similar to Earth, and the two are often referred to as sister planets; yet Venus has about 90 times as much atmosphere!

Our best explanation of why the atmospheres are so different results from an earlier discovery about the planet which was made with radio telescopes. When radio waves were first detected coming from Venus, they were about twice as strong as expected. Either some unknown phenomenon on or near the planet generated the extra radio energy, or else the surface had to be very hot, around 750K, as hot or hotter than the sunlit side of Mercury. The first successful space probe to visit another planet, *Mariner 2,* obtained close-up measurements of the radio waves and found they came from the surface,

Table 8–3
Table of comparative atmospheres

	Percentage Composition by Volume			
	Venus	Earth	Mars	Jupiter
Hydrogen (H_2)	~0%	5×10^{-5}%	~ 0%	**90%**
Helium (He)	~0%	5×10^{-4}%	~ 0%	**9%**
Nitrogen (N_2)	3.2%	**78.1%**	2.7%	~ 0%
Oxygen (O_2)	<0.1%	**20.9%**	0.13%	~ 0%
Neon (Ne)	small	2×10^{-3}%	~ 0%	small
Argon (Ar)	small	0.9%	1.6%	small
Carbon monoxide (CO)	10^{-3}%	10^{-4}%	0.07%	~ 0%
Carbon dioxide (CO_2)	**96.6%**	.03%*	**95.3%**	~ 0%
Methane (CH_4)	< 10^{-4}%	2×10^{-4}%	~ 0%	.07%
Ammonia (NH_3)	~0%	~ 0%	~ 0%	.02%
Water vapor (H_2O)	~0.1%	0.1 to 2.8%	0 to .03%	~ 10^{-4}%
Total surface pressure compared to Earth:	90	1.0	.007	No defined surface

*Approximately 250,000 times as much CO_2 exists in marine deposits and rocks.
< means "less than"
~ means "about"

definitely showing the high temperature to be there. Also, the effects of Venus' atmosphere on radio signals sent back to Earth as *Mariner 2* disappeared behind the planet gave the first direct indication of a high atmospheric pressure on Venus.

The Russian probes that successfully landed on the surface agreed with these results, although the first probes entering the atmosphere were crushed by the great pressure before reaching the surface. The fact that the later probes did reach the ground and survived for more than a few minutes at the extreme temperatures and pressures is impressive. The Russians refrigerated their capsules before dropping them into the atmosphere and depended on thick layers of insulation for protection from the 750K heat at the surface. One probe landed in an area of rocks with jagged edges, while another found a region of smoother (and hence more eroded and probably older) rocks.

The landers had a clear view of distant objects, so the cloud layers do not extend to the ground. The high temperatures near the surface must keep droplets from forming and, furthermore, would quickly evaporate any droplets that might fall from the clouds. The day side of the planet's surface is well lit, so a fair amount of sunlight manages to filter down through the thick clouds to heat the surface. The warmed surface then radiates energy at infrared wavelengths. However, the carbon dioxide in the atmosphere is opaque at many infrared wavelengths, so the energy emitted by the surface cannot escape very easily. Because of this trapped energy, the Venusian surface temperature builds up. This effect, creating high temperatures by admitting visible light but restricting outflowing infrared waves, is called the **greenhouse effect.** A greenhouse on Earth and a car with closed windows parked in sunlight both demonstrate this effect, since glass is opaque to infrared wavelengths. The glass also prevents the warm air from rising and being replaced by fresh cool air, so more than the greenhouse effect serves to keep a greenhouse or a closed car warm. The high temperature of Venus compared to Earth thus appears to be a result of the dense carbon dioxide atmosphere.

What happened on Venus to give it so much more carbon dioxide gas than Earth has? Actually, this turns out to be the wrong question. Instead we should ask, what happened on Earth to have so little carbon dioxide gas here? The answer is primitive Earth was cool enough to have some liquid water, and carbon dioxide gas dissolves readily in water. Almost all of the carbon dioxide released to our atmosphere ends up in the oceans. There it reacts with dissolved minerals to produce solid carbonates, forming such minerals as limestone and chalk. If the carbon dioxide trapped in rocks and sediment on the earth could be released again, we would have a dense atmosphere like Venus.

A planet not far from a sun cannot have both a dense carbon dioxide atmosphere and liquid water. A CO_2 atmosphere implies a strong greenhouse effect, which heats the surface enough to evaporate the water, thereby stopping the CO_2 from dissolving. Conversely, if liquid water exists, the CO_2 gas is rapidly removed from the atmosphere. As Venus and Earth were forming their atmospheres, Venus, being closer to the sun, was slightly warmer, preventing enough water from condensing to remove its CO_2 gas. Had the early Earth been slightly warmer, the carbon dioxide would have won out over the oceans, and Earth would be nearly as inhospitable as Venus. On the other hand, if Earth had been slightly farther from the sun, the water would occur mostly as ice. In this case, carbon dioxide would also remain in the atmosphere, but it would surround a cold planet. This is the case for the next planet, Mars. In hindsight, we are lucky that Earth formed where it did.

Mars

There are only two planets except Earth for which surface details are telescopically visible—Mercury and Mars. But Mars, being a superior planet, is much more favorably placed for observation, particularly when it is at opposition and is in the sky throughout the entire night. It is also closest to Earth at opposition, although because of its orbital eccentricity, its distance at opposition can vary from 56 to 100 million kilometers. It is small wonder that Mars has received more attention than any other planet. It has an atmosphere and shows permanent markings, seasons, polar ice caps, and occasional large storms. Some observers even thought there were fine lines running across the planet's surface which they interpreted as canals carrying water from the polar regions to the equatorial regions. We now know that this is not true, but we had to wait for the space probes to prove it. Its day is only slightly longer than a day on Earth, and its axis is tilted almost the same amount as Earth's.

The first close-up view of Mars was obtained by *Mariner 4* in 1965. The photos showed a nearly uniform gray. Only when the contrast was increased about 100 times were details on the surface clearly revealed. An astronaut flying by Mars would not notice the

Table 8–4
Mars: Vital statistics

Size: 6800 km = 0.53 Earth
Mass: 0.107 Earth
Density: 3.9 g/cm^3
Rotational period: $24^h37^m23^s$
Orbit
  Semi-major axis: 1.52 AU = 227.9 million km
  Eccentricity: 0.093
  Inclination: 1°.85
  Sidereal period: $686^d.98$

tremendous fine structure on the surface revealed by the computer enhancement of spacecraft photos.

Mariner 4 was followed four years later by *Mariners 6* and *7,* all photographing different latitudes on the same side of Mars as they flew rapidly by the planet. In the pictures returned, the surface appears desolate and cratered, as shown in Figure 8–9. The atmosphere is thin, with a surface pressure less than one percent of Earth's atmospheric pressure. The composition is mostly carbon dioxide, with small amounts of argon, nitrogen, and oxygen. The craters do show signs of erosion, but not the water erosion common on Earth. Instead, winds reaching speeds of hundreds of kilometers per hour and carrying small particles of sand have eroded the craters and made them much smoother than craters on the moon. The straight narrow canals which some people had reported are not in evidence, nor does anything else indicating the presence of life appear in the photographs from these space probes. *Mariners 6* and *7* also found some smooth plains

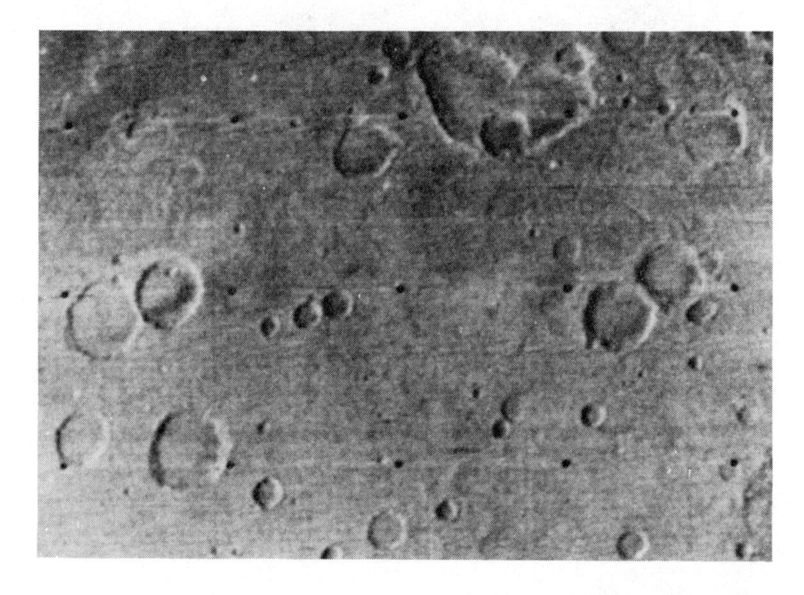

Figure 8–9 Mars photographed from *Mariner 6*.

and regions of particularly irregular terrain. The Mariners detected no magnetic field, which is not surprising because of Mars' small size. The space probes also improved our value of the mass of Mars, thus confirming that its mean density is only slightly greater than that of the moon. Thus, Mars also lacks a large iron core.

After having three probes flying past the planet, scientists were anxious for probes that could orbit continuously around the planet. This has two main advantages: (1) complete coverage of the planet is obtained as it rotates inside the spacecraft orbit, and (2) observations extending over many months reveal any changes taking place on the surface. *Mariners 8* and *9* were launched for this purpose, but unfortunately *Mariner 8* only got as far as the mid-Atlantic. *Mariner 9* was incredibly successful, however, and more than made up for the loss of her sister craft. The maps of Mars shown on page 147 resulted from *Mariner 9* observations. *Vikings 1* and *2* followed six years later, with each Viking containing both orbiter and a lander. The two orbiters examined in greater detail the features revealed by *Mariner 9*, and we have now pieced together a history for the planet far different than expected.

Figure 8–10 shows a huge canyon discovered by *Mariner 9*. At its widest point the canyon is about 100 kilometers across and 6 kilometers deep. It is 5000 kilometers long. The Grand Canyon, shown to almost the same scale in Figure 8–10, is dwarfed by comparison. The Martian canyon is big enough to span the United States completely, all the way from the Atlantic to the Pacific. It shows up as a large scar slightly to the left of center in Plate 5B, photographed by *Viking 1* as it approached the planet. The canyon, named *Valles Marineris* or Valley of Mariner after *Mariner 9,* apparently is the result of a spreading of the crust of the planet, similar to movements that can take place on Earth during major earthquakes. As the spreading continued, the canyon became deeper and wider. Occasional landslides have occurred when part of the canyon wall collapsed, helping to widen the canyon further. Three giant landslides can be seen in Figure 8–11,

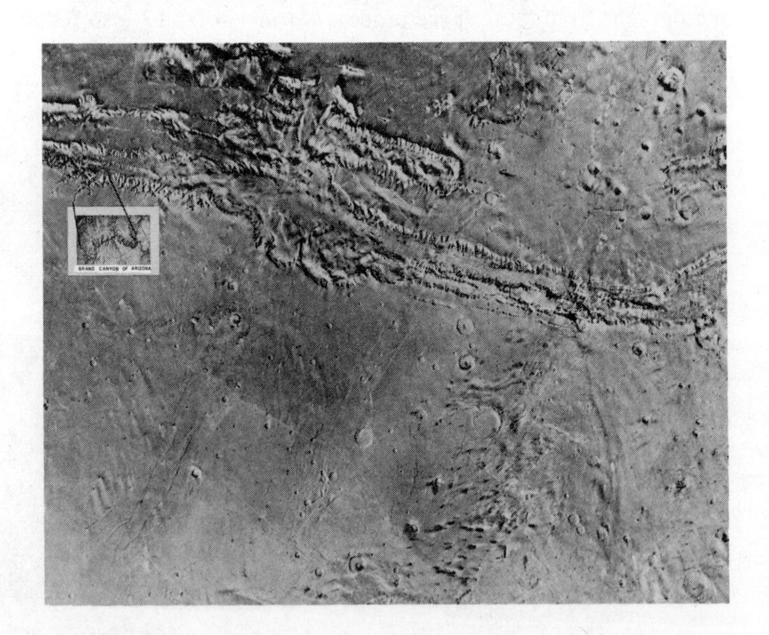

Figure 8–10 The large canyon system on Mars.

Plate 5A Venus.

Plate 5B Mars.

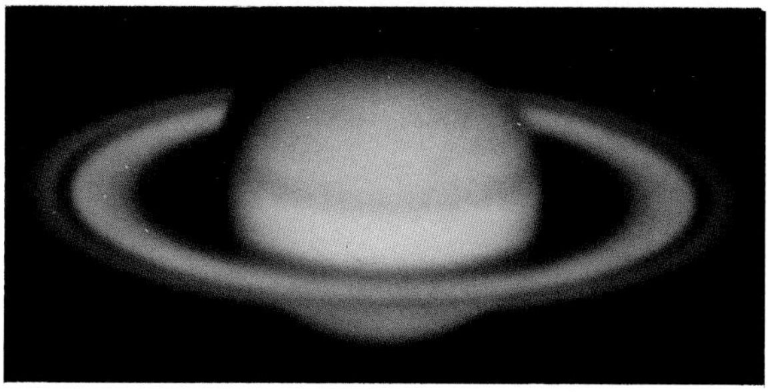

Plate 5D Saturn.

Plate 5C Jupiter.

Plate 6A Viking photograph of the surface of Mars.

Plate 6B Turbulence near the south pole of Jupiter.

where parts of the wall fell about 2 kilometers, then spread out over the floor. The slow spreading of the canyon walls and landslides may still continue to deepen and widen *Valles Marineris*.

Many photographs of other areas of Mars show that the planet has had extensive seismic activity. In Figure 8–12, the long, narrow straight valleys are characteristic of features formed on Earth by a spreading of the earth's crust.

Possibly the most exciting discovery by *Mariner 9* was the huge mountains; the biggest, Olympus Mons, is shown in Figure 8–13. Most of the Martian craters are thought to be impact craters, but the central craters on Olympus Mons and its companions are certainly volcanic. Olympus Mons is nearly 700 kilometers across at its base, big enough to cover the state of Nebraska. It is 27 kilometers high and has a crater that is 70 kilometers in diameter, big enough to contain the area of Rhode Island. Wind and dust storms

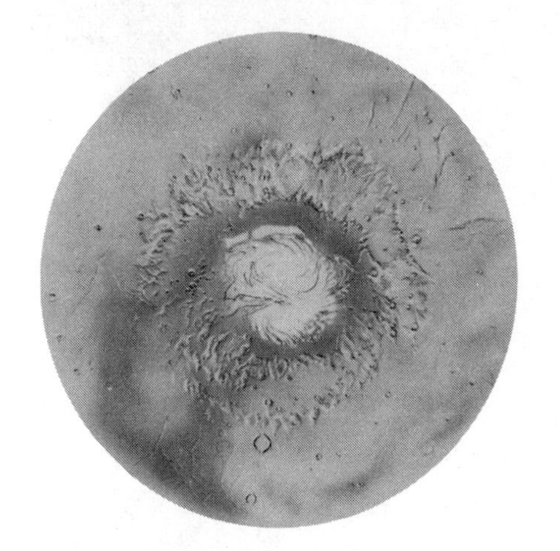

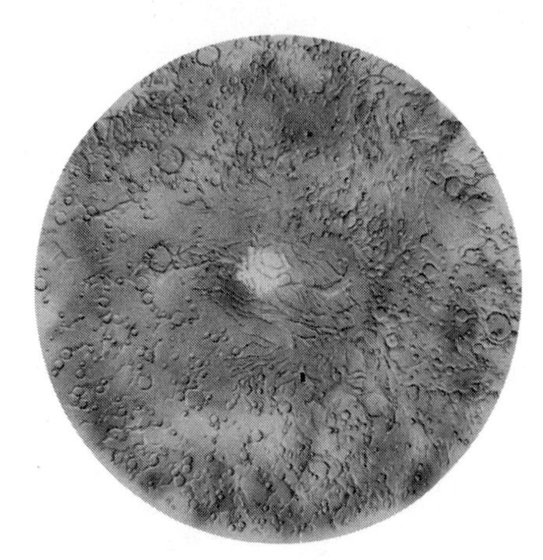

N 0°

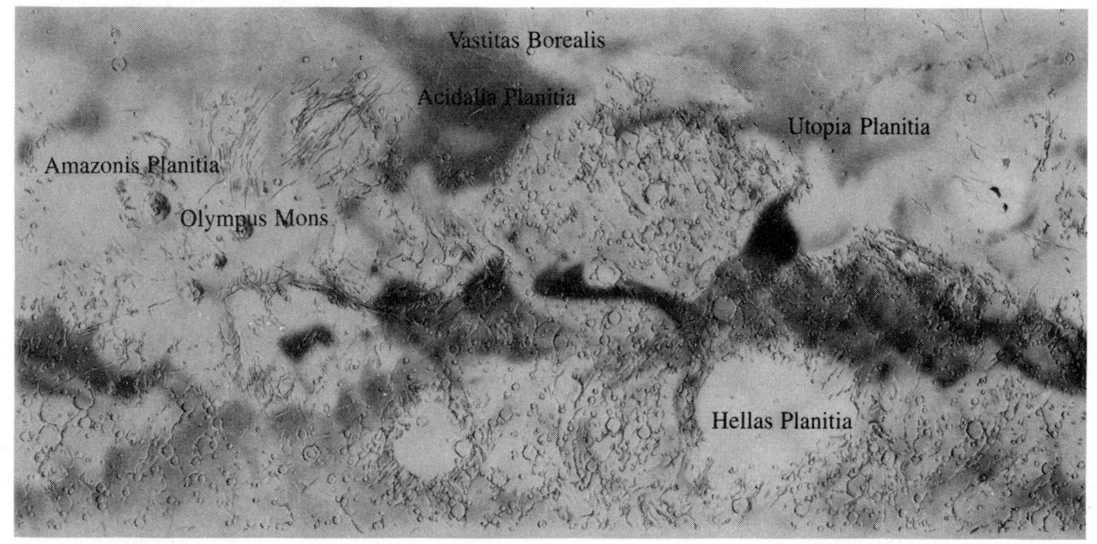

Mars

Figure 8–11 Detailed views of the great Martian canyon.

Figure 8–12 Faults in the Martian surface.

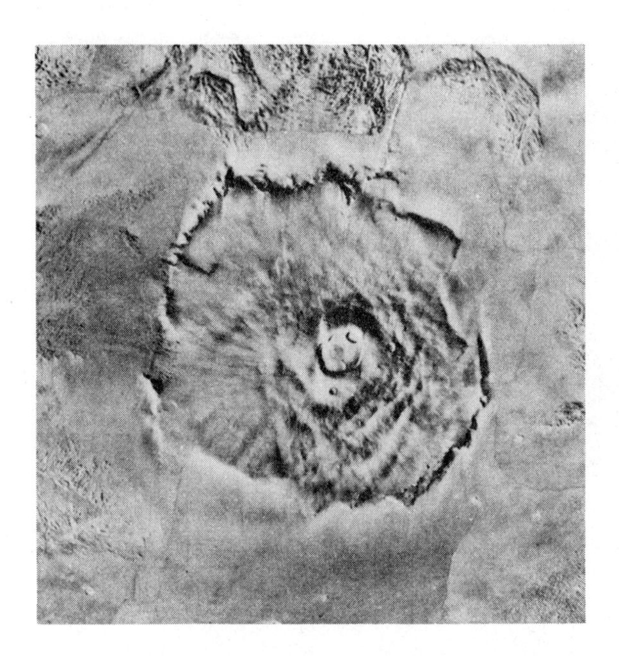

Figure 8–13 Olympus Mons.

presumably have cut away at the base of the mountain, forming a cliff 4 kilometers high which rings the base.

A view by *Mariner 9* (Figure 8–14) shows that after the cliff around the base was formed, a fresh lava flow occurred over a section of the cliff. This happened recently enough that the flow has not been eroded much by the wind. Evidently, lava flows have occurred fairly recently in the life of the mountain and presumably may happen again. Many other smaller volcanoes appear on Mars. A group of them are shown in Figure 8–15, where a tiny crater can be seen at the top of one.

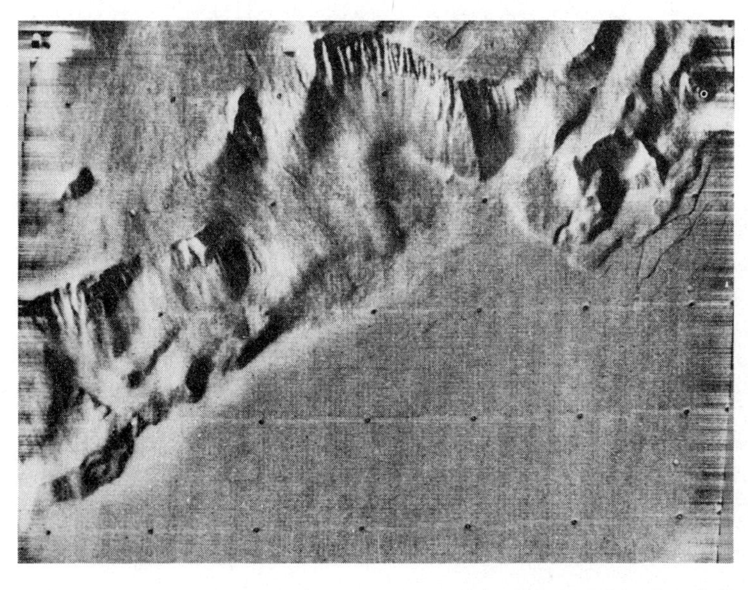

Figure 8–14 Detailed view of the cliff around Olympus Mons.

Figure 8–15 Small volcanoes on Mars.

Much of the gas in our atmosphere and the water in our oceans probably came from vapors released during volcanic eruptions, so the discovery of volcanoes on Mars suggests water vapor was released there also. So it may not be surprising that patterns like those shown in Figure 8–16 are found on Mars. Many tiny valleys are cut into the surface, coming together as they go downhill. This is exactly like the erosion patterns which occur on Earth when heavy rain falls on ground unprotected by vegetation. Apparently, at times in the past, it may have rained on Mars.

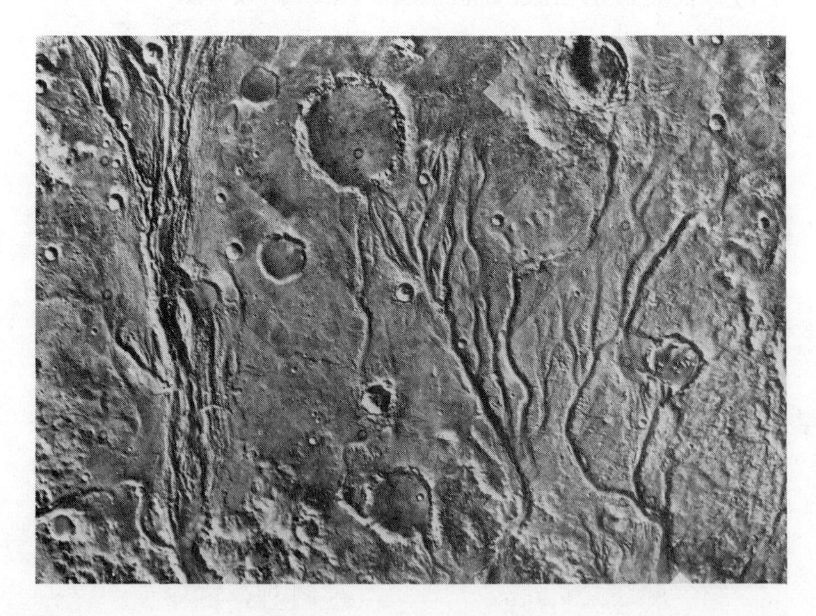

Figure 8–16 An erosion pattern on Mars.

Figure 8–17 shows another source of running water on Mars and, in the process, explains some of the irregular terrain found by *Mariners 6* and *7*. The ground slopes downward to the left, showing flow marks caused by large amounts of running water. The flow apparently came from the irregular terrain that appears to have collapsed. A great deal of frozen water must have existed below the surface, until it became heated. The ice melted, allowing the ground to collapse, with the released water gushing out to the left. A plausible suggestion of what provided the warmth is flowing hot rock rising from deeper layers. Of course, if the rock had actually reached the surface, a volcano would have been formed instead.

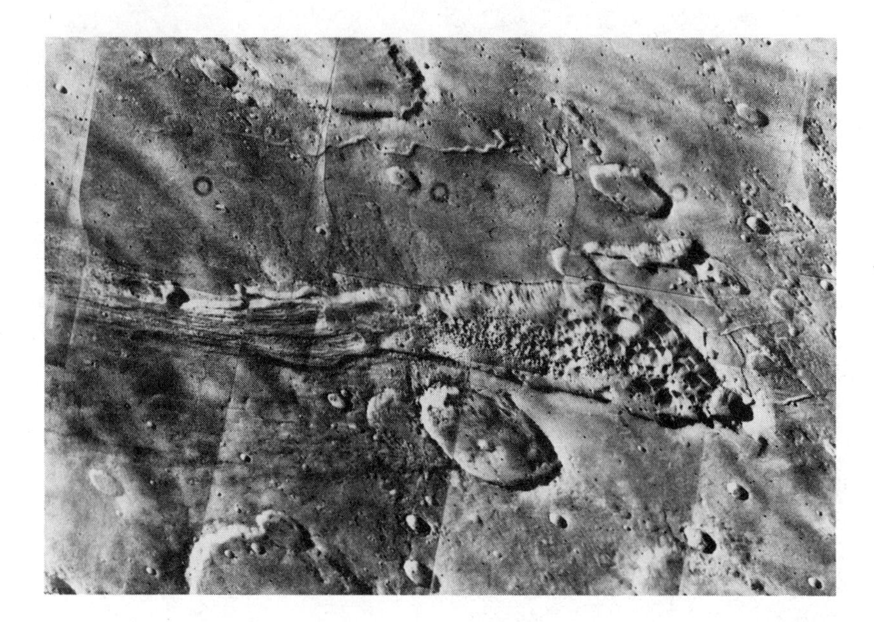

Figure 8–17 A collapsed surface region of Mars.

Details of many of the erosion patterns agree closely with patterns formed by rapid release of large quantities of water. Figure 8–18 shows a crater that is located in the middle of one channel. The wall of the crater formed a barrier, and as the water flowed around it, the teardrop-shaped island was formed. The deepest part of this particular channel is 100 meters below what was the shoreline, so the flood causing this feature must have been an impressive event. Eventually, these channels spread out as they reach a flat area, and the sign of running water disappears. But no evidence for standing bodies of water such as lakes or oceans has been found. Hence, liquid water did not exist on the surface long enough to remove much of the carbon dioxide from the atmosphere.

The Viking orbiters found other evidence that water existed beneath the surface when they photographed impact craters like that shown in Figure 8–19. Note the flow pattern in material ejected from the crater. These features, completely unlike anything found on the moon or Mercury, can be duplicated on Earth if an explosion takes place in waterlogged ground. Apparently, permanent ice below the surface, called *permafrost,* was instantly turned into a liquid by the impact which produced the crater, allowing this distinctive type of flow.

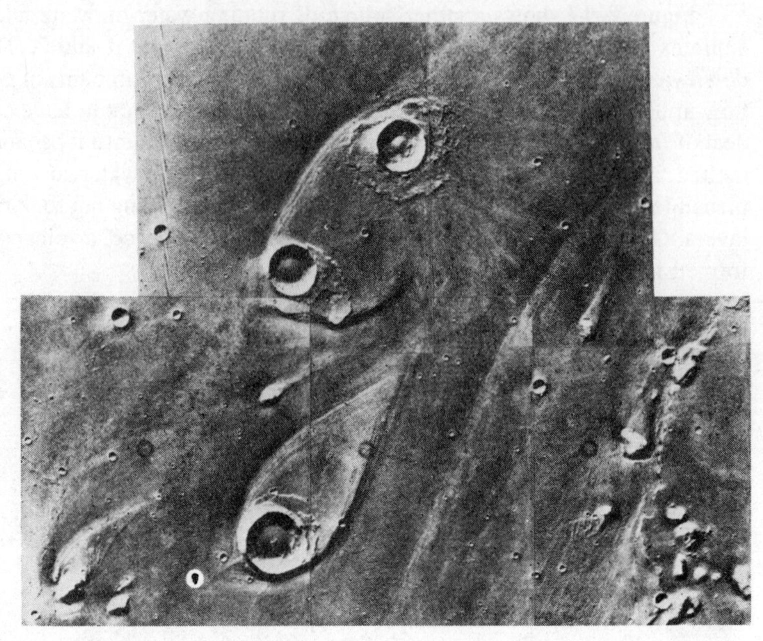

Figure 8–18 Islands in an ancient Martian channel.

Figure 8–19 At left, the flow pattern of material ejected from a Martian crater.

There is a reason to be surprised at signs of flowing water in Martian channels: because of the present low atmospheric pressure on Mars, water can exist only in the form of ice and water vapor. If you were to put a glass of water on Mars, it would begin to boil violently and then suddenly freeze solid. Presently, only tiny amounts of water vapor exist in the atmosphere of Mars (although most of that forms a frost on the surface each night). Then where did all the water go? Some could have seeped back into the surface, freezing into permafrost. Some could also be trapped in the polar caps. The caps grow so rapidly in the Martian winter and shrink so rapidly in summer that most of these areas must be a thin frost layer. The temperature of the winter cap equals the temperature at which carbon dioxide would freeze, forming "dry ice," at the pressure found in the Martian atmosphere. In other words, the CO_2 in the polar atmosphere freezes in winter. The Viking landers measured the changing atmospheric pressure on the surface as part of the atmosphere was condensed out first at one pole and then at the other, confirming that large amounts of CO_2 really do freeze in the winter polar cap.

But the central region of the north polar cap persists through the summer when the polar temperature gets high enough to evaporate solid carbon dioxide. This central region is at just the temperature expected for frozen water under the conditions occurring on Mars. Apparently a lot of water remains locked in the northern summer polar cap.

Detailed pictures of the residual caps show series of ridges in and around the remaining ice. Figure 8–20 shows the ridges of the northern cap which we interpret as dust that has been collected and left behind as the water ice melted or evaporated in the past. The inclination of Mars' axis of rotation to its orbit slowly oscillates with a period of about a million years due to perturbations which change the plane of the orbit of Mars. When the inclination is greater than it is now, enough radiation may reach the poles to evaporate the residual ice caps, leaving a new ridge. The ridges are not all concentric, indicating the possibility that the surface of Mars also moves relative to its rotational axis. Twenty or thirty ridges are easily identified, which suggests that the Martian ice ages may have been going on for at least several tens of millions of years. Also, layers of dust at the edge of the summer polar cap indicate that the Martian climate has changed periodically in the past. Perhaps Mars is presently in a temporary ice age. In a few hundred thousand years, when the tilt of the axis has once more increased, the water will be released; and Mars may once again have running water.

The close-up views of the Martian surface obtained by the Viking landers are strangely earthlike. Plate 6A shows the many rocks of varying types scattered across the

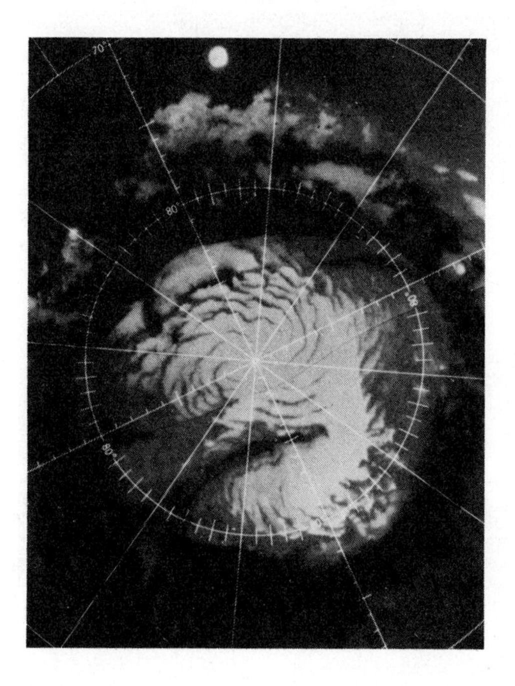

Figure 8–20 Ridges in the north polar cap of Mars.

landscape, thrown there by impacts which formed distant craters. Some rocks are filled with holes, a common occurrence in some volcanic rocks on Earth, where gases which were dissolved in the rock form bubbles as the rock solidifies. The soil is mostly a mixture of silicon oxide (basically sand) and oxidized iron but contains about one percent water, chemically bound in rock crystals.

Daily weather reports from the lander sites help us to understand the Martian climate and wind patterns. Although only one seismograph functioned on the landers, it has detected "Mars-quakes" which have characteristics closer to earthquakes than moonquakes. A major goal of the landers was to search for signs of life, but no unambiguous indications of life were found. The most important result in the search for life was that organic molecules, the building blocks for life as we know it, are just about totally nonexistent at the Viking lander sites.

Accompanying Mars (named for the god of war in Roman mythology) are two tiny moons, *Phobos* and *Deimos* (the Greek gods of fear and terror). The satellites are irregular in shape, with the largest dimensions being 27 and 15 kilometers for Phobos and Deimos, respectively. The orbits of both satellites are close enough to Mars that gravity has locked each moon's rotation to its revolution. Hence, one side of each Martian moon always faces Mars, just as our moon keeps one face toward the earth.

The irregular shapes of the moons are illustrated by the photo of Phobos in Figure 8–21. The surface is cratered, with strange parallel grooves covering part of it. The surface is coated with fine, dark dust, darker than the dust covering our own moon. One Viking orbiter came close enough to Phobos to permit measurement of its gravitational field. The weak field indicated that the moon's mass is small, so small that the density of Phobos is just a little over one-half that of Mars itself. The future of Phobos is limited. Because of perturbations in its motion, it is slowly spiraling in towards Mars. It is now so close that in only a few more tens of millions of years it will probably crash into Mars.

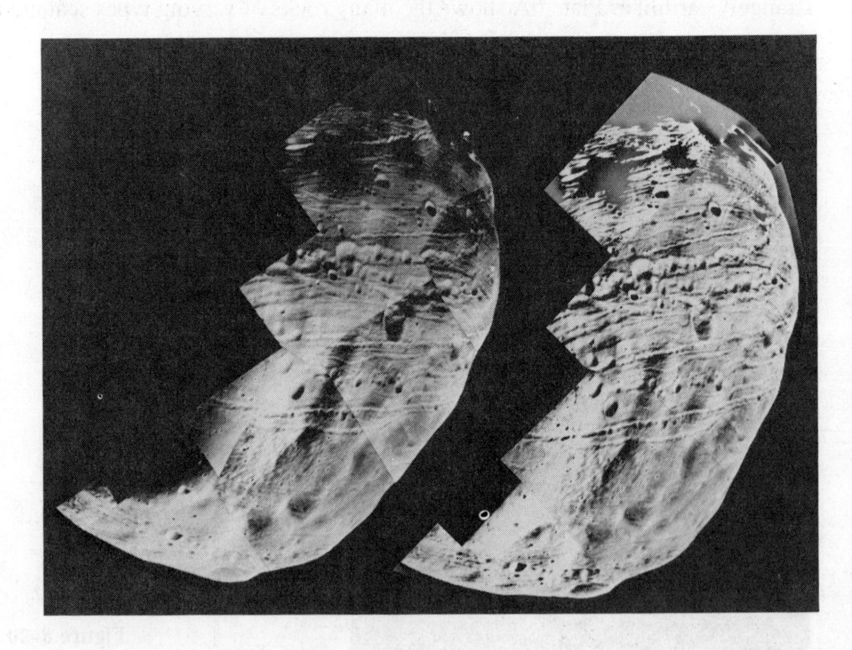

Figure 8–21 Phobos.

Jupiter

The four planets closest to the sun have high densities and consist mostly of rock and metals. They are often called **terrestrial planets** after Earth, the most massive of the group. With Jupiter, we come to the largest of the next four planets, which all have much lower densities. They are referred to as the **Jovian planets,** after Jupiter.

Except for the sun, Jupiter is the dominant member of the solar system. Its mass is more than twice that of all the other planets put together, and Jupiter usually has the largest apparent diameter of any planet in the sky. Its disc is visible with binoculars. Even the four brightest of its moons, the Galilean satellites, are visible with binoculars and occasionally have been glimpsed with the naked eye by sharp-eyed observers. Like Venus, Jupiter's atmosphere contains opaque clouds, as shown in Plate 5C.

Not only is Jupiter the largest planet, but it rotates the fastest of all. The regions near the equator rotate faster than the other parts of the planet, but all parts complete one rotation in less than ten hours. This very rapid rotation helps stretch the clouds into the bands that show up in Plate 5C. It also gives it an extreme rotational bulge at the equator, readily seen when the planet is viewed through a telescope.

Notice the oval feature to the left in Plate 5C. This is called the *Great Red Spot* because of its characteristic orange color. It has been seen with varying strength since 1831. Some people think it may even have been noticed as early as 1660 by Cassini, who used a spot on Jupiter to measure its rotation. The Red Spot is apparently a large, long-lived region of convection, which continually brings up material from deep in the Jovian atmosphere somewhat like a long-lasting hurricane. It appears to float at the top of the clouds and does not seem to be tied to anything firm below it. Over the decades it has speeded up and slowed down in its rotation around Jupiter, at one time gaining nearly one complete rotation compared to the clouds around it. It has strengthened and faded in intensity over the decades in which it has been observed.

Table 8–5
Jupiter: Vital statistics

Size: 143,000 km = 11.2 Earth
Mass: 317.9 Earth
Density: 1.3 g/cm^3
Rotational period: 9^h50^m to 9^h55^m
Orbit
  Semi-major axis: 5.203 AU = 778 million km
  Eccentricity: 0.048
  Inclination: 1°.3
  Sidereal period: 11.862 yr

Jupiter does not receive much energy from the sun at its great distance, and therefore it is quite cold. It was a surprise when it was found to be one of the brightest objects in the sky at short radio wavelengths. The signal at these wavelengths is steady; it comes from two regions on either side of the planet, not from the disc of the planet itself. The signal has the characteristics of radiation given off by electrons traveling near the speed of light in a strong magnetic field. Thus, as we would expect of a large, rapidly spinning body, Jupiter has a strong magnetic field. By a process we still do not understand, the magnetic field is able to accelerate some electrons to speeds close to the speed of light. These very fast particles were discovered by the first space probes to visit the giant planet, *Pioneers 10* and *11*. Some of these fast particles even reach the vicinity of Earth.

In addition to this steady signal at short wavelengths, Jupiter occasionally gives off strong outbursts of radiation at long wavelengths (around twenty meters.) The outbursts may be very intense for a few minutes and then disappear for hours or even days. These signals are the strongest long-wavelength radio waves we find in our sky, completely overpowering the radio emission from the sun. The outbursts are more likely to happen when certain regions of Jupiter face Earth; and these regions rotate at a very constant rate, with none of the wanderings of the Great Red Spot and other clouds. In addition, the likelihood of receiving these bursts is greatly increased whenever Io, the innermost Galilean moon, is in one of two preferred positions in its orbit relative to a line joining Jupiter and Earth. That a moon much smaller than Jupiter should exercise such control over this radio emission is surprising. The effect must depend on Io's effect on Jupiter's magnetic field. Jupiter and its magnetic field rotate very rapidly, and Io must exert a drag on the field as it tries to sweep by. Presumably, the magnetic field is generated in the core of Jupiter, which rotates at a very regular rate. By timing the regions of Jupiter that face Earth during outbursts, it is possible to find the rotation rate of the core of the planet, which is very close to the rotation rate of clouds far from the equator.

There was another surprise when the energy balance of Jupiter was studied. If the energy it radiates is compared to the energy it receives from the sun, we find that Jupiter gives off about twice as much energy as it receives. It must have an internal energy source, either heat left over from its formation, heat from a continuing slow gravitational contraction, or possibly both. The heat source causes the strong convection seen in its atmosphere. At low latitudes, the bright colors are regions where gases are rising in the atmosphere, while the dark belts occur where the material descends again. In the polar regions of Jupiter, convection occurs in small cells, giving rise to the granulated appearance shown in Plate 6B.

Spectroscopic analysis of the Jovian atmosphere has shown molecular hydrogen, methane, and ammonia to be prominent. In addition, *Pioneer 10* detected helium. Jupiter's atmosphere is vastly different than those of the terrestrial planets, as shown in Table 8–3, but the dominance of hydrogen and helium is remarkably similar to that of the sun. The only reason Jupiter did not itself become a star is that it lacked enough mass. Beneath the atmosphere, we do not expect to find a definite solid surface like the terrestrial planets. Instead, the pressures and densities increase toward the center of Jupiter until the atmosphere liquefies. The central regions are probably still fluid, but under such high pressure that the hydrogen behaves like a liquid metal.

Jupiter is such a massive planet that it is easy to overlook the importance of its moons. Two of the Galilean moons are as large or larger than Mercury, and the other two are about the size of Earth's moon. Like the moons of Earth and Mars, the four large satellites of Jupiter rotate so as to constantly keep one side toward their planet. Io, the innermost of these moons, is an excellent reflector of sunlight, reflecting about as well as white sand. Io has a slight reddish color, however, and is brighter on the leading side as it orbits Jupiter and redder on the trailing side. The surface of Io appears to be covered with various kinds of salts, much like salt flats on Earth. The elements in the salts are continuously being released into space, with the element sodium being particularly common. (Common table salt is sodium chloride.) About ten million sodium atoms are released from each square centimeter of Io every second, and the atoms have collected into a ring around Jupiter along Io's orbit. Hydrogen, potassium, and ionized sulfur are also present as large rings. Some of the sodium atoms become ionized while still near Io, providing Io with a thin ionized atmosphere. It is apparently this ionized atmosphere which pulls on Jupiter's magnetic field to cause the radio outbursts.

Both Io and the next large moon, Europa, are basically rocky objects, but most of Europa is covered with frozen water. The third Galilean moon, Ganymede, is the largest. Although it probably has a rocky core, it consists mostly of water, in either a liquid or frozen state. Part of its surface is frost-covered. The last large moon, Callisto, has even less rock and more water.

In addition to the four large moons found by Galileo, Jupiter probably has a very large number of small moons, many of which were most likely bodies that were captured by Jupiter as they were passing nearby. So far nine and possibly ten of these small moons have been found.

Saturn

Saturn is the most striking of all the planets because of its rings (*see* Plate 5D). The disc of the planet itself shows a banded structure similar to Jupiter's. And like Jupiter, it rotates faster at the equator than at the poles. Its most intriguing property is its low average density—so low that it would float on water (given a bucket of water 121,000 kilometers across). Saturn, like Jupiter, must be mostly hydrogen and helium. But being less massive than Jupiter, Saturn cannot provide enough gravitational force to compress much of these gases to the very high density of Jupiter's core. Hence, the average density of Saturn is very low.

Table 8–6
Saturn: Vital statistics

Size: 121,000 km = 9.5 Earth
Mass: 95.2 Earth
Density: 0.7 g/cm^3
Rotational period: 10^h14^m to 10^h38^m
Orbit
  Semi-major axis: 9.54 AU = 1427 million km
  Eccentricity: 0.056
  Inclination: 2°49
  Sidereal period: 29.458 yr

The rings of Saturn are quite interesting in themselves. They look solid in a telescope, but spectroscopic studies show Doppler shifts that decrease outward across the rings. This indicates that the velocities fall off with distance from the planet. Furthermore, the velocities decrease at just the rate predicted by Kepler's harmonic law. The rings must be composed of countless tiny objects, all orbiting in Saturn's gravitational field. The particle interpretation is confirmed when Saturn's orbital motion carries the rings across our line of sight to a bright star—the star dims but usually continues to shine through the rings. Radar reflections from the rings also show them to be composed of small bodies, a few centimeters to a few meters across. A calculation of the tidal force from Saturn at the distance of the rings indicates that the tides would be so great on a body with as much material as the rings that it would be disrupted. Furthermore, the same tides act to prevent any ring material this close to the planet from collecting into a single large body. The reflection of sunlight by the rings closely matches that expected from ice, indicating that at least the surfaces of the particles are ice-covered.

Saturn has at least ten moons. The largest of these, Titan, is about the size of Mercury and the largest moons of Jupiter, but Titan has the distinction of having a dense cloudy atmosphere, rich in hydrogen and methane. At its surface, the atmospheric pressure may be one-half the atmospheric pressure on Earth.

Among the more mysterious objects in the solar system is Saturn's ninth satellite, Iapetus. The side of the satellite that faces forward as it orbits Saturn is six times brighter than the trailing side. Hence, one side is made of white material (probably ice), and the other consists of very dark material.

Uranus

The planets we have discussed so far were all known to the ancients, since they are among the brightest objects in the sky. Uranus is the first planet to be "discovered." Because it is just barely visible to the unaided eye, it really requires a telescope to be seen properly. It was not identified until the eighteenth century, when Sir William Herschel noticed it while making a telescopic survey of star positions. Since it did not appear pointlike, he knew that it was not a star. By observing it over a period of time, he detected its orbital motion and knew that he had found a new planet. Uranus is so far away from the sun and from Earth that we can see very little detail. We must await the arrival of a very deep space probe to get detailed knowledge of Uranus.

We can observe one very interesting property of Uranus: its axis of rotation is tipped nearly into the plane of its orbit. Therefore, one pole points roughly to the sun for part of a revolution on its orbit; then the other pole points to the sun for the next part. Because the sidereal period of Uranus is about eighty-four years, the mid-latitudes of the northern and southern hemispheres take turns in complete darkness and sunlight, lasting about twenty years. These intervals are separated by periods when the sun rises and sets, also lasting about twenty years. In other words, summer and winter on Uranus are each twenty-one years long, separated by a twenty-one-year autumn and spring.

The prominent ammonia absorption features seen in Jupiter's spectrum almost disappear in the spectrum of Uranus, but the methane features continue to be strong. This is because Uranus is so far from the sun that its atmosphere is cold enough to freeze out the ammonia. When in its solid state, ammonia can no longer contribute lines to the spectrum. But the methane still exists as a gas. Because the density of Uranus is greater than that of either Jupiter or Saturn, it cannot contain as much hydrogen or helium as the two largest Jovian planets. Uranus is probably more of an icy object than the gaseous planets Jupiter and Saturn.

Table 8–7
Uranus: Vital statistics

Size: 52,000 km = 4.06 Earth
Mass: 14.6 Earth
Density: 1.2 g/cm^3
Rotational period: about 14^h
Orbit
  Semi-major axis: 19.18 AU = 2870 million km
  Eccentricity: 0.047
  Inclination: 0°77
  Sidereal period: 84.01 yr

In 1977, Uranus passed between the earth and a faint distant star. Observers measured the star's brightness as it disappeared behind the planet, hoping to find out more about the planet's atmosphere. The experiment paid unexpected dividends. About thirty-five minutes before Uranus covered the star, and again thirty-five minutes after the star was uncovered, the star momentarily dimmed several times. The dimming was due to several thin rings circling the planet in the plane of the equator. The rings, however, contain much less material than those of Saturn; but apparently, having such debris in orbit around a planet is not such a rare occurrence as was once thought.

Five satellites have been found around Uranus. The largest is about 1000 kilometers in diameter, much smaller than the largest satellites around Jupiter and Saturn. They move in the equatorial plane of Uranus, at nearly right angles to the planet's orbit.

Neptune

After the motion of Uranus had been studied for a while, it was found slightly off from the position which would be predicted by Newton's laws. In 1843 a young Englishman, John Adams, concluded that the irregularities were due to the perturbing influence of an even more distant planet. He calculated where the perturbing planet should be in the sky and sent his predictions to the Astronomer-Royal at Greenwich, who chose not to search for the planet. In 1846 a French mathematician, Leverrier, independently calculated the position of the same object. He eventually sent his predictions to the Berlin Observatory. Galle, the astronomer there, promptly found the new planet, Neptune, within one degree of the predicted position. Adams' prediction, by the way, turned out to be less than two degrees in error. The discovery of Neptune is considered to be a great triumph for celestial mechanics, the application of Newton's laws to celestial objects.

Neptune is almost exactly the same size as Uranus, although Neptune is somewhat more massive so it has a higher density. Methane has been observed in its spectrum. Of the two moons known to orbit Neptune, one is rather small. But the other, Triton, is unique. It is the only large satellite in the solar system that orbits in the opposite direction to that in which its planet rotates.

Table 8–8
Neptune: Vital statistics

Size: 48,600 km = 3.81 Earth
Mass: 17.2 Earth
Density: 1.7 g/cm^3
Rotational period: about 18^h
Orbit
Semi-major axis: 30.06 AU = 4497 million km
Eccentricity: 0.009
Inclination: 1°77
Sidereal period: 164.8 yr

Pluto

After allowances were made for the influence of Neptune, there were still some small discrepancies in Uranus' motion. Several astronomers proposed a ninth planet and, as Adams and Leverrier before them, calculated the mass and location of the disturbing body. The most persistent was Percival Lowell. Lowell searched for the new planet at his observatory in Arizona from 1906 until his death in 1916. After his death, the search was continued there. Lowell's brother donated a special photographic telescope to aid in the search in 1929, because the predicted position of the object was in a portion of the sky with many stars. Finally, in February, 1930, Clyde Tombaugh, then a graduate student, located the planet by comparing two plates which had been taken the previous month. The new planet was named Pluto. Figure 8–22 shows the motion of Pluto relative to the stars near it in the sky and shows how Tombaugh found it.

Figure 8–22 The discovery photographs of Pluto.

Was this another triumph for celestial mechanics? Unfortunately not, because more modern computations show that Pluto is not massive enough and is too far away to perturb Uranus' motion significantly. The small deviations in Uranus' motion, after allowance is made for Neptune, are now known to be small errors of observation. It is simply a happy circumstance that Lowell's calculations, with bad data, led him to predict the correct region of the sky in which to search. More important, of course, was his obstinate determination to find the ninth planet.

Pluto has the most eccentric and most inclined orbit of any of the major planets. Because of its eccentric motion, it sometimes passes inside the orbit of Neptune, but there is almost no chance of a collision because of the large inclination of Pluto's orbit.

Table 8–9

Pluto: Vital statistics

Size: less than 6000 km = less than 0.47 Earth
Mass: 0.01 Earth
Density: ?
Rotational period: $6\overset{d}{.}39$
Orbit
Semi-major axis: 39.53 AU = 5912 million km
  Eccentricity: 0.249
  Inclination: $17\overset{\circ}{.}17$
  Sidereal period: 248.5 yr

Pluto is difficult to classify as either a terrestrial or a Jovian planet, for we do not even know its density with any reasonable accuracy. Its surface appears to have a layer of frozen methane on it. There are objects in the outer solar system that Pluto does resemble, however, namely the large moons of the Jovian planets. Considering that Pluto sometimes travels inside Neptune's orbit, and that Neptune's large moon has the strange retrograde orbit, it is tempting to conclude that Pluto was once a satellite of Neptune and that a close encounter between the two large satellites caused one to be ejected and the other to have its orbit altered drastically.

Some people argue that we will never know the true character of Pluto until a space vehicle is sent to its vicinity. But recently, in June 1978, James W. Christy, while examining photographs of Pluto taken with the 61-cm astrometric reflector of the Naval Observatory at Flagstaff, Arizona, noticed something peculiar about the image of the planet. It was slightly elongated. He reexamined three plates taken earlier in 1978, five taken in 1970, and two more taken in 1965. These earlier plates also showed the same effect. The elongations of the image show a systematic shift in time which Christy suggests can be explained if Pluto has a satellite with an orbit that has a maximum separation from Pluto of 0.9 second of arc as seen from the earth. If this is true, the satellite has an orbit with a semi-major axis of about 20,000 kilometers.

The changes in position of the elongation appear to have a period of 6.4 days, the same as Pluto's rotational period. Of course the amount of data currently available is quite limited, so this cannot be considered a well-determined fact as yet. If the 6.4-day value is the period of the satellite, then Newton's form of Kepler's harmonic law can be used to calculate the mass of Pluto. For the above values of the semi-major axis and period, the mass of Pluto turns out to be only about 0.0024 times the mass of the earth, that is, around 0.2 percent. This is much smaller than most values proposed for Pluto's mass. If this is confirmed, Pluto is even more perplexing and difficult to explain than before.

Christy has suggested a name for the satellite, if it is verified by future observations. He would call it Charon, the name of the mythical boatman who ferried the souls of the dead across the River Styx to the underworld domain of Pluto.

No matter what we eventually find out about Pluto's size or mass, we know that it is quite distant from the sun. An observer on Pluto would see the sun as an extremely bright star. Other stars would be easily visible in the Plutonian daytime sky. Receiving so little energy, Pluto is very cold. On a "hot" summer day, at noon, it would only be about forty degrees above absolute zero. The name chosen for Pluto is still quite appropriate, the god of the underworld.

KEY TERMS

terrestrial planets
Jovian planets
solar wind
greenhouse effect

Review Questions

1. Which number best represents the distance from the sun to Saturn compared to Saturn's diameter?
   (1) 2
   (2) 25
   (3) 180
   (4) 25,000
2. What new devices used for planetary observations became available after World War II?
3. Why is Mercury very rarely seen?
4. Why can't we map the surface of Venus using optical telescopes on Earth?
5. How were the rotation periods of Mercury and Venus found?
6. What factors determine whether or not a planet retains an atmosphere?
7. Why has Earth been unable to retain much hydrogen and helium in its atmosphere when it can hold on to oxygen and nitrogen?
8. How was the high temperature of Venus' surface measured?
9. What conditions are necessary for the greenhouse effect?
10. Name the terrestrial planets and the Jovian planets.
11. Of the terrestrial planets, which has the most atmosphere and which has the least?

12. On Earth, regions of active geological faults are often accompanied by volcanoes. Is there a similar association of faults and volcanoes on Mars?
13. What evidence is there that liquid water once flowed on Mars?
14. What differences are there between craters on Mars and craters on Mercury and the moon?
15. How are the circular ridges around the Martian poles interpreted?
16. How are the Jovian planets different from the terrestrial planets?
17. What are the results of Jupiter's rapid rotation?
18. What results of Jupiter's magnetic field have we observed from Earth?
19. What is the generally accepted explanation of the Great Red Spot?
20. What two planets show peculiarities in their rotations?
21. How were Uranus, Neptune, and Pluto discovered?
22. How were rings around Uranus discovered?
23. When Pluto passes within the orbit of Neptune, is there any danger of a collision?

Discussion Questions

1. Which planet is more likely to have a strong magnetic field, Saturn or Pluto?
2. What are some similiarities and some differences between Mercury and the moon?
3. What stops the solar wind from hitting Earth?
4. Why is the atmosphere of Venus so different from that of Earth?
5. What are the main constituents of the polar caps of Mars? How have they been identified?
6. The Viking landers found that as summer progressed in the northern hemisphere of Mars, the atmospheric pressure at the surface decreased. Considering the composition of the atmosphere, how might this decrease be explained?
7. Why is Jupiter much larger than Earth? If it is so much larger, why does it have a lower density?
8. Based on the amount of heat supplied by sunlight, the temperature of Jupiter's cloud tops is expected to be 105K. But the temperature is measured to be about 130K, twenty-five degrees warmer. How could this twenty-five-degree difference indicate Jupiter radiates about twice as much energy as it receives from the sun?
9. What are two possible origins of Saturn's rings?
10. What can be learned about Saturn's rings by observing the sunlight reflected from them?

9

Comets, Meteors, and Asteroids

In 1766 Johannes Daniel Titius, a professor of mathematics and physics in Germany, devised a numerical rule for finding the distances of the planets from the sun. For the known planets of the time—Mercury, Venus, Earth, Mars, Jupiter, and Saturn—the rule works quite well. However, the rule predicts that there should be a planet between the orbits of Mars and Jupiter at a distance of about 2.8 AU from the sun. Why had this planet never been seen?

Herschel's discovery of Uranus in 1781 showed that a planet could exist and yet escape detection, at least for a while. The distance of Uranus from the sun also agreed remarkably well with the distance predicted by the rule of Titius for a planet beyond the orbit of Saturn.

One astronomer in particular, Baron von Zach, became convinced that there was also a hidden planet between Mars and Jupiter. He initiated a telescopic search in hope of discovering it, but soon found the undertaking too much for one individual. In 1800, at an informal gathering of German astronomers, von Zach proposed that an invitation be sent to astronomers throughout Europe to join in the search for the hidden planet. Specified areas along the ecliptic were to be assigned to individual astronomers as their particular regions of study.

Among those to whom letters were sent advising of the search for the hidden planet was Father Guiseppe Piazzi, a Sicilian astronomer. Even by today's standards, mail delivery was extremely slow. Piazzi was unaware that von Zach's letter was on its way as he conducted routine observations on New Year's Eve of 1800. During that night, Piazzi saw a faint uncatalogued star in the constellation Taurus. The next night he noticed that the position of this faint star had shifted somewhat. The next night, still another shift.

Piazzi was not thinking in terms of hidden planets, however. He thought he had discovered a comet. Anxious for confirmation of his discovery, in late January of 1801 he sent a letter to Johann Elert Bode, the director of the Berlin Observatory. Rather than thinking in terms of comets, Bode thought in terms of a hidden planet;

Halley's comet

for he, like von Zach, strongly believed in the validity of the rule of Titius. As a matter of fact, the numerical rule for the distance of the planets is frequently called Bode's law.

Bode discussed Piazzi's letter with von Zach. The two of them decided that Piazzi had indeed found the hidden planet. Unfortunately, the letter had not reached Bode until late March, 1801; by that time the constellation Taurus, and the new planet along with it, had moved into the daytime sky. To compound the situation, by the time it would again be in the night sky, no one would know precisely where to look for it. Piazzi had had to stop his observations in the middle of February due to illness. The methods for calculating orbits available then needed more data than Piazzi had accumulated.

Luckily, von Zach announced Piazzi's discovery during the summer of 1801 in his monthly astronomical magazine. A brilliant (and even that adjective is an understatement) young mathematician, Karl Friedrich Gauss, noticed the announcement. By coincidence, he had been working on a new method for calculating planetary orbits. He turned his attention to the problem of Piazzi's object and calculated its orbit. He then predicted where observers should look to find it when it reappeared in the night sky.

Today there are mathematicians who regret that Gauss ever picked up von Zach's magazine. He went on to spend many years further developing the mathematics of orbital calculations, spurred by his interest in the hidden planet and also by his respect for Newton and his laws of motion and gravitation. To a few pure mathematicians, the application of mathematics to any work of nature is a complete waste of time.

It is not so much that Gauss wasted his own time on astronomy and physics that disturbs these mathematicians as that he did not turn full attention to developing theorems of pure mathematics, particularly in the theory of numbers. You see, there are pure mathematicians who piddle around merely refining the ideas of Gauss. Had he developed still more ideas, they would have more with which to piddle.

There are some who argue that, Newton notwithstanding, Gauss had the greatest intellect in the annals of humanity. Gauss, however, modestly considered Newton to stand apart from all other geniuses. What a pity that most people have never heard of Gauss!

At any rate, on the night of December 31, 1801, precisely one year after its discovery, Wilhelm Mathais Olbers located the new planet exactly where Gauss said it would be. So the hidden planet was found. And its distance from the sun? 2.8 AU!

Piazzi named the new planet Ceres in honor of the mythical patron goddess of Sicily. Ceres' reign as the "no longer hidden" planet did not last long, however, for on March 28, 1802, Olbers found a second planet at about 2.8 AU from the sun. On September 1, 1804, another planet was found at a distance of about 2.8 AU from the sun; and on March 29, 1807, yet another. By 1850 thirteen new planets had been found. Like rabbits, they had multiplied to a total of 450 by the end of the nineteenth century.

Clearly, there is not a single planet between Mars and Jupiter, but many minor planets. These objects, much smaller than even the smallest of the planets (Ceres, the largest, has a diameter less than 1000 kilometers), are called the asteroids. Today the detected asteroids number in the thousands.

One or many, asteroids swing about the sun, their distances averaging about 2.8 AU. In this respect, Bode's law or the rule of Titius remains applicable. It is in the cases of Neptune and Pluto that it breaks down, predicting distances that do not agree with the actual distances. Note that Bode's law is not a law like Newton's laws. It is simply a good way to remember the distances of the planets and the asteroids, with the exceptions of Neptune and Pluto.

The asteroids, even though small in size, are probably the source of at least a few meteors which at times cause fantastic spectacles in the earth's atmosphere. They are but one kind of debris orbiting the sun. In this chapter we concern ourselves with this debris in the solar system.

Minor Planets

The object discovered by Piazzi, now known as Ceres, turned out to be the first of many similar objects. However, they were first called **asteroids** because of their starlike appearance in a telescope. The preferred name is now **minor planets,** although they are occasionally called *planetoids*.

The discovery of Ceres was particularly intriguing because the semi-major axis of its orbit, 2.77 AU, placed it at the correct distance from the sun for what a number of people had come to believe was a missing planet. The speculation about a missing planet was based mainly on a numerical progression developed by Titius. This progression is frequently called Bode's law because Bode, while director of the Berlin Observatory, publicized it among German astronomers. Figure 9–1 graphs the smooth progression of planetary distance with the number of each planet counted outward from the sun. Ceres fits nicely between Mars and Jupiter.

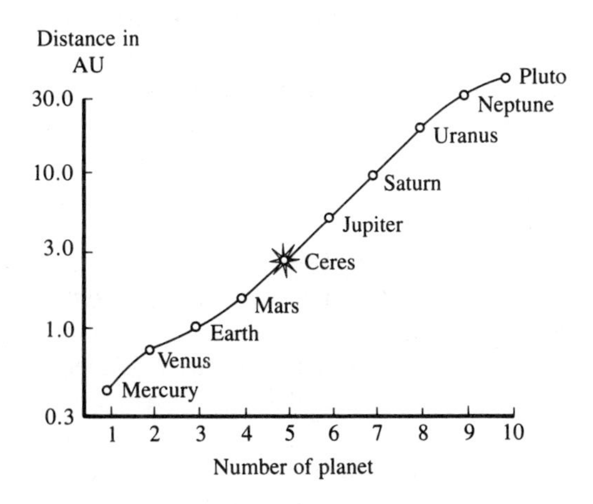

Figure 9–1 Planetary distances.

Bode's law (or the progression of Titius) is a convenient way to remember the approximate values of the semi-major axes of the planets in astronomical units. The progression is developed by first writing down the numbers 0, 3, 6, 12, 24. . . . Except for zero and three, each succeeding number is found by doubling the preceding number.

Next, add four to each number and divide by ten. The resulting progression gives the approximate value of the planets' semi-major axes with surprising accuracy. Table 9–1 compares the values found from Bode's law with the actual values.

Table 9–1
Bode's law

Planet	Bode's law	Actual value (AU)
Mercury	(0 + 4)/10 = 0.4	0.387
Venus	(3 + 4)/10 = 0.7	0.723
Earth	(6 + 4)/10 = 1.0	1.000
Mars	(12 + 4)/10 = 1.6	1.524
Ceres	(24 + 4)/10 = 2.8	[2.77]
Jupiter	(48 + 4)/10 = 5.2	5.203
Saturn	(96 + 4)/10 = 10.0	9.539
Uranus	(192 + 4)/10 = 19.6	18.18
Neptune	(384 + 4)/10 = 38.8	30.6
Pluto	(768 + 4)/10 = 77.2	39.4

The relationship breaks down for Neptune and Pluto. The predicted size of Neptune's orbit agrees quite well with the observed value for Pluto's orbit but not with that for Neptune itself. Ceres, however, fits the prediction, falling between Mars and Jupiter. When Uranus was discovered in 1781, Bode was quick to note that it obeys the progression of Titius. This convinced him that the predicted value of 2.8 AU must be the distance of another undiscovered planet. When he heard of Piazzi's discovery, Bode immediately proclaimed it the missing planet required by Titius.

But in March, 1802, just a little more than a year later, another body was accidently discovered at nearly the same distance from the sun as Ceres. Instead of a missing planet, astronomers realized that there might be many minor planets and began a concerted search for them. By 1890 more than 300 had been discovered visually in telescopic searches. In 1891 photography was introduced into the search. Figure 9–2 demonstrates how a minor planet is discovered on a photographic plate. The telescope is accurately guided on a starfield during a relatively long exposure. Any bright minor planets will show up as short trails because of their motion relative to the stars.

In the 1975 catalogue of minor planet orbits, there are 1861 objects listed. Many more have been seen on photographic plates but have not been tabulated because their orbits have not been calculated. For example, there are several thousand minor planet trails on the Palomar Sky Survey photographs, most of which have not been catalogued. It is estimated that we could detect well over 100,000 with the 200-inch telescope if we systematically searched for them.

The known orbits of the minor planets all lie close to the plane of the ecliptic. The average of the orbital inclination is 9°.5. Only about two dozen minor planets' orbits have inclinations greater than 25°. The semi-major axes of most of the orbits lie between 2.3 and 3.3 AU. The smallest known orbit is 0.84 AU; the largest, 5.8 AU. Both of these orbits are very eccentric (0.45 and 0.66). The average eccentricity for all known orbits is only 0.15.

Figure 9–2 A minor planet trail.

It is no surprise that Ceres, the first asteroid to be discovered, is also the largest. It is about 1000 kilometers in its largest dimension. The sizes fall off rapidly after Ceres. There are about 200 minor planets larger than 100 kilometers. Around 500 are between 50 and 100 kilometers, while the remaining ones are smaller than 50 kilometers. The smallest minor planets seen from the earth are only about 150 meters across. The first space probe to pass through the *asteroid belt,* the region between Mars and Jupiter where most minor planets are found, detected over 100 small bodies orbiting around the sun. These bodies ranged in size from 20 centimeters down to a grain of sand. Extrapolating this count to the entire asteroid belt gives estimates of billions of small objects.

The minor planets are known to be irregularly shaped because of the way they reflect radar signals and sunlight. When we analyze the light received from an asteroid over an interval of one night, we often find the light to be variable, growing brighter and fainter over a period of hours. This results from the rotation of an irregularly shaped body. Depending upon what part of the body is facing us at any moment, it reflects either more or less sunlight. When its largest dimension is seen broadside, it reaches a peak brightness. When its smallest dimension faces the observer, there is less effective reflecting surface and it appears fainter. By timing the variations in brightness, we can find the rotational period. Careful analysis over many nights even reveals whether the asteroid rotates direct or retrograde and the direction in space in which the rotational axis points.

The irregular shapes are consistent with the sizes and masses of the minor planets. The gravity of even the largest asteroids is not great enough to cause them to be spherical. Ceres is estimated to have an escape velocity of only one-half kilometer per second, less than one-twentieth that of the earth. Even the largest of the minor planets is therefore incapable of retaining an atmosphere.

At the distance of the asteroids, the sun can only heat bodies to a temperature around 200K. While this is quite cold by ordinary standards, ices of water or carbon dioxide would slowly evaporate. The minor planets are made of mostly rocky materials with some metals mixed in, but with little or no ices on their surfaces. Their ability to reflect sunlight confirms this composition, because they would appear much brighter if coated with ice.

Comets

Well beyond the orbit of Jupiter, the temperatures of bodies heated by sunlight are so low that an object consisting mostly of solidified gases in the form of ices could exist for extremely long times. But at these great distances small bodies would be difficult to see. In 1977 one such object was discovered; staying mostly between the orbits of Saturn and Uranus, even at its closest approach to the sun it would be no brighter than the planet Pluto. Occasionally, however, a cold, icy object from the outer reaches of the solar system approaches the sun. The sun heats it, evaporating some of the ices and giving the body a temporary atmosphere. But the small body cannot retain the atmosphere, so the gases escape into space.

Some of the escaping atoms and molecules are ionized by ultraviolet light from the sun. The ions, unlike the neutral atoms of the gas, are swept up by the solar wind and stream away from the sun, forming a **gas tail**. Figure 9–3 shows two photographs of one of these objects taken 21 minutes apart. The feature marked by the arrow has moved away from the sun with a speed near 400 kilometers per second, the speed typical of the solar wind. These objects are called **comets**, which means "long-haired," because of the appearance of their tails. A color photograph of a bright comet is shown in Plate 7B.

Figure 9–3 Photographs of a comet taken 21 minutes apart.

Comets have been seen since the beginning of civilization. Some of our oldest records describe the unexpected appearance of these strange objects. They were usually taken as evil omens and caused great terror among the superstitious. Even as recently as 1973, some people interpreted the appearance of Comet Kohoutek as a warning of the end of the world or some similar disaster. While we do not know everything there is to know about comets, we do know that they cannot cause catastrophes on the earth unless we were actually to collide with one.

Comets contain more than just ices. Some solids which do not evaporate are imbedded in the ice. This material is in the form of dust particles. As the ice evaporates, the solid particles are released from the body of the comet and follow along in orbit about the sun. The small particles respond to the pressure of the solar radiation and are slowly pushed away, forming a **dust tail**. Figure 9–4 shows both the gas tail and the dust tail issuing from a comet's head. The dust tail generally points away from the sun and lags behind the more or less straight gas tail. This happens because the dust is pushed gently away by the **radiation pressure**, in contrast to the rapid motion induced in the gas by the solar wind.

Figure 9–4 The dust and gas tails of a comet.

The spectrum of a comet usually includes a faint continuous spectrum crossed by absorption lines. This spectrum is nothing more than sunlight reflected by the dust particles. The gas around the comet head and in the tail, however, is excited by absorbing sunlight. Upon de-exciting, it radiates in spectral lines characteristic of the atoms or molecules in the gas. This causes emission lines, superimposed upon the spectrum of reflected sunlight. The emission features come from simple molecules or fragments of larger molecules broken up by the ultraviolet radiation from the sun.

The comet's head expands to sizes of roughly 100,000 kilometers as it nears the sun. The density of the matter within the head is much lower than in a good laboratory vacuum, however. The tails can stretch over millions of kilometers. When the comet is close to the sun, the head is surrounded by a huge hydrogen cloud approximately ten times the diameter of the visible head. These hydrogen envelopes were first detected by artificial satellites, since the cold hydrogen in them emits only in the ultraviolet part of the spectrum which cannot penetrate the earth's atmosphere. The hydrogen comes from water molecules and hydrocarbon molecules broken up by sunlight.

The tail of a comet always points approximately away from the sun because of the radiation pressure and the solar wind. Thus, instead of always streaming behind the comet, the tail can actually precede the comet in some parts of its orbit. Figure 9–5 demonstrates how the tail of a comet would behave as the comet makes its pass about the sun. When it is close to the sun, the increased heating feeds more material into the tail. Therefore, the size of the tail gets larger as the comet sweeps in closer to the sun and diminishes as it moves outward from the sun.

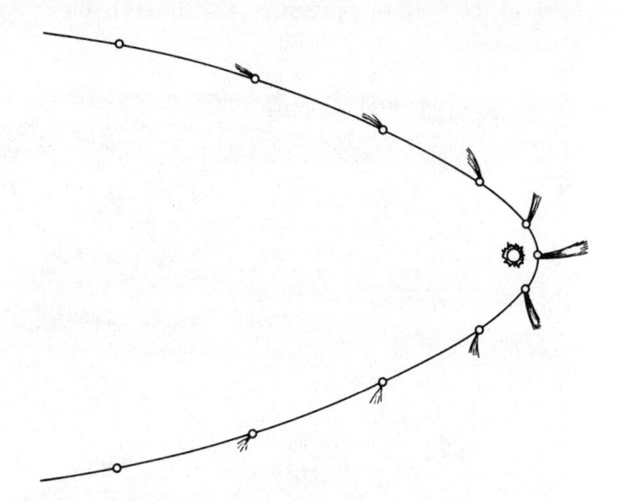

Figure 9–5 Orientation of a comet tail.

The brightness of a comet usually increases rapidly as its distance from the sun decreases, both because more material is released by evaporation and because the sunlight which excites the gas and reflects from the dust becomes much more intense. Consequently, the brightest comets are those that get closest to the sun. Some become bright enough to be seen in the daytime sky, but we see these extremely bright comets only once every few decades. Figure 9–6 shows such a comet, Ikeya-Seki, as it traveled rapidly through the outer reaches of the solar atmosphere in 1965. The apparent brightness of a comet, however, also depends upon where the earth is located in its orbit relative to the comet when it is seen. For example, Halley's comet, which is on an orbit which brings it back inside the earth's orbit once every seventy-six years, was very striking in its last appearance in 1910. Its next appearance in 1986 will probably be disappointing because the earth will be on the wrong side of the sun as the comet comes into the inner part of the solar system. It may not even be visible to the unaided eye.

With each pass close to the sun, a comet loses material by evaporation. Sometimes the effects are so strong that the comet fractures into two or more pieces when it passes its perihelion. Ultimately, after a number of solar passages, the comet will completely

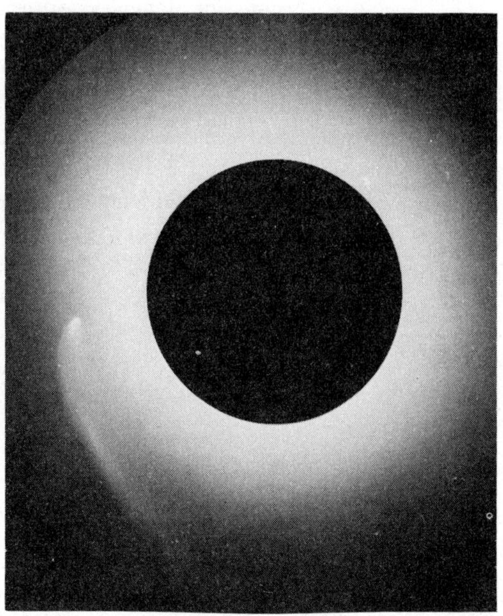

Figure 9–6 Comet Ikeya-Seki near the sun.

evaporate and be destroyed. But practically every year unexpected comets are discovered. There must be a supply of comets to replenish the ones that are inevitably destroyed.

Most comets, when first discovered, are on very eccentric orbits, approaching the sun with speeds close to that needed to escape the sun's gravity. No comet has yet been found to approach the sun with a velocity great enough to be definitely on a hyperbolic orbit. Thus comets, on elliptical orbits, are bound to the sun and belong to the solar system. But comets, particularly those with orbital planes close to that of the ecliptic, can be subjected to strong perturbations by the planets (mainly Jupiter) as they come into the inner region of the solar system. Some comets speed up, which throws them onto a hyperbolic orbit. These comets have made their last trip about the sun; they depart from the solar system. Other comets have been slowed down by planetary perturbations, causing them to move on smaller orbits of shorter periods. More frequent evaporation as they pass the sun hastens their final destruction.

The orbits of newly discovered comets have semi-major axes of close to 25,000 AU, indicating that they spend most of their time quite far from the sun. Furthermore, the orbits of comets can have any inclination, so the comets are not restricted to the flat system occupied by the planets and the asteroids. Newly discovered comets are also as likely to have westward as eastward revolutions about the sun. A model to explain these observed characteristics was presented by the Dutch astronomer Jan Oort. A diagram of his model is shown in Figure 9–7. A cloud of dark, cold comets, called the **Oort comet cloud**, surrounds the sun in a huge spherical shell. The comets are only loosely bound to the sun. Perturbations by other stars passing near the sun disturb some of the comets enough to deflect them onto highly elliptical orbits. It takes millions of years for a deflected comet to make its trip to the sun from the cloud. Ultimately, it spends a month or so shining because of the intense sunlight before it swings back out to the bitter cold from which it came. Of course, because of planetary perturbations, some comets go into even smaller orbits and have quite short periods—several years to thousands of years. This guarantees their ultimate destruction by the sun.

The frozen ices and dust of a comet, like a dirty iceberg, are thought to be only a few kilometers in size. For billions of years, before being perturbed into smaller orbits about the sun, the iceberg exists unchanged in space. The comets therefore are among the best examples of the matter from which the solar system formed. Some day a space probe may rendezvous with a comet to examine it in detail and return the first direct analysis of this early matter of the solar system.

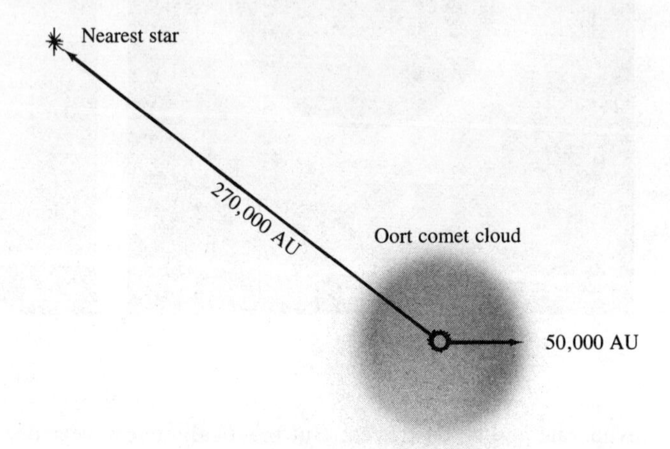

Figure 9–7 The Oort comet cloud.

Meteors and Meteorites

Some of the solar system debris, material left behind by comets or particles associated with the asteroids, is swept up by the earth. When these particles are on a collision course with the earth, they hit the earth's atmosphere. They enter the atmosphere with high speeds, between 12 and 72 kilometers per second, which is between 35 and 220 times the speed of sound in the atmosphere. At these speeds air friction is very important, and the objects are heated enough to vaporize most materials. The air through which the object passes is also heated. The hot gas along the path, made up of both the vaporized gases from the object and the air of the atmosphere, briefly gives off light before it cools again. An observer on the ground, seeing this event at night, would see a point of light streaking through the stars. He would see a shooting star, properly called a **meteor.** Figure 9–8 is a photograph of a meteor tail. The solid particle that caused the trail was never visible; what was photographed was the luminous gas heated by the particle's passage.

Whether or not the particle, or more properly some part of it, will reach the surface of the earth depends upon its initial mass. Most of the objects which cause meteors have masses smaller than one gram and completely vaporize before they penetrate our atmosphere very far. A large object, on the other hand, does not have enough time to vaporize completely during its short trip through the atmosphere.

Figure 9–8 A meteor trail.

Meteors are very common. About 25 million meteors, bright enough to be seen with the unaided eye at night, occur every twenty-four hours over the entire earth, although from any one spot on earth an observer might see an average of only about 100. If we consider the meteors down to one-hundredth the brightness that can be seen without a telescope, there are about 8 billion each day. While the mass of a particle causing a bright meteor typically is only about one-quarter gram, the total number of meteors implies that the earth picks up between 10 and 100 tons of solar system debris each day.

Once or twice a day, a piece of matter large enough to make it to the ground strikes our atmosphere. The piece that makes it to the ground is called a **meteorite**. There are three general groups of meteorites: **stony, stony-iron,** and **iron.** Stony meteorites are the most common. Over ninety percent of meteorites recovered after they were seen falling are stony. But a stony meteorite looks much like any other rock, especially to an untrained observer; and most which have come to earth unseen are overlooked. But iron meteorites, because of their peculiar appearance and resistance to erosion, are more easily identified, even when they were not seen to fall. Consequently, over half of the meteorites which have been recovered but were not seen to fall are irons.

The iron meteorites are particularly interesting. They usually contain a rich amount of nickel, anywhere from five to fifteen percent. When we cut, polish, and etch an iron meteorite with acid, we often see a pattern like that shown on the left in Figure 9–9. The pattern results from the crystalline structure of the metal. The form and size of the crystals in a metal depend upon its initial temperature and rate of cooling. The structures seen in the iron meteorites indicate that most were formed at a temperature near 800K and cooled between one and ten degrees every million years. Obviously, they could not have come from the icy-cold comets. They do fit well with our concept of the asteroids, however.

If a body between 300 and 600 kilometers across had a rocky outer shell and an inner core of nickel-iron, the rock would act as an insulating blanket, keeping the heat stored in the metallic core. Calculations show that if the core were at a temperature around 800K, the rock would not let it cool off by more than a few degrees or so each million years. This calculation fits well with the iron meteorite patterns. In addition, the fine bands shown on

Figure 9–9 Crystalline structure and bands in iron meteorites.

the right in Figure 9–9 are similar to what occurs in metal crystals when they are subjected to a sharp collisional shock.

The meteorites, therefore, most probably come from asteroids which have been shattered in collisions. They could not have come from a single big planet which broke up, because the cooling rate in a big planet would be too slow to fit the observed crystal pattern.

Meteorites are always associated with particularly bright meteors. The brightest meteors are called *fireballs* or *bolides*. Some fireballs have been bright enough to see in full daylight. The largest meteorite ever found weighed about 45 tons and was discovered in South Africa. The largest known single meteorite found in the United States weighs 13 tons; it was found in Oregon. But several craters thought to have been formed by even larger meteorites have been identified in different parts of the world. One of the most obvious, the first to be identified, is the Barringer Meteorite Crater in Arizona. The crater is 1.3 kilometers across and 180 meters deep, with a rim that rises 45 meters above the surrounding land. Over 25 tons of meteoritic material have been found in and around the crater, but borings into the crater bottom have not revealed any large buried fragments. Apparently, the meteorite blew up on impact, scattering and burying pieces over a wide area. There are no known cases of the killing of a human being by a meteorite, but a woman in Alabama was struck and injured by one in her home in 1954.

A most remarkable case of an object that almost became a meteorite occurred in 1972, when a body calculated to be about the size of a house entered the earth's atmosphere. Its path carried it parallel to the earth's surface, about 60 kilometers above the ground over the Rocky Mountains. It left the atmosphere over southern Canada and continued in orbit around the sun. If its orbit had been slightly different, it could have impacted on the earth with the force of an atomic bomb, and a new meteorite crater would be added to our list of such formations. A photograph of this meteor over the Grand Teton Mountains is shown in Plate 7A.

It is fortunate that most meteors are caused by very small particles and are not the larger particles associated with asteroids. Because meteors, on the average, enter the earth's atmosphere with random inclinations to the plane of the ecliptic, most meteors are probably associated with comets. This assumption is confirmed by the phenomenon of the **meteor shower**. Figure 9–10 shows a photograph of a meteor shower, when a large number of meteors are seen in a short time. They appear to radiate from one region of the sky, which implies that the particles causing a shower are traveling on essentially parallel orbits about the sun. Studies of meteor showers show that the particles are associated with the orbits of former comets and most likely are dust particles released when the comets were evaporated by solar heating.

Figure 9–10 A meteor shower.

The trajectories of meteors, from which the orbits of the particles causing them can be calculated, are found by photographing meteor trails from two locations. They can also be found by radar. The particles themselves are too small to be detected by radar, but the column of ionized gas left by the intense heating in the meteor trail is much larger and is also a good reflector. The radar measurements have one great advantage; they work in the daytime as well as at night. The positions and speeds of the meteors indicate that they occur mostly around 100 kilometers above the ground and that the particles causing them rapidly slow down in the thin upper atmosphere. Thus, the particles are not solid rock or iron, but fluffy material like that expected to be released by a comet. In contrast to the case of the usual meteor, the trajectories for meteors yielding meteorites lead to orbits related to those of asteroids, with low inclinations to the ecliptic and semi-major axes consistent with the distance to the asteroid belt. The trajectories and hence the orbits of the fluffy meteors, however, are characteristic of comets, with large eccentricities and frequently large inclinations.

There is one particular meteorite of note: the great Tunguska meteorite of 1908, which occurred in a remote region of Siberia. It was accompanied by a brilliant daytime fireball. Trees were knocked down for a distance of 30 kilometers and over 1500 reindeer were killed. The shock was detected as far away as England. From its great effect, the mass of the meteorite has been estimated to have been about 100,000 tons. But no actual meteorite or single large crater has ever been found. The most probable explanation is that the body was actually a small comet which violently broke up before it hit the ground. Any pieces reaching the ground would have simply evaporated because they would have been mainly ice.

Finally, there is another way the solar system debris makes its presence known. Under proper observing conditions (the absence of bright moonlight or city lights), a faint band of light can be seen just after dusk in the west or before dawn in the east. The band falls along the ecliptic and is named the **zodiacal light**. It originates from the scattering of sunlight by dust orbiting the sun mostly inside the orbit of the earth. The tiny particles of the zodiacal light cannot remain in this region of the solar system indefinitely. Hence, they must be continuously replenished by bits of dust, presumably contributed by the asteroids and comets.

Origin of the Solar System

The theories of the origin of the solar system fall into two classes. The planets may have formed as a by-product of the formation of the sun, in which case we would expect many other stars to have solar systems also. A second possibility is that planets formed later as a result of some unusual and catastrophic event in the sun's life, such as a collision or near-collision with a passing star. In 1939 Lyman Spitzer at Harvard College Observatory showed that if hot gas from the sun were ejected into space either by a collision with some large object or by huge tides raised by a passing star, the high temperature of the gas would cause it to quickly disperse. Planets, asteroids, and comets could not have originated in such a hot ejection of gas. So we are left with the theories in which the sun and its companions formed together.

The most widely accepted theory of the formation of the solar system is the nebular theory, which has been around in various forms since it was proposed by the philosopher Immanuel Kant in 1755. Actually, Kant put forth his idea before there was much evidence by which it could be tested. At first, new evidence from the observations of the solar system seemed to refute the theory, or at least not to favor it much. But today most of the objections have been countered, and this theory probably represents the birth of the solar system quite well.

Some of the nebulae—regions of diffuse matter between the stars—are seen to have dark globules, actually huge clouds of cold gas and dust, like those in Figure 9–11. We think these dark globules are contracting under their own gravity out of the original material of an immense interstellar cloud. As the globules shrink, the matter in them is compressed, causing them to become steadily hotter. The turbulence in the huge cloud would impart some rotation to the condensing globules. The globules, as they contract, would rotate ever faster. So a globule which starts out as essentially a large diffuse spherical cloud begins to distort in shape as it contracts. It develops a marked equatorial bulge. With even more contraction, most of the material becomes concentrated in a rapidly spinning core. The remaining material is left in a flat, disclike arrangement, slowly rotating in the equatorial plane of the core. The central condensation heats up and

Plate 7A   Earth-grazing meteor over the Grand Tetons.

Plate 7B   Comet.

Plate 8C The sun in X-ray radiation.

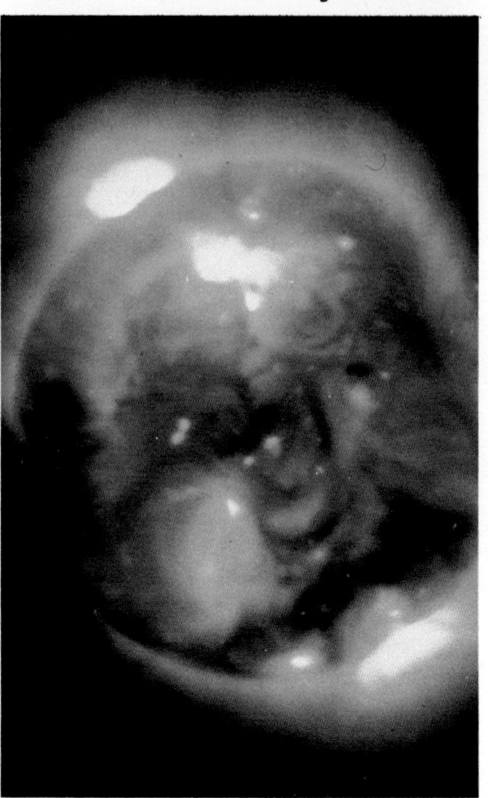

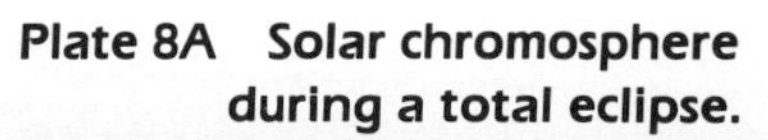

Plate 8A Solar chromosphere during a total eclipse.

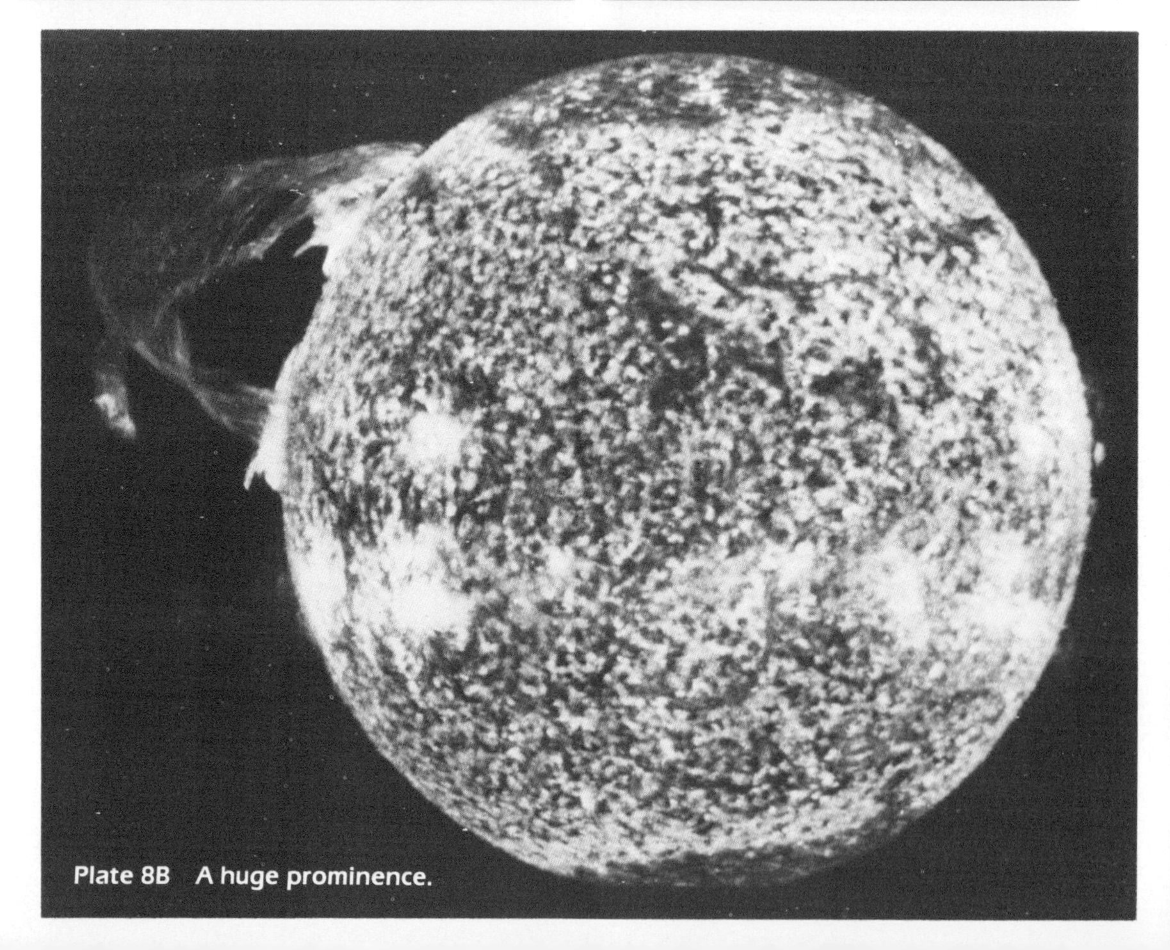

Plate 8B A huge prominence.

becomes a star—that is, a sun. The disc contraction, and the resultant faster rotation, continues until the material in various parts of the disc reaches the appropriate orbital velocity about the sun. The concentrations of gas forming in the disc then revolve around the sun in circular orbits, with the periods given by Kepler's harmonic law. At this point, gravity can no longer make the disc contract. With the contraction ended, the gravitational heating of the disc ceases. But the sun keeps the inner regions of the disc from cooling off rapidly, while the outer regions become colder by radiating away heat which cannot be replaced by the distant sun.

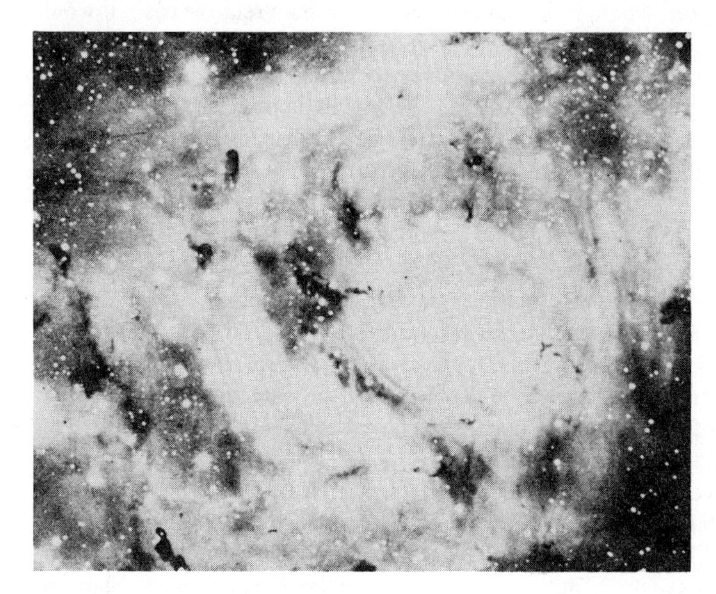

Figure 9–11 A nebulosity with dark globules.

The nature of the solar system now depends upon the chemistry of the cooling gases. Three facts about these gases are important to remember:

1. The gases are almost all hydrogen and helium, and at the temperatures which existed in the disc, hydrogen and helium remain as gases.
2. The rest of the matter in the disc consists predominantly of light atoms such as oxygen, carbon, and nitrogen. These generally form solids by reacting with hydrogen, but only at comparatively low temperatures.
3. The even less common and heavier atoms such as magnesium, aluminum, silicon, calcium, and iron, however, are capable of forming solids at higher temperatures.

In the cold, outer parts of the disc, the temperatures soon become low enough for water, methane, ammonia, and other simple molecules to form into ices. (Note that water, methane, and ammonia result from the most common element, hydrogen, reacting with oxygen, carbon, and nitrogen, respectively.) However, in the inner part of the disc, because of the sun's heat, it does not become cool enough for the light molecules to condense. The temperature remains high enough that only the heavier atoms like iron and molecules of various minerals are able to condense. The small condensation particles are probably fluffy, more or less like heavy snowflakes. As the fluffy flakes orbit about the sun, some occasionally bump into each other and stick together to form bigger particles. These particles pick up even more of the flakes and, in turn, join with other particles, forming even bigger pieces. Eventually, a few pieces become large enough to gravitationally influence smaller particles which come near them. The largest pieces therefore sweep up most of the remaining matter, ultimately growing to the size of planets.

In the inner regions, where it is too hot for the lighter elements to condense, the rocky and metallic planets like Mercury, Venus, Earth, and Mars are formed. Since Mercury formed in the hottest region, where iron could still condense with relative ease, it would have a much higher iron content, thus accounting for its great density. Since the elements that form these rocky and metallic planets are not very common but are relatively heavy, these planets end up being fairly small but dense. The Jovian planets, on the other hand, formed in the cold regions, where the ices condensed and collected the more abundant light elements as well as the heavier elements. They became much more massive but less dense. Jupiter and Saturn became massive enough that their gravity could also capture large amounts of hydrogen and helium gases. So much of these gases were present that Jupiter and Saturn became much larger and more massive, but with even lower densities.

This same pattern of lower density objects having larger orbits occurs in the four large satellites of Jupiter. The inner two are primarily rocky, while the outer two are primarily liquid or frozen water. Presumably, Jupiter was quite warm soon after it formed, and this warmed up the inner two moons enough to vaporize the water and ices that were in them. So the formation of Jupiter and its moons may have been a miniature version of the formation of the sun and planets.

In the region now occupied by the asteroid belt, the particles did not quite accumulate to a planet. The collecting of smaller bodies to form larger ones requires that the bodies come together gently. When the original particles were moving on essentially circular orbits, this condition was satisfied. But, most likely before the matter between Mars and Jupiter could completely coalesce into a planet, Jupiter had grown large enough to perturb the nearby pieces of matter. The particles went into more eccentric orbits, and their collisions became more violent. Instead of tending to stick together, the larger pieces now tended to break up. These violent collisions could account for the fine banded structure in the metallic crystals of the iron meteorites. In the first billion years of the solar system, most of the matter that occupied the asteroid belt was probably swept up by Jupiter, Mars, Earth, and the moon.

Since the major and minor planets were formed from the flat, rotating disc, they would all revolve about the sun in a flat system, with the same direction of motion about the sun, as indicated in Figure 9–12. Their rotations would also be direct, and their axes of rotation would tend to be perpendicular to the equatorial plane of the original condensing globule. The retrograde rotation of Venus and the large inclination of Uranus' axis probably resulted from the way they happened to collect their material. Possibly Venus and Uranus started rotating normally until they collided with some smaller bodies, which literally knocked them off their axes and modified their rotation. Only Pluto does not seem to fit this model well. It may be an escaped moon of Neptune and may not have formed like the other planets.

The comets need some special explanation. There are two ideas about how they may have formed and found their way to the Oort cloud. In one scheme, they are pieces left

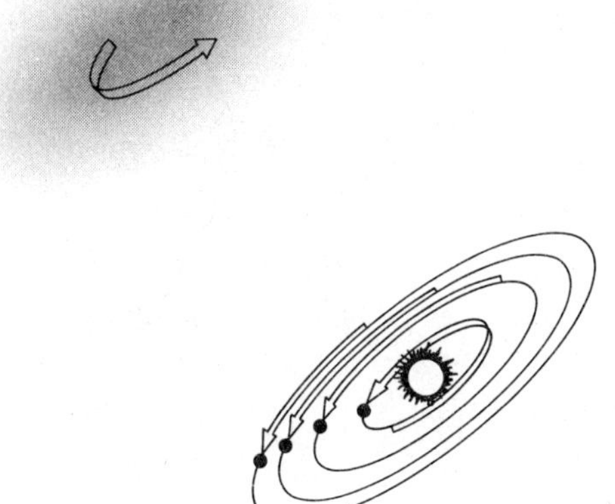

Figure 9–12 The early solar system.

over from the original condensing cloud. The pieces fragmented from the cloud as its rotation increased with its initial contraction. Some of the smaller pieces left behind were not large enough to form into separate solar systems; instead, they merely condensed into billions of dirty icebergs that became comets. Some of these would be at the extreme limits of the solar system, barely bound to it by the sun's gravity.

The other scheme proposes that huge numbers of dirty snowballs formed in the regions of Uranus and Neptune. Most collected to form these two planets, but a number were not captured. Instead, they were perturbed by the newly forming planets and sent to the outer reaches of the solar system. The choice between these two theories must await more definite evidence on the chemical composition of the comets, which would differ depending upon just where the comets originally formed.

The slowness of the sun's rotation had been the major block to the acceptance of the nebular hypothesis. When we calculate the sun's rotation rate expected from the contraction of the original cloud, we find that the sun should be rotating once every few hours. But it only rotates about once a month. In fact, much more of the original rotational energy of the cloud is now in the orbital motion of Jupiter than in the rotation of the sun. The phenomenon of the solar wind, which was not definitely established before the launching of artificial satellites, could reduce the sun's rotation. This finding defeats the strongest argument against the nebular theory. As indicated in Figure 9–13, the magnetic field of the sun, rotating with it, drives the ionized particles in the solar wind to greater and greater velocities around the sun as they move outward. At about 0.1 AU, the velocities of the particles in the solar wind will become great enough that the spreading magnetic field will no longer accelerate them. The more or less radial lines of force in the field will bend and start to trail behind the sun's rotation. The acceleration of solar wind particles is accomplished by removing rotational energy from the sun. While it is a gentle brake on the sun's rotation, working over the 4.5 billion years of the solar system, it could have removed most of the rotation of the sun, leaving it with its present rather leisurely spin.

It would be rash to say that we know all there is to know about how stars and solar systems form or how cosmic gas condenses to form the ices and silicate and iron particles.

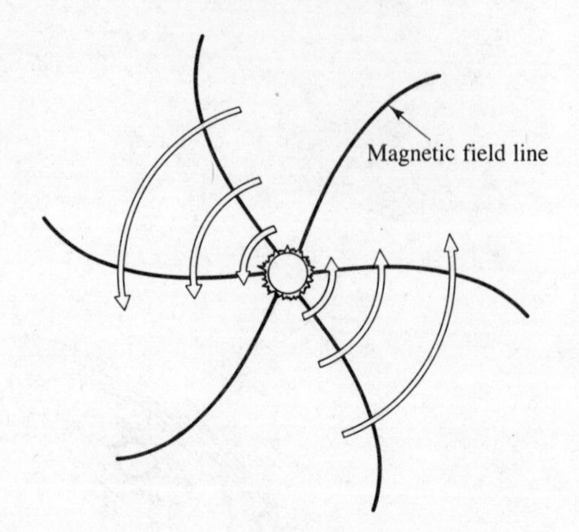

Figure 9–13 Rotation of the sun's magnetic field.

But our current ideas are probably on the right track. The formation of stars, with many having planetary systems, appears to be quite natural and common. Of the incredible number of stars known from telescopic observations, it seems impossible that there are not other stars with solar systems like ours. In fact, there are probably so many other solar systems in the universe that it is extremely unlikely that the life on our earth is the only life in the cosmos. The circumstantial evidence for life elsewhere grows each day—so much so that some astronomers are proposing and have even begun to search for positive evidence. They are looking for radio signals sent by intelligent life in other parts of space.

KEY TERMS

minor planet	meteor
asteroid	meteorite
comet	stony meteorite
gas tail	iron meteorite
dust tail	stony-iron meteorite
radiation pressure	meteor shower
Oort comet cloud	zodiacal light

Review Questions

1. *True or false:* A minor planet and an asteroid are the same thing.
2. Where in the solar system are most of the minor planets found?
3. If asteroids look like stars through a telescope, how could you tell which points of light seen with a telescope are asteroids?
4. How does a large asteroid compare in size with the visible head of a comet? With the "dirty iceberg"—that is, the solid core of a comet?

5. If an asteroid looks like a point of light through a telescope, how can we measure its period of rotation?
6. What determines whether a small solid body orbiting the sun will be an asteroid or a comet?
7. Why are comets generally farther from the sun than the asteroids are?
8. Why are neutral (un-ionized) gases not swept into the gas tail of a comet?
9. What causes the dust particles from a comet to form the dust tail?
10. Why do comets generally brighten rapidly as they get closer to the sun?
11. What evidence is there that there is a cloud of comets roughly 50,000 AU from the sun?
12. Why is the Oort comet cloud believed to be spherical, rather than a flat disc?
13. What is the difference between a meteor and a meteorite?
14. *True or false:* The best time to see a meteorite arrive at the earth's surface is during a large meteor shower.
15. When do meteor showers occur?
16. Which of the following would have a mass fairly representative of the mass of a meteor?
    (1) A few atoms
    (2) A few postage stamps
    (3) A golf ball
    (4) A bowling ball
17. What is the zodiacal light?
18. If two small objects orbiting the sun should collide, what determines whether they will stick together to form a larger object or break up?
19. Why does it appear unlikely that the planets condensed from material thrown off by the sun during a collision or near-collision with a passing star?

Discussion Questions

1. What lines would you expect to see in the spectrum of light from an asteroid?
2. Suppose an asteroid has its axis of rotation lying in the plane of its orbit; and when the asteroid appears at opposition in February or August, we see the asteroid pole-on. When opposition occurs in May or November, we would view the asteroid in the plane of its equator. How would the asteroid's brightness change in one rotation about its axis if viewed at opposition in February or in May? Could observations of the amount of variation be used to discover the asteroid's rotation axis?
3. Some comets have two separate tails, one fairly curved and diffuse and the other streaming straight out from the head. What are each of the tails made of? Why is one so much straighter than the other?
4. Each year we discover some comets thought to be making their first pass close to the sun. What leads to this conclusion?

5. If you see a meteorite land close to you in a field and you pick it up, what kind of meteorite is it likely to be? If you just happen to notice a meteorite that fell years ago, what kind is it likely to be?
6. How does the study of meteorites indicate that asteroids did not originate in a single planet which subsequently broke apart?
7. What evidence is there indicating that most meteorites were once parts of asteroids and that most meteors originated from comets?
8. If comets are sometimes ejected from our solar system by planetary perturbations, occasionally the ejected bodies should pass close by another star. Similarly, if comets are ejected by other solar systems, sooner or later one of these would pass close to our own sun. How would we recognize such an interloper, and what impact might its discovery have on our ideas of the origin of the solar system?
9. Why did the terrestrial planets form from generally heavier atoms than those which formed the Jovian planets?
10. Several of the outer moons of Jupiter have retrograde orbits. Why might these bodies be different from the other Jovian satellites?
11. What role may the solar wind have played in changing the rotation of the sun?
12. What bearing do theories about the formation of our solar system have on our estimates of the chance of detecting life elsewhere in the universe?

10

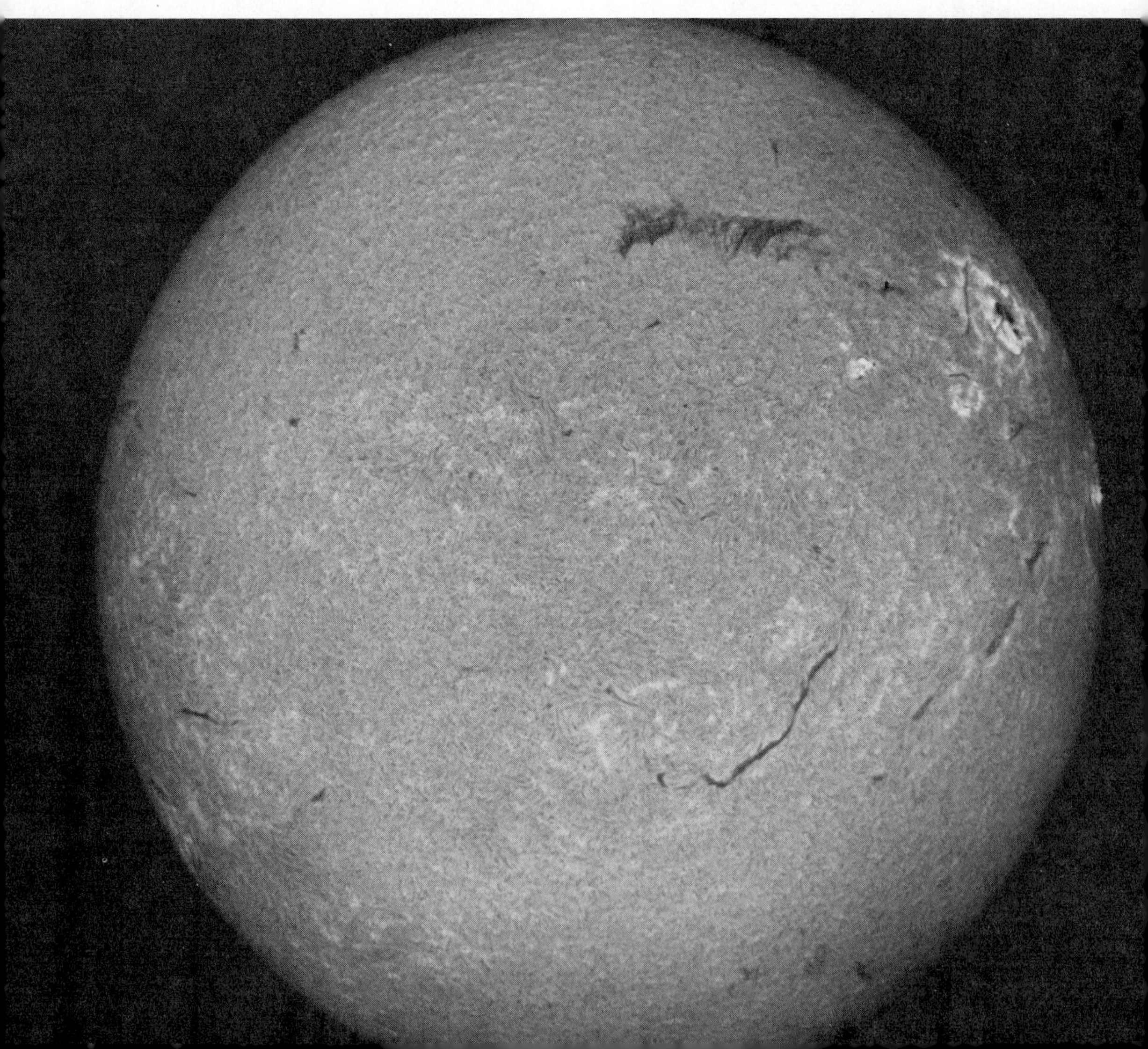

The Sun

During the early years of World War II, Germany conducted massive air strikes on England. Antiaircraft radar, which warned of imminent attack, was almost as important to Britain's defense as the famous Spitfire and the men who flew it so magnificently in aerial battle against the German air force.

The value of radar to the British was not lost on the Germans. During 1941 and 1942 they made a concerted effort to jam radar operation. Early in 1942, two German warships slipped through the English Channel under the cover of jamming from the French coast. As a result, the British War Office stepped up their counter-efforts to deal with radar jamming. They requested the Army Operational Research Group, which consisted in part of several distinguished scientists, to join in the task.

One member of the group, J. S. Hey, was asked to assume responsibility for advising the army on antijamming measures. Hey, although a physicist, had little knowledge of radio technology when he entered the British Army Radar School for six weeks of intensive training early in the war. However, his enthusiasm was fed by the urgency of the war effort as well as by the competence of his colleagues, and he soon became regarded as a radio expert. During the Battle of Britain, Hey's advice was sought frequently and at times frantically by the Anti-Aircraft Command and the War Office.

In late February of 1942, he received reports from radar sites located in many parts of England describing static so severe that radar operation was impossible. Had the Germans developed a device so powerful that transmission from occupied France could confuse Britain's entire radar network? No chances could be taken, and a full alert was ordered to meet the suspected arrival of a German air armada. However, no aircraft appeared.

Hey contemplated the problem of the new type of radar jamming and recognized that the direction of maximum interference recorded by each radar operator appeared to follow the sun. He immediately contacted the Royal Greenwich Observatory to find out if any unusual solar activity had been seen. Indeed, as it turned

The sun seen in the light of H-alpha

out, there was an extraordinary sunspot situated on the central meridian of the sun during the time of maximum interference. So quite inadvertently, during the frenzy of war, the outbursts of radio-wave emission from the sun associated with the magnetic activity of sunspots were discovered.

Over two thousand years earlier, on the Anatolian Plain of modern-day Turkey, a battle between Medes and Lydians had ended as a result of a solar eclipse. In that case, the sun's effect was purely emotional, but nevertheless powerful enough to cause warriors to lay down their arms. In the case of the battle between the British and Germans, the sun's effect was real; but it was understandable and not emotionally powerful enough to cause an end to the battle.

Today great efforts are being made to predict solar activity which can affect radio communications on earth. Not only does the direct radio-wave emission from the magnetic upheavals associated with sunspots cause interference, but so does the emission of ions and electrons which accompany this activity. These particles alter the reflecting properties of the earth's upper atmosphere, at times knocking out short-wave radio communications.

The worldwide defense systems of the United States and the Soviet Union rely so heavily on radio-wave communications that each country recognized its vulnerability during periods of extreme solar activity. As a result, each country has constructed satellite microwave communications networks to avoid the effects of these disturbances.

Some of the particles emitted by the sun occasionally burst through the earth's magnetic field, particularly in the north and south magnetic polar regions. Their interaction with air molecules high in our atmosphere causes the aurora, known to those in the Northern Hemisphere as the northern lights.

The Eskimos, a people who live in a part of the world where survival alone is so difficult that they have never had much time to develop the art of war, have been influenced in a quite different way by the particles emitted by the sun. One of their favorite games is an unusual form of football. The ball is a walrus skull. The object of the game is to kick the ball so that it falls with the tusks pointing downward and sticks in the snow. According to legend, the departed souls of dead Eskimos dwell in the "Land of Day," where they have unlimited pleasure. Laughing and singing, they kick a walrus skull across the snow of heaven; and the moving spirits of the dead can be seen as the northern lights.

According to the Eskimos, on a quiet night in the Arctic, when the northern lights are particularly vivid, you can hear their whistling cracking sound. They say that if you whistle in return, the lights draw closer to see the source of the sound better. On the other hand, if you spit, the lights draw back in disgust. Wrong as they may be about the nature of the northern lights, the Eskimos seem to have a lot more fun with the effects of the sun than the Medes, Lydians, English, or Germans!

The sun is the most important star in the sky—at least for the inhabitants of the earth. It supplies the light and warmth which makes the earth livable. It is the most obvious object in the sky, having the greatest apparent brightness of anything visible to us. Yet, if it were placed at the distance of the nearby stars, it would fade into relative obscurity and would be just visible to the unaided eye.

But it is quite close at hand, and it therefore supplies the astronomer with the only star for which details can be directly observed. All other stars, larger and smaller and more or less energetic, are so distant that they are mere points of light in even the most powerful telescopes. The sun is the starting place for understanding the stars. It has supplied many of the ideas about stellar behavior which have been expanded and modified to fit other and often widely different stars.

The energy of the sun is generated in its incredibly hot central regions, where the temperature is around 15,000,000K. We will discuss the specific energy generation process in a later chapter. For now we will concern ourselves with the outer layers of the sun from which the energy appears to emanate after its journey from the hot but invisible interior.

The Photosphere

To the naked eye, the sun appears as a large, smooth, luminous ball. Figure 10–1 is a direct light photograph of the sun. It shows what the sun looks like in a telescope with moderate magnification. The apparent surface from which the light appears to come is called the **photosphere,** which is derived by combining the Greek word *photos* with *sphere*. Photosphere literally means "ball of light." The parts of the sun from the photosphere outward are commonly called the *atmosphere* of the sun. They comprise the minute portion of the matter in the sun that can be directly observed.

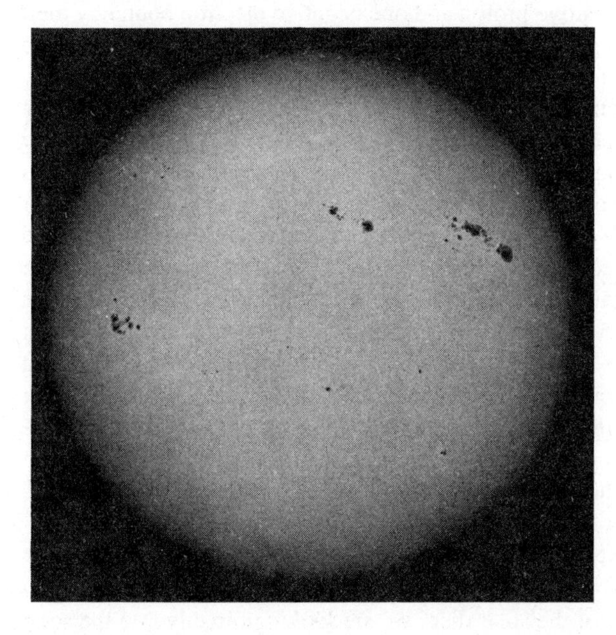

Figure 10–1 The solar photosphere.

The radius of the sun measured at the photosphere is 696,000 kilometers, or about 109 earth radii. It contains 1.99×10^{33} grams, which is equivalent to 333,000 earth masses. The mean density of the sun is consequently 1.4 grams per cubic centimeter, almost the same as that of Jupiter. The total energy radiated by the sun is given off at a rate of 3.90×10^{33} ergs per second. (An *erg* is the energy of motion possessed by a two-gram mass moving at a speed of one centimeter per second.) When the solar energy rate is converted to horsepower, it is equivalent to 5.23×10^{23}, or 523 thousand million million million horsepower. If a spherical black body having a radius equal to that of the photosphere were substituted for the sun, it would have to have a temperature of about 5800K to radiate the same total energy as the sun. The effective temperature of the photosphere, frequently called the "surface" temperature of the sun, is therefore given as being 5800K.

The dark markings on the photosphere seen in Figure 10–1 are called **sunspots,** the same phenomenon observed by Galileo when he trained his telescope on the sun. Relative to the north pole of the sun, the spots are carried in a counterclockwise direction by the sun's rotation, making them appear to move across the visible disc. It takes roughly a month for a sunspot to complete one full rotation with the photosphere. The spots at higher latitudes on the sun take longer to complete one rotation than do the equatorial regions, however; the farther from the equator, the longer it takes. This observation constituted the first evidence that the sun does not rotate as a rigid body. It is a huge ball of hot gas held together by gravity. The variation in the rate of rotation with latitude will prove important in understanding many of the phenomena observed in the solar atmosphere.

The photosphere is actually not a surface; instead, it is a gaseous layer with a density less than one-thousandth the density of the air at sea level on the earth. But the deep gases of the photosphere are much less transparent than the denser gases of the earth's atmosphere. The opacity of the photosphere—that is, the ability of its gases to impede radiation—arises because of the negative hydrogen ion, symbolized by H^-. The normal hydrogen atom consists of one positive proton and one negative electron bound by their mutual attraction. It therefore appears to have a total electrical charge of zero. The photospheric regions of the sun are hot enough to ionize some of the metal atoms, thereby providing a supply of free electrons. These electrons can attach to hydrogen atoms, giving each one an apparent negative charge and forming an **H-minus ion.** The system is weakly bound together, each electron being slightly more attracted by the proton than it is repelled by the other electron. Because of this weak binding, a low-energy photon (visible or infrared wavelength) has enough energy to eject one of the electrons when it is absorbed by an H-minus ion. The ejected electron does not have to move on any particular hyperbolic orbit about the normal hydrogen atom that is left behind. Thus, the negative hydrogen ion absorption can occur over a wide range of wavelengths and not just at specific energies in specific lines. There are enough H-minus ions in the photosphere at any instant to make it difficult for visible radiation to pass through this part of the solar atmosphere without being absorbed and re-emitted.

A close examination of Figure 10–1 reveals that the sun is not as bright near the limb of its disc as it is in the center. This dimming toward the edges is called **limb darkening.** The limb darkening becomes more pronounced when observed in shorter wavelengths. The schematic slice of the solar photosphere shown in Figure 10–2 can help explain the cause of limb darkening.

When we look at the center of the solar disc, we are looking directly into the solar atmosphere. The light we receive has originated at different depths in the photosphere; but

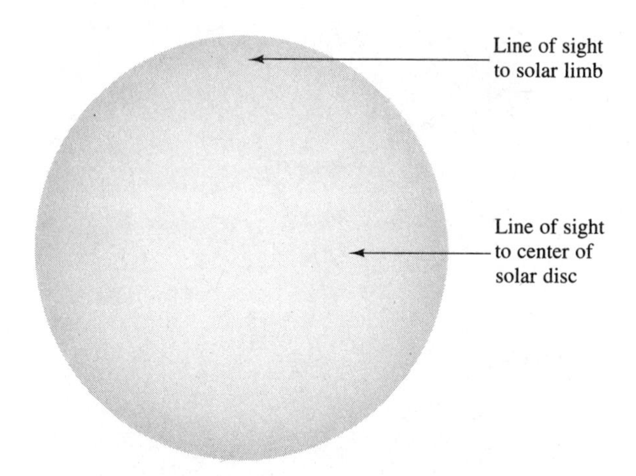

Figure 10–2 Limb darkening.

the deeper the layer from which the light has come, the more overlying atmosphere it has had to traverse and the more it has been blocked by absorption. The deeper layers are at the higher temperature and give off more light, but more of it is absorbed. Thus, the light that finally leaves the photosphere has a weak contribution from the hot deeper layers because of strong absorption. The largest contribution comes from intermediate layers, where the temperature is still fairly high and the absorption is reduced. Finally, a weak contribution is received from the cooler outermost layers. Even though there is no absorption to speak of, the lower temperature and density reduces the contribution of these layers.

Toward the limb, as shown in Figure 10–2, we look obliquely into the atmosphere—that is, we must look through more gas to see into the same depth in the photosphere. Therefore, the light from the hotter deeper layers is even more reduced by absorption and the light we receive is consequently more representative of that given off by higher, cooler layers. The brightness of the disc decreases as we look closer and closer to the limb because the light is more characteristic of the cooler and fainter upper layers. The enhancement of limb darkening at shorter wavelengths is also explained by this model. The black body radiation laws tell us that as the temperature decreases, the light radiated at shorter wavelengths falls off more rapidly than that at longer wavelengths. Therefore, since the light near the limb corresponds more to the higher, cooler layers, the limb gets darker more rapidly in blue light than it does in red light.

We can, in essence, invert this process. By measuring the solar limb darkening at several wavelengths, we can find how the temperature of the photosphere must change with depth to give the observed darkening. Such studies indicate that the deepest photospheric layers directly contributing to the light received from the sun are at about 8000K. The temperature rapidly falls off to about 4400K at the top layers. The total thickness of the layers contributing to the photospheric light is only about 400 kilometers. The photosphere is thus a relatively thin skin, covering a sphere over one million kilometers in diameter. It is this small relative thickness which causes the solar disc to have such a sharp apparent edge when viewed from the great distance of the earth.

When the photosphere is photographed under high magnification, it no longer appears uniform in brightness, but has a mottled appearance. Figure 10–3 shows the pattern seen away from the region of sunspots. The pattern has the appearance of a handful of rice grains thrown randomly on a piece of paper. It is commonly referred to as the solar **granulation.** The best resolution photographs, like Figure 10–3, reveal that the indi-

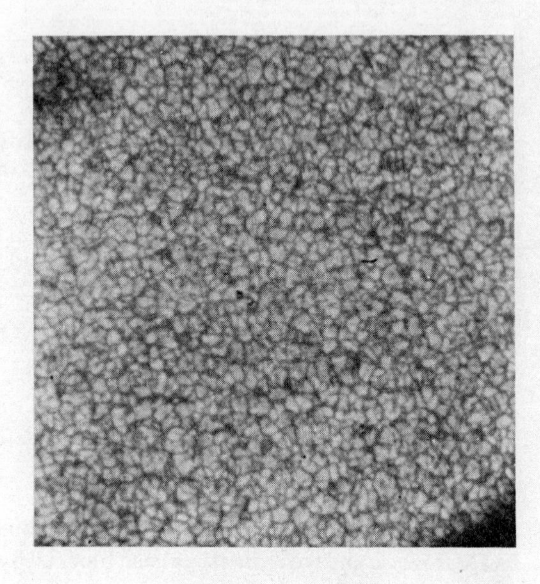

Figure 10–3 Solar granulation.

vidual granules are bright regions surrounded by dark lanes. On the average, any particular granule has a lifetime of about eight minutes. It measures about 1000 kilometers across, although granules as small as 300 kilometers can be seen. The bright regions of the granules appear to be about 50 to 100K hotter than the surrounding dark regions.

The granulation is a result of convection just below the photospheric layers. Hot gas rises in cells, marked by the granules, radiating away heat as it reaches the lower photosphere. The cooled gas then flows back into the subphotospheric region in the dark intergranular lanes, only to become heated and rise again. Thus, the gas stays in a constant state of turbulence, carrying heat energy from below the photosphere up into the photosphere.

The gases in the granules boil up and sink down with velocities around a few tenths of a kilometer per second. Some of this movement is transmitted to the overlying material in the photosphere. Very high resolution spectrograms of the sun show that the absorption lines have a wiggly appearance induced by Doppler shifts of the moving gases in which the absorption occurred. The sizes of the wiggles correspond to the granulation size and indicate that the higher portions of the photosphere above the granules have matter moving up and down with even greater velocities, about one kilometer per second.

The Chromosphere

Above the photosphere lies the region of the solar atmosphere called the **chromosphere.** It was given this name by Sir Norman Lockyer in the 1800s because of the characteristic red color which appears at the edge of the solar disc during a total eclipse when the moon just covers the photosphere completely (*see* Plate 8A). An instant or two later the color disappears, when the moon obscures this region of the solar atmosphere. Lockyer derived the name by combining the Greek word *chroma,* which means color, with *sphere.*

If a thin prism or a diffraction grating is placed over the objective of the telescope used to observe an eclipse, we obtain an emission spectrum just as the chromosphere flashes momentarily into view. Such a spectrum (shown in Figure 10–4) is called a *flash spectrum* because of its transitory nature. Normally, when we observe the sun out of eclipse, we see absorption lines caused by the cooler gases overlying the hot photosphere. In an eclipse, however, when the moon has just covered the photosphere, the thin tenuous gas of the chromosphere, still appearing beyond the lunar limb, no longer has a bright source of continuous light behind it. We see it in emission. In essence, what we see are a series of images of the thin crescent of the chromosphere in the characteristic wavelengths emitted by its gases. The strongest emission occurs in **H-alpha,** the brightest line of hydrogen in the visible part of the spectrum. Since hydrogen is the most abundant element in the sun, it is not surprising that this line predominates, giving the chromosphere its reddish color.

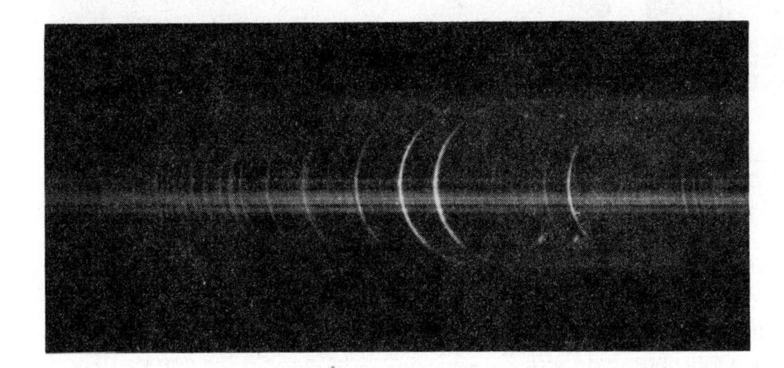

Figure 10–4 The flash spectrum.

The chromosphere extends from the top of the photosphere to the height where we no longer see the H-alpha emission. The total thickness defined in this way is usually 2000 to 3000 kilometers. At times it extends to 8000 to 10,000 kilometers in restricted regions.

If we were to observe the disc of the sun in the light of a strong absorption line, we could not see deeply into its atmosphere; in fact, we would see radiation originating in the chromosphere. Color filters and a device called a *spectroheliograph* have been developed which allow us to photograph the sun in an extremely restricted wavelength region. Consequently, we are able to view the chromosphere face on; we do not have to wait for solar eclipses to carry out chromospheric studies.

In Figure 10-5, you see two *spectroheliograms,* the name for photographs of the sun taken in the light of essentially one wavelength. The photograph on the left was obtained using H-alpha light, while that on the right used the light in a strong line of ionized calcium. The bright areas in the photographs are called **plages,** from the French word for beach. Notice also that dark threadlike markings called **filaments** are seen in the H-alpha photograph. In the ionized calcium spectroheliogram there is also a **network** of fine bright lines. A bit of the network has been sketched in to emphasize the cell-like structure. The cells are much larger than the cell structure seen in the solar granulation, being twenty to fifty times greater. Also, the chromospheric cells last for a day or two instead of a few minutes. Another contrast between the two kinds of cells is the direction of flow of the gases. In the chromospheric cells, the flow is mostly horizontal, from the center of a cell to the edges; but in the photospheric granulation, the gas motions are predominantly vertical.

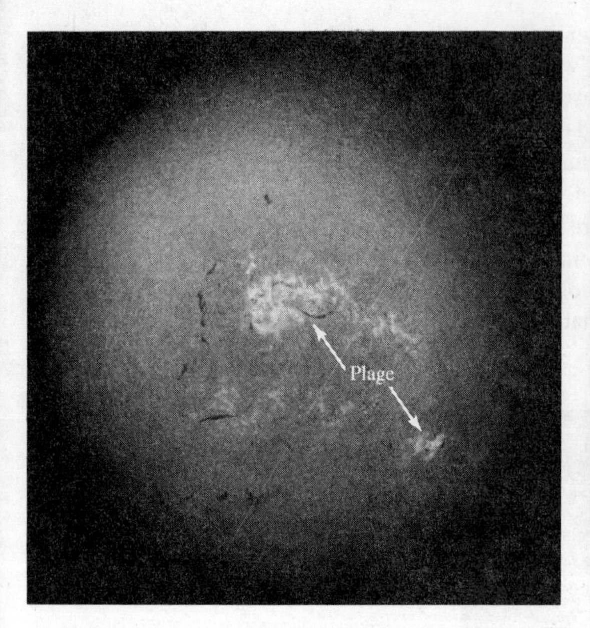

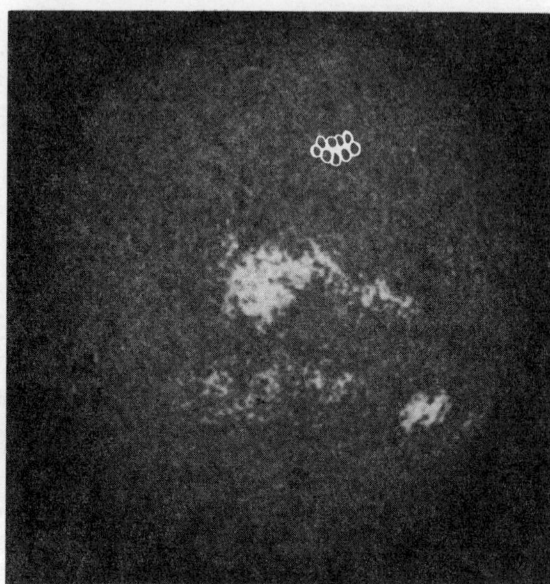

Figure 10–5 H-alpha and calcium spectroheliograms.

The H-alpha photograph of the limb of the sun (Figure 10–6) shows that the top of the chromosphere is not smooth, but that there are spikes protruding into the corona. These spikes, called **spicules,** shoot upwards with speeds measured in tens of kilometers per second. Individual spicules take between five and ten minutes to appear and disappear. In Figure 10–7 we see the spicules on the disc of the sun, but now dark, silhouetted against the deeper, brighter layers. The spicules do not occur uniformly across the surface of the sun, but rather originate at the borders of the chromospheric network cells.

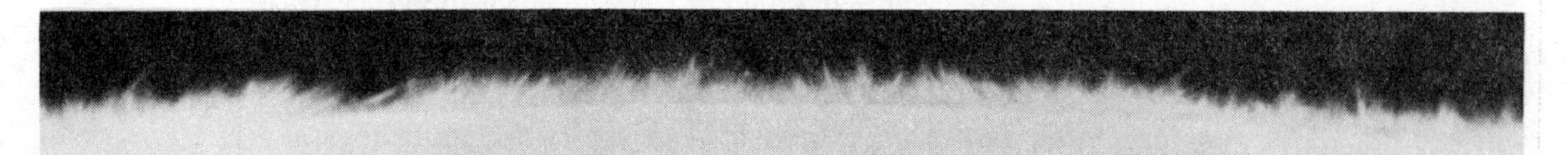

Figure 10–6 Spicules seen at the solar limb.

The spicules occur at network boundaries because of the behavior of the solar magnetic field in the network cells. Since the temperature of the chromosphere is high, much of its gas is ionized. The electrically charged particles of the ionized gas tend to move along magnetic field lines and can cross them only with great difficulty. Any vertical field lines appearing in a cell are rapidly swept to the boundary of the cell by the horizontally flowing ions. At the borders, where the horizontal gas flow stops and most of the matter appears to flow back to the photosphere, the vertical magnetic field piles up and is enhanced. The spicules, then, are columns of gas shooting upward along the vertical field at the cell walls. This conclusion is confirmed by comparing maps of the surface magnetism of the sun to the ionized calcium network. The magnetic field map shown on the left in Figure 10–8 follows the same pattern as the network shown on the right. Note how the field intensifies at the boundaries of the cells. The vertical field channels the gases in the spicules and keeps the material from simply dispersing.

Figure 10–7 Spicules at the network boundaries.

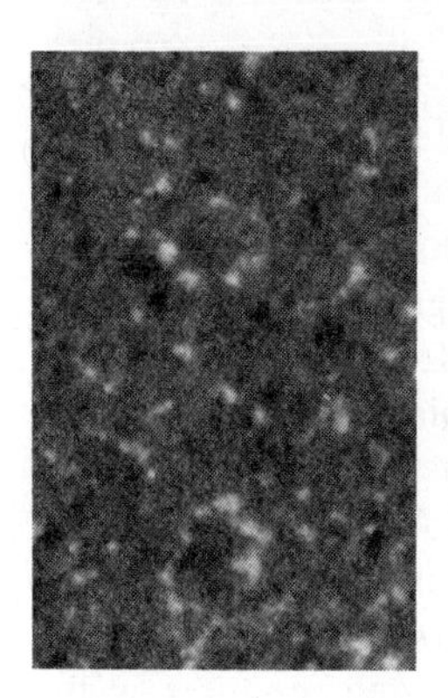

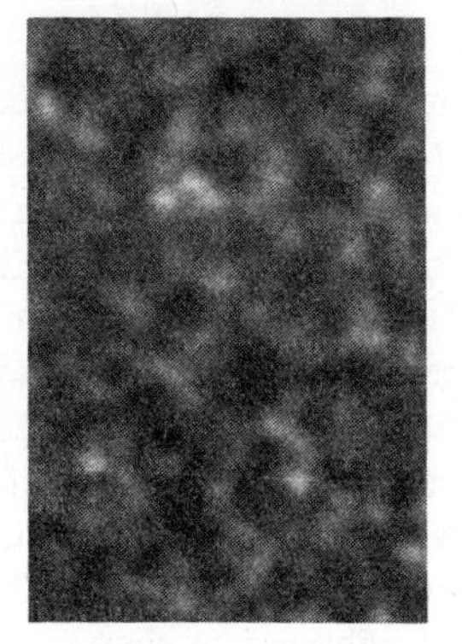

Figure 10–8 Section of a calcium spectroheliogram and the corresponding magnetic field map, shown on the left.

When the temperature of the chromosphere and photosphere are plotted against height, as shown in Figure 10–9, we get a startling result. Through the photosphere the temperature decreases as the energy flows outward from it—just what is expected to happen. But in the transition region between the photosphere and the chromosphere the trend reverses; throughout the chromosphere, the temperature increases outward, reaching values much higher than in the photosphere. Since heat energy always flows from hot to cold, some of the heat of the chromosphere must flow down toward the photosphere. The upper regions should cool off unless the heat energy is resupplied. We think the energy to heat the chromosphere comes from the turbulence causing the solar granulations. One proposed mechanism suggests that the motion of the gas in the granules, being partly transmitted to the thinner gases of the chromosphere, generates sound waves which dissipate in the chromospheric regions and above, thereby heating the gas. The very sharp rise in temperature at the upper end of the chromosphere happens because the gas loses its main cooling process: the emission of light by hydrogen atoms. When the hydrogen has become hot enough to be completely ionized, it also loses its ability to cool by radiating spectral lines, so the temperature rapidly increases to about 500,000K at the top of this part of the atmosphere.

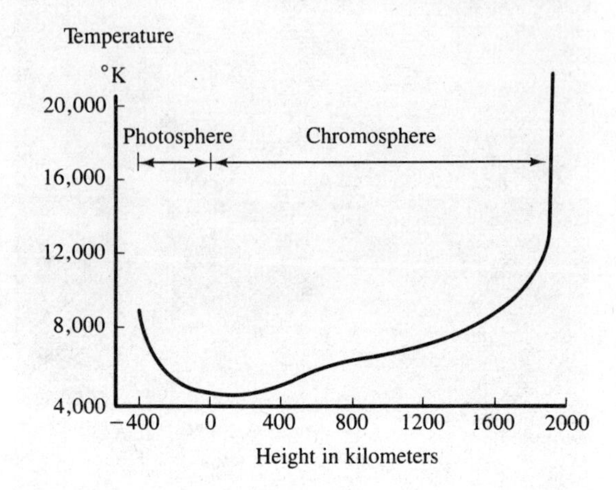

Figure 10–9 Temperatures in the photosphere and the chromosphere.

The Corona

Above the top of the chromosphere, where hydrogen is ionized so we no longer see H-alpha light emitted, lies the outermost part of the solar atmosphere. It is seen as the pearly white crown of light surrounding the darkened sun during a total solar eclipse. This region is called the **corona,** from the Latin word for crown. Figure 10–10 shows the corona as it appeared in the 1970 total solar eclipse. It extends out from the sun more than ten solar radii; in fact, some traces of it are detected hundreds of solar radii away. The gas of the corona is extremely diffuse; at its bottom it has the density of the top of the chromosphere—100,000 times less dense than the photosphere. At two solar radii the density is another one million times lower. The most striking thing about the corona, however, is its extremely high temperature: 1,000,000K to 2,000,000K. The reason that it does not swamp the photosphere in brightness at these high temperatures is the extremely low density—there is comparatively little matter to radiate energy. The combined light of the corona in the visible part of the spectrum is about equal to that of the full moon, and except at the time of an eclipse, it is completely submerged in the bright sunlight scattered in the daytime sky.

The very high temperature of the corona was proven, although it had been suspected earlier, when emission lines in its spectrum were shown to come from atoms such as iron ionized twelve and thirteen times. To strip twelve or thirteen electrons from iron atoms, the atoms must be moving with extremely high speeds when they collide with other atoms. These high speeds correspond to the 1,000,000 to 2,000,000K temperatures.

The escape velocity for particles near the sun, as with any other body, falls off with distance from the surface. At the base of the corona, the sun's gravity is still strong enough to prevent the gas particles from escaping. But at a distance near four solar radii, the

Figure 10–10 The corona.

escape velocity has decreased to a value equal to the average speed of the particles in the hot coronal gas. From this height upward, the atmosphere literally boils away, thus forming the **solar wind.**

While the corona is faint in the visible spectrum compared to the photosphere, it gets brighter than the photosphere in the very short ultraviolet and X-ray regions of the spectrum. This occurs in spite of its extremely low density because its temperature is so very much greater than the solar surface temperature. We have obtained pictures of the corona, looking directly at the sun with ultraviolet and X-ray radiation, by using satellites high above the absorbing atmosphere of the earth (*see* Plate 8C). These kinds of observations, not possible with earth-based telescopes, should help us to gain new insights into the complexities of the outer regions of the solar atmosphere.

Sunspots and Solar Activity

The granulation in the photosphere, the chromospheric network, spicules, and the corona all show variation with time, but they are always present on the sun. There are other features, however, which are sometimes present and at other times are almost entirely absent. The most obvious of these features are the sunspots.

A German amateur astronomer, Heinrich Schwabe, systematically observed the sun with a small telescope for nearly twenty years. He counted the number of spots visible in his telescope for each day that he could see the sun. In 1843 Schwabe announced that by his count the number of spots varied in cycles of 10 years. From records of sunspot observations over the last two centuries, we find the average length of one cycle of variation in the number of sunspots to be 11 years. Figure 10–11 shows the plot of sunspot numbers from 1900 through 1978. The time interval from one maximum to the next varies

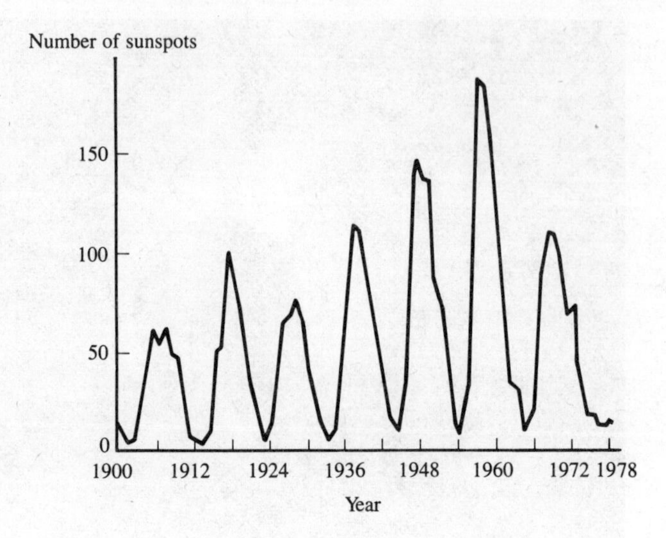

Figure 10–11 The sunspot cycle.

from cycle to cycle. The shortest interval ever noted is 7.5 years; the longest, 16 years. The 11-year **sunspot cycle** is the basic pattern of solar activity.

Recently, solar astronomer John Eddy rediscovered and verified some historical research by Gustav Sporer and E. W. Maunder at the end of the nineteenth century. They had found that although the sunspot cycle could easily be traced back to the year 1715, sunspots were very rare from 1645 to 1715. The "Maunder Minimum," as it is now called, was accompanied by other remarkable phenomena. For example, observers at total solar eclipses during this interval did not report either the corona or the chromosphere, although they were easily seen in more recent total eclipses. Apparently, there have been other extreme periods in sunspot activity. Oriental records of sunspots visible to the naked eye indicate that the sun had an unusually high number of sunspots during the twelfth and thirteenth centuries and relatively few spots in the late fifteenth and early sixteenth centuries. While we think we are beginning to understand the structure of the sun's atmosphere and the origin of sunspots, it is sobering to realize the regularity we presently see on the sun may not be a permanent feature.

More things change during a cycle than just the number of spots. Figure 10–12 shows the *butterfly diagram*. A mark is placed on the plot for each group of sunspots, showing its solar latitude and the time of formation. The outline of the plot for each cycle resembles the outspread wings of a butterfly—hence the name of the diagram. The spots in a cycle start out as a small number of spots forming and disappearing at latitudes around 30 to 35 degrees. As the cycle progresses, more and more spots appear and disappear, but the average of their latitudes grows progressively smaller. Then, as the cycle passes its maximum, the spots are seen in smaller and smaller numbers as the average latitude continues to drift toward the equator. Finally, the cycle ends, with the last few spots appearing around five degrees away from the solar equator. Before the cycle has completely ended, the next cycle has already started, with its first few spots starting to show in high latitudes. During a maximum, as many as a hundred spots may be visible at one time; while at the minimum of a cycle, days may go by with no spots at all being seen.

When the spots are examined in detail, they show considerable structure, but they divide into two general regions. As shown in Figure 10–13, the inner darker region is

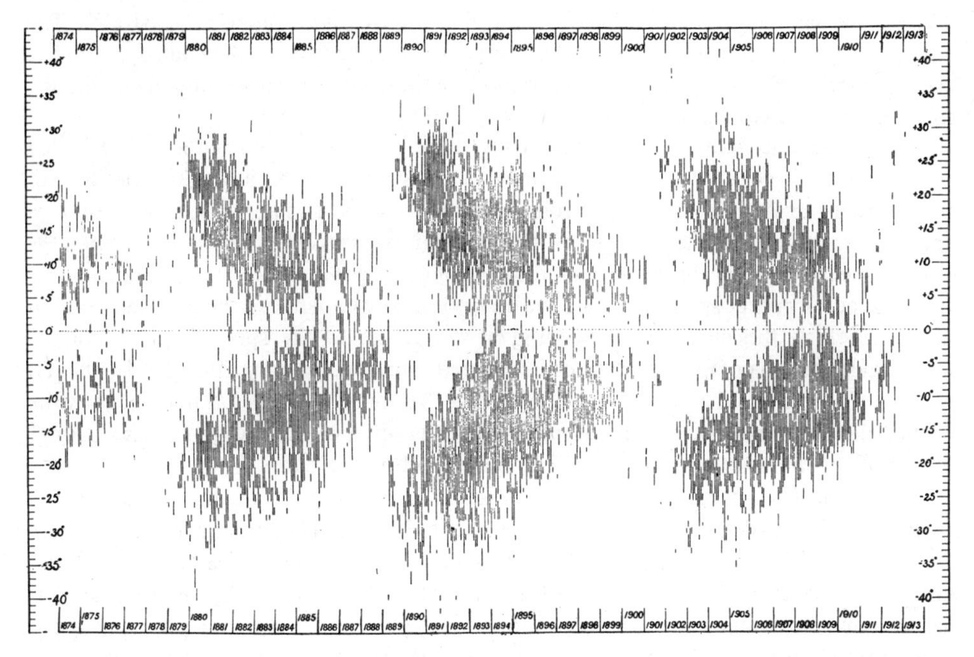

Figure 10–12 The butterfly diagram.

called the *umbra* and the less dark surrounding areas are called the *penumbra*. The apparently dark sunspot is actually quite bright—it just appears dark against the much brighter photosphere. If we could see a large sunspot isolated from the rest of the sun, it would give enough light to cast strong shadows on the earth. It would be tens of times brighter than the full moon. If we analyze the amount of light given off by the spot using Planck's black body radiation law, we find that the umbra must have a temperature around

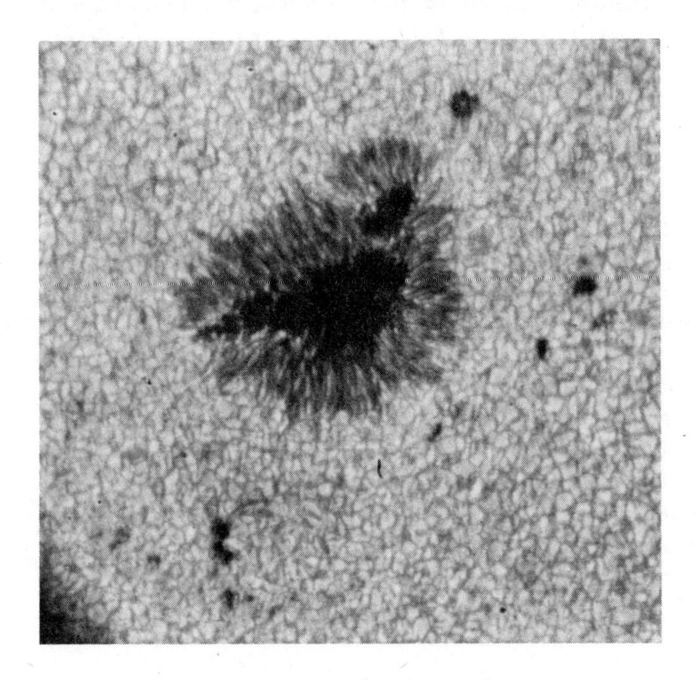

Figure 10–13 A sunspot.

4300K, the penumbra about 5700K, while the nearby photosphere is at 6100K. The sunspot is a cooler or, in one sense of the word, a refrigerated area in the photosphere. The cooling mechanism is not yet fully understood, but it must have something to do with the magnetic fields associated with the spots.

The magnetic fields are measured using the *Zeeman effect,* named after the Dutch physicist who discovered it in 1896. Whenever a gas is placed in a strong magnetic field, the normally single lines in its spectrum split into multiple components. The amount of splitting increases with the strength of the field and can be used to measure it. When the slit of a spectrograph is placed across the image of a sunspot, you see a single spectral line from the nearby photosphere. Across the sunspot it splits apart into the Zeeman components, reaching the maximum spreading at the center of the spot. By studying the individual Zeeman components, we can even determine the polarity of the field—that is, whether the field has the direction of the field at the north pole of a bar magnet or at the south pole. The magnetic fields required to give the observed splittings are very strong, thousands of times stronger than the magnetic field on the earth.

Sunspots often occur in isolated pairs. In sunspot groups, there are two subgroups, each with a principal spot. The spot pairs lie essentially parallel to the equator. The spot in the direction of the sun's rotation is called the *leading spot;* the other is called the *following spot.* The Zeeman effect shows that the leading spot (or the leading principal spot) always has the opposite magnetic polarity to the following spot (or the following principal spot). Furthermore, all leading spots in one hemisphere of the sun have opposite polarity to the leading spots of the same cycle in other hemisphere. Thus, during the 1957 sunspot maximum, the leading spots in the northern hemisphere were all north magnetic poles while the following spots were south magnetic poles. In the southern hemisphere, the reverse was true: the leaders were south poles while the followers were north poles. In the next sunspot cycle, the polarity in the two hemispheres reverses. The solar sunspot cycle can be considered as a 22-year cycle, when we include the magnetic reversal along with the variation in the number of spots.

The discovery of the magnetism associated with the sunspots raised a number of questions. Why does the sun have such magnetic fields? Why do the fields reverse every 11 years? What is the connection between the number of spots and this 11-year interval? And finally, why do the spots tend to appear at lower latitudes as a given cycle progresses? To try to answer these questions, we will use the model proposed by Horace Babcock in 1960. This model does a reasonably good job of accounting for the observed phenomena associated with the sun's magnetism, although it does not help us yet to understand why the sunspot cycle essentially stopped for 70 years during the Maunder Minimum.

We begin by considering how the magnetic fields make their appearance during the formation of a sunspot pair. The magnetic fields suddenly emerge from beneath the photosphere, taking about half an hour to rise in an arch above it. The left half of Figure 10–14 shows a photograph of a dark filament arch which traces out the magnetic arch, and the right half is an artist's drawing of what we think the magnetic field looks like. The magnetic field lines are wound or twisted like a rope. The magnetic rope floats to the surface, then bends upward into an arch extending above the photosphere. The two places where the rope pierces the photosphere are the regions of strong magnetic fields. Where the rope leaves the surface we have one polarity; where it reenters, the opposite polarity.

To generate the magnetic ropes, assume that the sun begins with a magnetic field like that of the earth. As shown in the upper left diagram of Figure 10–15, the magnetic field

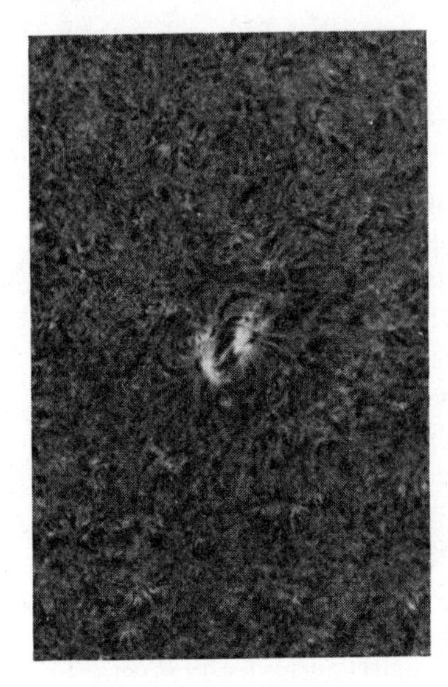

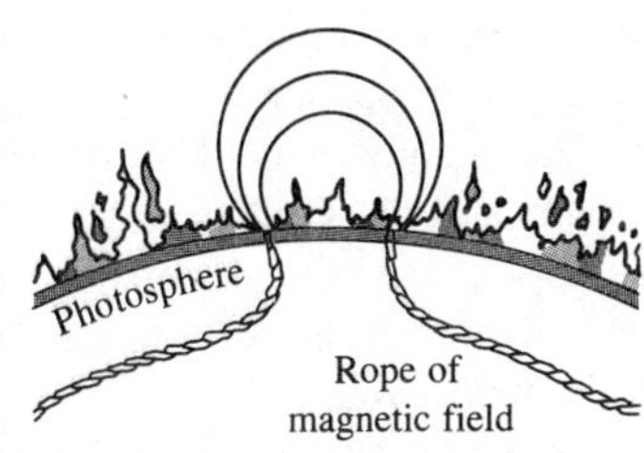

Figure 10–14 An arch filament and a diagram of a magnetic rope.

lines form large loops from the north to the south pole; then they make a closed path by continuing beneath the sun's surface. But the equator of the sun rotates faster than the polar regions. The magnetic lines are therefore distorted because they advance faster in the equatorial regions. As shown in the succeeding diagrams in Figure 10–15, the field lines get wrapped around the sun more and more as time progresses. After several years and many rotations, the lines are tightly wrapped. The turbulence below the photosphere twists the wrapped lines into the magnetic ropes. Since the wrapping is greatest at high latitudes, the ropes are formed there first. Consequently, the first spots begin far away from the equator.

The different polarity in the two solar hemispheres is a natural consequence of the model. Assume the lines go from the south pole of the sun to the north pole below the photosphere. As the wrapping proceeds, the lines in the northern hemisphere will be

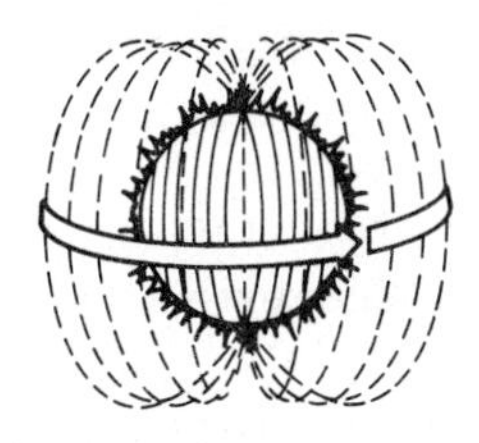

Start

One month later

Six months later

Two years later

Figure 10–15 Solar magnetic field wrapping.

distorted until they point right to left; but in the southern hemisphere, they will point left to right. Thus, the ropes in the two hemispheres will have different directions, causing the spot pairs formed whenever they erupt through the photosphere to have opposite polarity arrangements in the two hemispheres.

After the spots form, their magnetic field spreads out and the spots slowly decay. As the field spreads through the chromosphere, the bright plages seen in spectroheliograms appear about the spots. The magnetic field slowly disperses with the decay of the particular spot or group pair, and the field from the following spot slowly migrates to higher latitudes in the vicinity of the nearest pole of rotation. The fields from the many following spots in a cycle collect near the poles to form a new general magnetic field that will be needed to begin the next cycle.

In roughly eleven years, all of the ropes have erupted and dispersed and the general field returns to its original condition. There is one exception: The fields drifting from the following spots are in the opposite direction to the general field that started the cycle. The next cycle, therefore, comes from an oppositely directed field. The magnetic ropes formed by the rotation and turbulence of the sun are of necessity reversed in direction from those of the last cycle. The leader-follower polarities are consequently reversed in accord with observations.

When the fields about the two spots in a spot pair begin to spread, they approach each other along a line between the two spots. Any gas in the corona caught between the two fields is trapped and protected from the hot coronal gas, which cannot flow across the magnetic fields. The trapped gas therefore cools down, permitting some of the hydrogen ions to return to neutral hydrogen by capturing electrons. The recombined neutral hydrogen is once again capable of emitting and absorbing spectral lines. When we see this gas silhouetted against the chromosphere, we see dark filaments like the one shown in the H-alpha photograph in the left half of Figure 10–16. The right half of Figure 10–16 is a magnetic map of the same region, where bright and dark areas mark regions of north and south magnetic fields. Note how the filament seen in the H-alpha photograph threads its way along the border of the north and south magnetic fields.

When the cool gases are seen at the limb of the sun, they appear to be suspended above the edge of the disc and appear bright against the blackness of space behind. They are then called **prominences.** Figure 10–17 shows that the same gas cloud can produce both a prominence and a filament. The upper part of the cloud looks bright against the background and is called a *prominence,* while the rest of it appears darker than the sun behind it and is therefore called a *filament*. During the early observations of the sun, filaments and prominences were not recognized as being the same phenomenon viewed differently and were given entirely different names.

Prominences or filaments can last for long periods, up to several months. Occasionally, they erupt; they are best seen at this time when viewed as prominences. The material rapidly rushes outward from the sun. The Skylab astronauts observed the sun with an opaque disc covering the photosphere to give an artificial eclipse. This let them study the corona. Fortunately, they carried out several of the observations during eruptive prominences. The material violently ejected into the corona by a prominence formed a huge bubble moving through the corona with sufficient velocity to escape the sun's gravity and spew out into space. The Skylab experiments also produced the beautiful photograph of a huge prominence shown in Plate 8B.

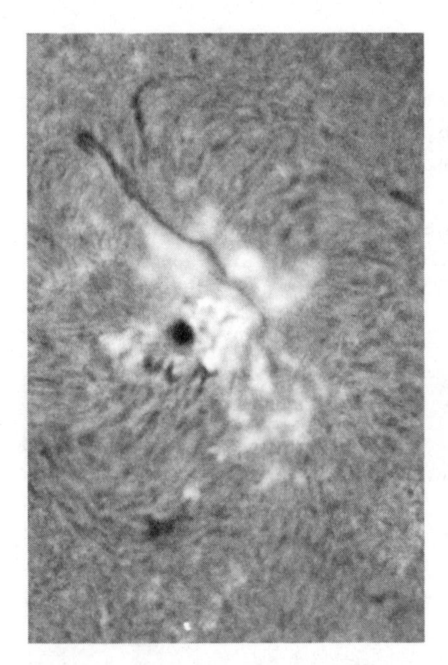

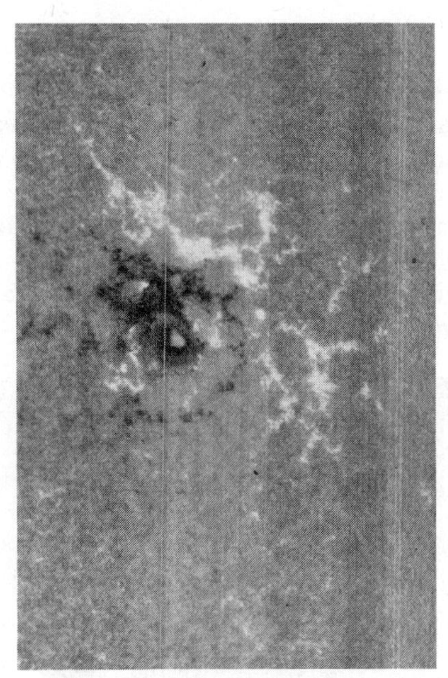

Figure 10–16 A filament and the corresponding magnetic field map.

Figure 10–17 A long-exposure photograph of a prominence superposed on a photograph of its continuation seen as a filament.

Flares

Possibly the most spectacular single events on the sun are the solar flares. A **flare** is a sudden and tremendous eruption in the chromosphere. Flares are best seen in H-alpha photographs, where they appear as a rapid brightening in a plage area. The two H-alpha photographs in Figure 10–18 were taken only twenty minutes apart. The large flare in the right-hand picture demonstrates how bright the H-alpha emission can become. Energies at all wavelengths—X-rays through radio waves—are emitted. The most intense flares give off so much light that they become visible in all wavelengths against the photosphere. These are called *white-light flares*.

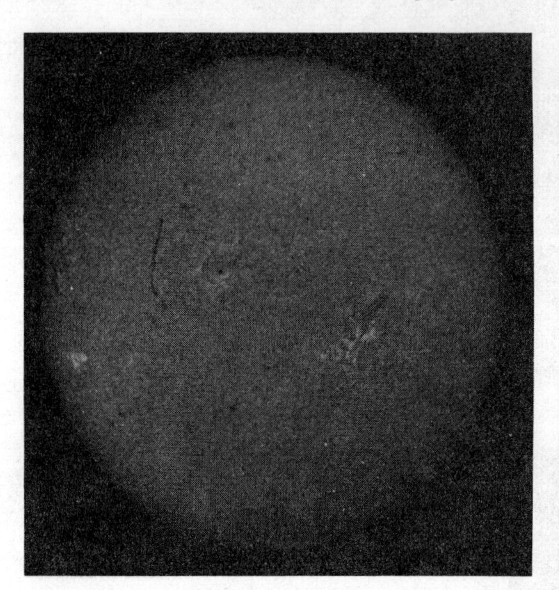
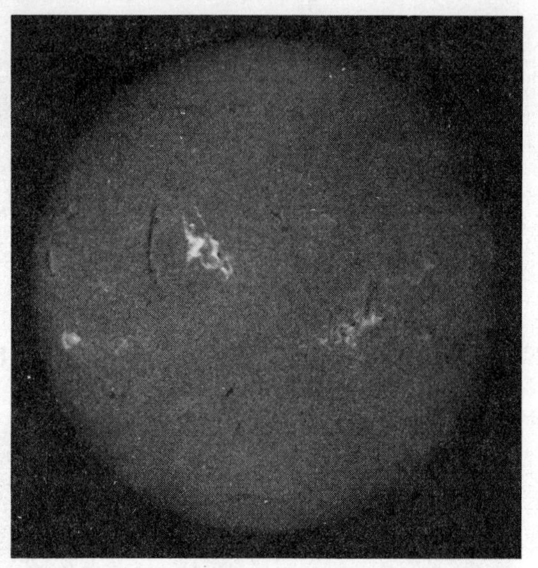

Figure 10–18 The sun before and during a flare.

The flares are relatively short-lived phenomena, brightening to full brilliance in a few minutes. The larger flares can remain visible for an hour or more before fading away, but the small ones subside in a few minutes. While flares as yet are rather unpredictable, their occurrence rate goes along with the eleven-year activity of the sun. At the sunspot maximum, small flares occur on the average about once an hour and the very large flares about monthly.

During its brief existence, a large flare can give off enough energy to supply the total energy consumption of the United States for 60,000 years. Surprising as this is, the total energy of such a flare is still only about one percent of the energy given off normally by the sun each second. We do not yet understand the mechanism of a flare or even where the disturbances that cause them originate.

Besides the large amount of radiant energy released during a flare, matter is also ejected. An example of more or less slowly ejected matter is shown in Figure 10–19. Some matter is ejected violently, however, streaming much more rapidly through the corona. This causes radio waves to be emitted along the way, permitting the matter to be followed on its mad rush upward. Some of the matter comes off at a speed around 1000 kilometers per second, but other parts of the matter speed outward nearly one hundred times faster—about one-third the speed of light. If the ejected matter is directed toward the earth, not only do we see the light emitted from the flare, but some of the ejected matter impinges upon our atmosphere. The light from the flare takes about eight minutes to make the 150-million-kilometer trip from the sun. This is followed less than half an hour later by the high-speed particles, while the slow particles take a relatively leisurely day or two to arrive.

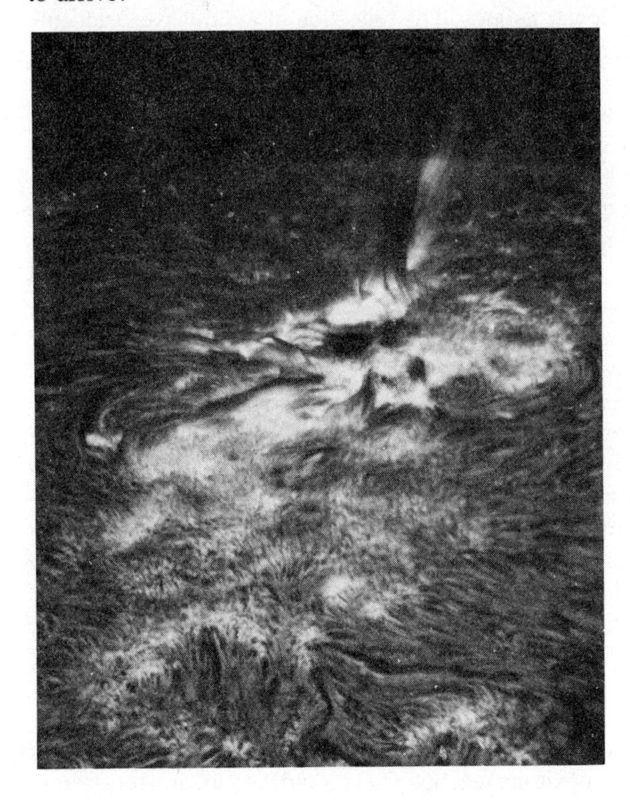

Figure 10–19 An eruption on the sun.

When the highly ionized matter from a flare approaches the earth, it interacts with the earth's magnetic field. If the flare is strong to moderately strong, the entire field of the earth is upset and a so-called *magnetic storm* takes place. Sensitive magnetic compasses suddenly go into peculiar fluctuations, appearing to lose their normally steady direction-finding property. Also, ions that had been trapped in the earth's field can suddenly dump into the atmosphere, causing the atmosphere to glow in the **northern** (or southern) **lights** (*see* Plate 9A). While this phenomenon of the aurora is best seen at high geographical latitudes, the northern lights can be easily seen from mid-latitudes during sunspot maxima and after particularly strong flares.

Not surprisingly, reports of aurora were extremely rare during the 70-year Maunder Minimum. The reappearance of this phenomenon at the end of the prolonged sunspot minimum caused considerable excitement since most people alive then had never seen the northern lights. Edmund Halley, a steady and tireless observer of the night sky, was 60 years old at this time, and he was most impressed with his first view of the aurora.

The ultraviolet light and X-rays from the flares also ionize the earth's upper atmosphere. Under normal circumstances there always exists a rather well-behaved region high in the earth's atmosphere called the *ionosphere,* generated by the normal short-wavelength radiation from the sun. The ionosphere reflects short-wavelength radio waves. It is the reason why short-wave communications can be carried out around the curvature of the earth. Without the reflection, you could not communicate with anyone beyond the horizon because the radio waves normally travel along more or less straight lines. When the flare radiation hits the atmosphere. the ionization suddenly increases. The ionosphere also pushes down to lower heights. The reflection of radio waves suddenly changes. What had been a good signal from a distant transmitter can suddenly weaken or even disappear completely. Sometimes the ionosphere is so enhanced that the radio waves are completely absorbed by it, and short-wave radio communication simply stops.

Effects of solar activity on lower layers of the earth's atmosphere are more difficult to measure. Powerful though solar flares may be, the energy reaching the earth from even a large flare is much less than the energy contained in weather patterns on the earth. It is therefore difficult to understand how a flare or related disturbances on the sun can have much effect on our day-to-day weather. However, several scientists have reported changes in weather patterns occurring when changes in solar activity take place. Others have reported that changes in the annual growth patterns of tree rings vary with the number of sunspots over the decades. The world climate was rather warm throughout a period of increased sunspots during the twelfth and thirteenth centuries. (The "Middle Ages warm epoch" actually lasted from the twelfth through fourteenth centuries.) On the other hand, the climate was unusually cold during the fifteenth and sixteenth centuries and later in the seventeenth and eighteenth centuries when sunspots were extremely infrequent. The connection between sunspots and weather or climate is far from proven, but the possibility must be taken seriously and is of considerable practical importance. For example, another cold spell like that accompanying the Maunder Minimum would seriously cut world food production. Perhaps the ultimate test of whether changes in the sun affect our climate will be to find out whether the climate on Mars varies at the same time.

Many questions about solar activity and its effect on the earth are still unanswered. So far, all we can say is that solar activity (including sunspots, plages, flares, and filaments) depends on the presence of strong magnetic fields, and these in turn result from the simple fact that the equatorial regions of the sun rotate faster. If the sun merely rotated as a solid body, long-distance radio communication would be more reliable, but the sun and its interactions with the earth would not be nearly so interesting.

KEY TERMS

- photosphere
- chromosphere
- corona
- H-minus ion
- limb darkening
- granulation
- H-alpha spectra line
- spicule
- plage
- chromospheric network
- solar wind
- sunspot
- sunspot cycle
- prominence
- filament
- flare
- northern lights

Review Questions

1. Why is the sun's appearance in the sky different from that of other stars?
2. *True or false:* The photosphere of the sun includes the entire sun except for the chromosphere and corona.
3. What are the granules of the photosphere?
4. What evidence is there that the photosphere is made up of a gas and not a solid?
5. What is an H^- ion?
6. What causes the wiggly lines in the photospheric spectrum?
7. How are photographs of the chromosphere taken?
8. What are spicules?
9. Why do spectral lines from chromospheric gases change from absorption to emission lines when we look at the center of the disc and then at the limb during a total solar eclipse?
10. What is the chromospheric network?
11. How does the temperature vary from the lower photosphere through the chromosphere to the corona?
12. How was the very high temperature of the corona finally determined?
13. If the hotter of two black bodies radiates more brightly, why is the corona fainter than the photosphere?
14. Which region of the sun's atmosphere is the strongest continuous emitter of X-rays?
15. *True or false:* A sunspot that appears early in the sunspot cycle will probably last for the entire cycle, moving from higher to lower latitudes as the cycle progresses.
16. In a sunspot, is the temperature of the umbra hotter, cooler, or the same as the penumbra?
17. How can the strength of the magnetic fields in the photosphere be mapped?
18. What concentrates the magnetic fields inside the sun to build up a field strong enough to make sunspots?
19. If a recently formed sunspot group has only two spots, how can we explain why magnetic polarities of the spots are opposite?
20. What was different about the sun and its effect on the earth during the Maunder Minimum?
21. What is the difference between a solar prominence and a filament?
22. What are the major differences between a plage and a flare?
23. What causes the northern lights?

Discussion Questions

1. Why are H^- ions so much more effective than normal atoms in increasing the opacity of the photosphere?
2. Some X-ray spectral lines are emitted from the lower corona, and photographs of the

sun taken with radiation at the wavelengths of these lines show the sun with limb brightening, not the more familiar limb darkening. Why?

3. The gases rising in photospheric granules are hotter than their surroundings until they reach the photosphere. Why do these temperature differences cease when the gases reach the photosphere, and not at deeper or higher layers?
4. The apparent size of the sun can be measured in photographs taken using both white light and the H-alpha spectral line. Which image of the sun will look slightly larger, and why?
5. Why do the spicules appear only at the borders of the cells of the chromospheric network?
6. Why is the sun's corona so hot?
7. What causes the solar wind?
8. If you observe the sun near sunspot minimum and you notice two adjacent sunspots, what two methods could you use to decide if the spots are the tail end of the old cycle or some of the first spots of the new cycle?
9. What observations about sunspots can be explained by the model in which the different rotational periods on the sun wind up the magnetic field?
10. What does the lack of sunspots during the Maunder Minimum imply about the sun's magnetic field?
11. What might support a prominence or filament, so that the sun's gravity doesn't cause it to fall to the chromosphere?
12. What are some of the various ways that a flare on the sun or its aftereffects can be detected on the earth?

11

The Distances and Motions of the Stars

Isaac Newton's formulation of the laws of mechanical motion and gravity left little room for doubt that the earth was not the center of the universe. It is merely one of several planets orbiting the sun. Newton's descriptions of motion and gravity were published in 1687 under the title **Philosophiae Naturalis' Principia Mathematica,** often called simply **The Principia.** Many consider it to be the greatest work on natural science ever produced by a human being.

This greatest work came dangerously close to never existing, at least in completed form. That it did come into existence, many believe, is a result of the efforts of Newton's contemporary, Edmund Halley. In 1686 Halley was appointed by the Royal Society to supervise the printing of Newton's treatise. The Royal Society, however, was financially broke at the time and ordered that Halley foot the bill for the printing costs. Besides supervising the project and paying the costs, Halley also collected much of the material for the book and corrected the proofs.

All of this still would not have been enough to ensure completion of the project. Before work on the third and most important volume of the set was begun, Newton, in a fit of anger, demanded that all work be stopped. His wrath was incurred by a colleague's claims that Newton had learned of the inverse square law of gravitation from him.

Halley managed to persuade Newton to go on with the project by gathering sufficient evidence to counter the allegations. Ultimately, hurt feelings were salved and **The Principia** was published in its entirety.

There are problems of human emotion, and there are problems of science. Serious scientists of Newton's time, including Halley, were confronted by the problem of stellar parallax. In view of Newton's work, one would have been hard put to insist that the earth did not move about the sun. Because of this motion, the annual shifting of positions of stars, that is, parallax, occurs. But at that time, methods of observations were not sufficiently refined for parallax to be detected.

A rich star field

Halley and others realized that the distances to stars would become known once their parallaxes were determined. The amount of annual shifting of a star's position due to the earth's motion around the sun depends on the star's distance and the size of the earth's orbit. But in Halley's time, not even the size of the earth's orbit was known accurately. A first step in determining the distances of stars in terms of terrestrial distances is accurately measuring the sun's distance from the earth.

It might appear that you could measure the sun's distance simply by measuring its position from two well-separated observatories on earth. However, the sun is far enough away that precise measurements of its position were and are difficult to make.

Since the distances in the solar system were known to scale, Halley developed a scheme to measure the sun's distance indirectly by measuring the closer distance to Venus as it passed between the earth and sun. French astronomers applied Halley's method. They obtained a value of 153 million kilometers for the sun's distance—that is, the astronomical unit, which is quite close to the modern value of about 150 million kilometers. Of course, Halley's method is no longer in use. Today we can use radar techniques to obtain the presently accepted value for the astronomical unit.

Nevertheless, Halley's method had led to a value for the size of the earth's orbit that could have been used to determine stellar distances if only parallax had been observed. And indeed, Halley did analyze the accuracy which might be attained with instruments of the day and did develop a method for measuring parallaxes. But this method bore no fruit. Parallax was not to be observed for nearly 100 years after Halley's death. In 1838 Bessel, Henderson, and Struve detected the parallaxes of the stars 61 Cygni, Alpha Centauri, and Vega, respectively.

So Halley never lived to see the observational proof of what he knew was there, stellar parallax. On the other hand, he stumbled onto proof of what no one, including himself, had ever considered seriously—proof of the motions of stars. While examining Ptolemy's catalogue of star positions, Halley noticed that the positions of certain stars were different than those measured in the seventeenth and eighteenth centuries. At first he concluded that the differences were due to inaccuracies in the earlier positions. Finally, after thinking it over, he realized that the differences in position were so large that he had to face the fact that the "fixed" stars are indeed not fixed at all—they have motions of their own.

This view was quite revolutionary in the eighteenth century. But eventually astronomers came to accept the notion of moving stars. At present, much of what we know about the structure of various types of star systems comes from rather routine studies of stellar motions.

Although he contributed much in so many different ways, Edmund Halley is not a famous man. He forever lies in the shadow of a giant, Isaac Newton. Only for a few short weeks every 76 years is Halley a household word—when the comet bearing his name passes near the sun.

We begin to understand the true nature of the stars only after their distances from the sun become known. In the attempts to measure stellar distances it was found that what early astronomers called the *fixed stars,* as opposed to the planets, or wanderers, actually were not immobile objects attached to the sky. They do indeed move relative to each other, but because of their great distances, their apparent motions are always small, normally requiring observations taken over many years to detect. Just as we have seen that the earth is not the immobile center of the solar system, we will, from a study of stellar motions, see that the sun is not the immobile center of the stars that surround us in the sky.

Stellar Parallax

The distances of the stars from the sun are all very large. The nearest star is nearly 250,000 times more distant from the earth than the sun is. Because of these large distances, no direct process of length measurement can be applied to the determination of stellar distances. Instead, we use the process of parallax measurements. Even this method can be used accurately for only a relatively small number of stars—about 700—which are comparatively near the sun. The distances to these stars and to the sun form the basis for the calibration of even more indirect distance determinations of more distant objects. The method of parallaxes is so fundamental in astronomy that astronomers will often quote the parallax of a star when asked about its distance.

Parallax is defined as the angular, or apparent, change in the position of an object due to a change in position of the observer. As shown in Figure 11–1, a relatively nearby star will appear to change its position in the sky relative to more distant stars as the earth moves on its annual orbit about the sun. The angle between the two lines of sight is called the *parallax angle*. Figure 11–1 is grossly exaggerated to illustrate the principle of parallax. If it had been drawn to scale, for even the nearest stars, the two lines of sight could not be distinguished from each other in the diagram. Even the nearest neighbor of the sun, for example, has a parallax which is less than one second of arc, the angle that would be subtended by a penny when viewed at a distance of nearly three kilometers.

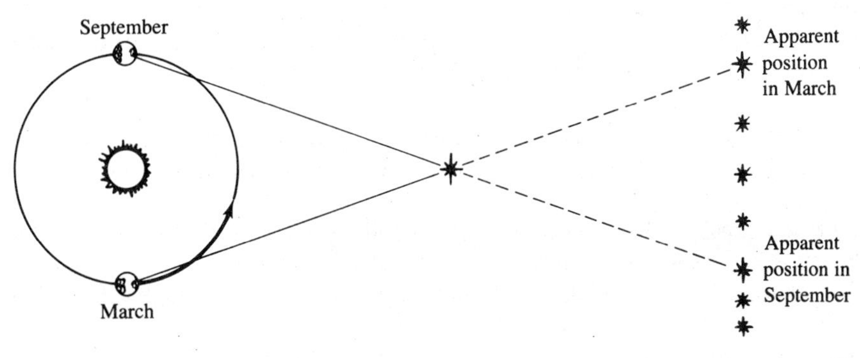

Figure 11–1 Annual parallax.

The actual parallax angle observed will depend upon both the distance between the two points of observation and the distance to the object being observed. Notice in Figure 11–2 that the nearer star will have a larger parallax angle than the more distant one. The parallax angle of a star, therefore, indicates the distance to the star. But to compare the parallax angles for different stars, and hence their distances, we must insure that the same baseline, or separation in the two observation points, is used for all measurements. Consequently, astronomers have agreed to define the parallax of a star as the parallax angle that would be measured from two points separated by exactly one astronomical unit and with the line between the points perpendicular to the direction of the star being observed. In the course of a year, any star can be observed from two points on opposite sides of the earth's orbit which, therefore, have a separation of two AUs. By choosing the times of observation correctly, the line joining the two points will be perpendicular to the direction of the star. Parallax measurements for a given star are always made when the earth is near the two optimum positions for that star, but the result of the measured parallax angle is always converted to the value which would have resulted from a separation of precisely one AU. The value determined this way is called the **trigonometric parallax** of the star. This term is always reserved for parallaxes determined by the method just outlined.

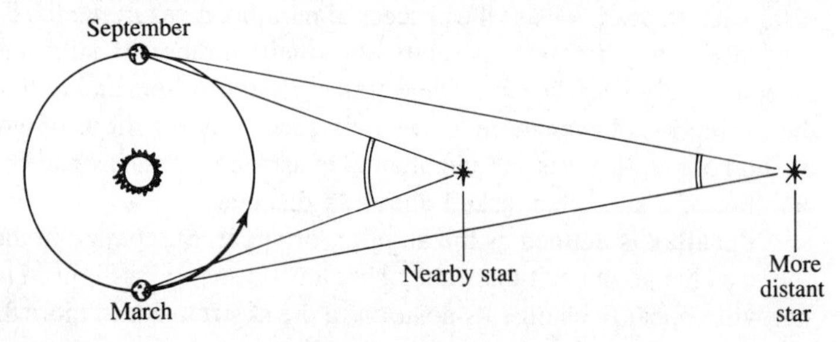

Figure 11–2 The relation of parallax to distance.

Measuring the small angles involved in stellar parallaxes is difficult. It is usually done by measuring the small changes in the positions of the star's image on photographs taken about six months apart. The image displacements of the parallax star are measured relative to the locations of the background stars on the photographs. This reduces errors in positioning the photographs in the measuring device, errors which otherwise could completely mask the small parallax shifts. Even with the long focal length telescopes used for parallax work, the stellar image positions must be measured to accuracies of a micron (one-thousandth of a millimeter) to obtain reliable parallaxes. For example, the star with the largest known parallax would only show a total parallactic movement of 73 microns—less that the thickness of a thin piece of paper—when using a telescope with a 10-meter focal length.

For the very small parallax angles observed in stellar parallaxes, the distance to a star is simply related to the star's parallax. As the distance to a star increases, its parallax decreases by the same amount. Thus, if the parallax of one star is one-tenth that of another star, it is ten times farther away. Alternatively, if the distance to a star were increased five times, its parallax would be decreased to one-fifth its original value.

Because of this simple relationship, it is convenient to define a new unit of length to express stellar distances. The unit is called the **parsec** and is defined to be the distance at which a star would have a parallax of one second of arc. No star is known to be that close to the sun. Proxima Centauri is the closest star found so far, at a distance of about 1.31

parsec. Since the parallax of a star in seconds of arc is equal to one divided by its distance in parsecs, Proxima Centauri has a parallax of roughly 0.75 second. As another example, a star with a parallax of 0.001 second would have a distance of 1000 parsecs.

The parsec is a large unit of distance, at least by terrestrial standards. It is equal to 206,265 AU or about 30 million million kilometers. Another comparison is that one parsec is equal to 3.26 light-years, that is, the distance light traveling at 300,000 kilometers per second would travel in 3.26 years.

The smallest value of trigonometric parallax that can be measured is around 0.001 second, although reliable measurements are usually no smaller than 0.01 second. Thus, for stars at distances greater than 100 to 1000 parsecs, trigonometric parallaxes fail, and other means must be used to find a star's parallax or distance.

Stellar Motions

Even before the first parallax had been measured in 1838, astronomers had noticed that some stars appear to move slowly across the sky. For a particular star this apparent motion, although small, is always in the same direction. It can build up to a fairly large angular position change over a period of many years. This change is different than the parallactic shift in a star's position because the parallax effect is oscillatory and thus does not accumulate over a period longer than one year.

The progressive angular motion of a star across the sky is called the **proper motion** of the star. It is expressed in terms of seconds of arc per year. Because of the possible proper motion of a star, we cannot obtain its parallax from only two photographs; this is because we do not know how much of the motion of the star's image resulted from parallax and how much resulted from proper motion. Three or more photographs allow us to sort out the effects, however.

Figure 11–3 shows the angular change in position—that is, the proper motion μ—which would result in one year for two stars moving relative to the sun. It is, of course, a greatly exaggerated drawing, since most proper motions amount to fractions of a second of arc per year. Although the two stars have the same proper motion, the more distant star must move faster to appear to sweep through the same angle in the sky as the nearby star. Consequently, once the proper motion of a star has been measured, we still do not know whether it is a nearby star moving slowly or a distant star moving rapidly. In general, however, it usually turns out that stars with large proper motions are among the closer stars to the sun. For example, Barnard's star, which has the largest known proper motion (10.3 seconds of arc per year), also turns out to be the second closest star to the sun.

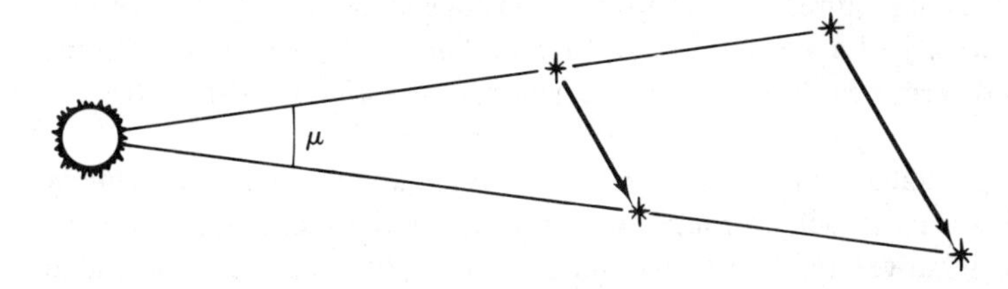

Figure 11–3 Proper motion.

Of the thirty-seven stars known to be closer to the sun than five parsecs, only four have proper motions less than one second of arc per year.

Only the part of a star's motion that is perpendicular to the line of sight to the star can contribute to its proper motion. If the distance to the star, as well as its proper motion, can be measured, it is possible to calculate the perpendicular speed of the star. Since this speed is at right angles to the line of sight, which is a radius of the celestial sphere, it appears to be tangent to the celestial sphere and is called the **tangential velocity** of the star. In Figure 11–4 it is obvious that the larger the proper motion, the larger the tangential velocity must be. Also, for a given proper motion, the tangential velocity must increase with increasing distance to the star. Combining these two facts, the tangential velocity of a star must be proportional to the product of the star's proper motion and its distance. The tangential velocity, in kilometers per second, can be calculated by the expression

$$T = 4.74\ \mu d \text{ km/s}$$

where T = the tangential velocity

μ = the proper motion (seconds of arc per yr)

d = the distance in parsecs

The number 4.74 arises from converting parsecs to kilometers, years to seconds of time, and seconds of arc to radians to obtain the velocity in kilometers per second.

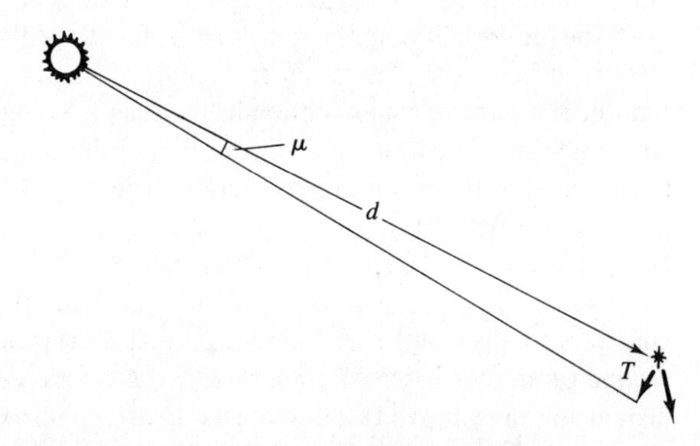

Figure 11–4 Tangential velocity.

As well as moving across the line of sight, a star may be moving along the line of sight at the same time. Again, this movement along a radius of the celestial sphere is called the *radial velocity* of the star. It can be found by measuring the Doppler shift in a star's spectrum and does not require that we know the distance to the star. By convention, the measured radial velocity of a star is corrected for the motion of the earth when the star's Doppler shift is determined. The result is the *heliocentric radial velocity* of the star or, in other words, the velocity of the star toward or away from the sun.

Once both the tangential and radial velocities of a star are known, its **space velocity** may be calculated. This is its velocity through space referred to the sun, that is, how fast and in what direction it moves relative to the sun. Because the tangential and radial velocities are perpendicular to each other, the rule of Pythagoras can be used to calculate the size of the space velocity. Thus, we form the following equation:

$$V^2 = T^2 + \overline{RV}^2$$

where V = the space velocity

T = the tangential velocity

$\overline{RV}$ = the radial velocity

Since we also find the direction in which the star appears to move across the sky when we find its proper motion, we can determine the direction of the space velocity, but this requires the application of trigonometry. Note that there are three independent measurements which must be made before the space velocity of a star can be found: its proper motion, its radial velocity, and its distance, or parallax.

Because the space velocity is referred to the sun, it is made up of a combination of the sun's motion and the motion peculiar to that particular star. As shown in Figure 11–5, to find the **peculiar velocity** of a star—that is, its velocity relative to a hypothetical stationary observer—we must add the effect of the sun's motion to the space velocity of the star. But, unfortunately, the sun's motion is not known beforehand. It must be determined from a knowledge of the motions of the other stars in the vicinity.

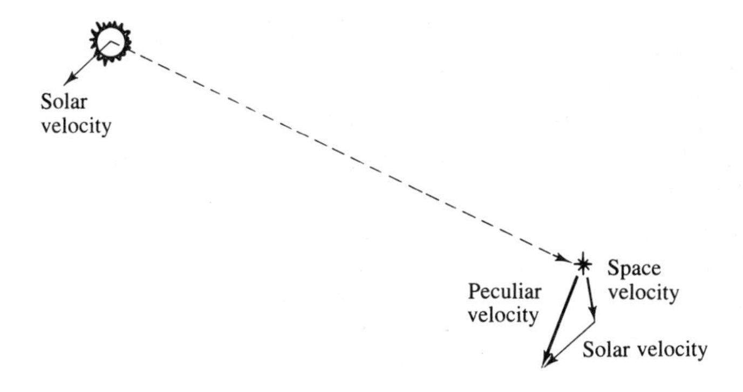

Figure 11–5 Peculiar velocity.

To see how the sun's motion is found, we begin by assuming that all stars except the sun are stationary. If this were the case, each star would appear to have a space velocity equal to that of the sun but in just the opposite direction to the solar motion. Figure 11–6 shows what would happen. The direction toward which the sun is moving is called the **solar apex**, and the point just opposite in the sky is called the **solar antapex**. It would appear to the observer being carried along with the sun as if the stars were streaming across the sky from the apex to the antapex. Stars lying in the direction of the apex would show radial velocities of approach, while those toward the antapex would show radial velocities of recession. They would, of course, show no tangential velocity and hence no proper motion. Stars in a direction perpendicular to the solar motion, on the other hand, would appear to have no radial velocity but would show a tangential velocity. Depending on their distances, they would have varying degrees of proper motion. In betweeen these special directions, the stars would appear to have both radial and tangential velocities reflecting the actual movement of the sun.

But the other stars are not standing still; they have their own peculiar velocities. To the first approximation, in the neighborhood of the sun, the peculiar velocities of the stars

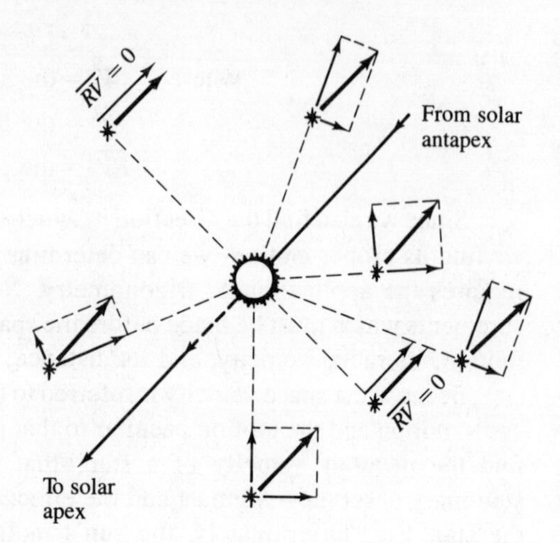

Figure 11–6 Apparent motion of stars due to the solar motion.

are random. That is, as many stars are moving in one direction as another, and furthermore, the range in the sizes of the peculiar velocities is the same in all directions. Thus, in a large sample of stars, the average of their peculiar velocities will be zero. Since the space velocity of each star consists of its own peculiar velocity plus the reflection of the solar velocity, the average of the space velocities will, in fact, be simply the reflected solar motion. Measuring the space velocities of the bright nearby stars and averaging them leads to the conclusion that the sun is moving with a speed of about twenty kilometers per second roughly in the direction of the star Vega. If you look in the direction of Vega, you see the largest average radial velocity of approach; if you look in the opposite direction, you see the largest average velocity of recession. The largest average proper motion is measured along a circle on the celestial sphere midway between the apex and antapex. This is what we expected from our analysis based upon the assumption that only the sun is in motion, but now we must take averages to cancel out the effects of the peculiar velocities of the stars.

We have already noted that good parallaxes can only be obtained when they are one-hundredth of a second or larger. Thus, we can only get good distance measurements by trigonometric parallaxes for stars within 100 parsecs, or 325 light-years, of the sun. Since we need the distance to a star to find its space velocity, our analysis of the solar motion by the method just described is based upon the very close sample of stars within a sphere of radius 100 parsecs surrounding the sun. Figure 11–7 schematically represents the sun and its neighbors within a distance of 100 parsecs. These stars are more or less moving through space as a group. They are like a swarm of bees, each having its own peculiar motion in the swarm as the whole swarm moves through space. If an observer wished to study the peculiar motions of the bees, he or she could run along parallel to the swarm and would thus cancel out the swarm motion. For any bee, the motion observed would be just its peculiar motion. Astronomers invoke a similar idea in studying stellar motions. We define a point in space, called the **local standard of rest**, which has a velocity equal to the average velocity of all stars, including the sun, lying within a distance of 100 parsecs from the sun. The solar motion determined from this sample of stars is really the motion of the sun referred to the local standard of rest. In a later discussion of the total

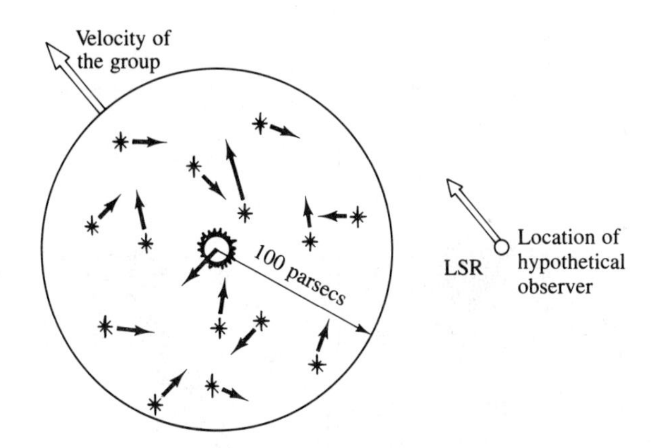

Figure 11–7 The local standard of rest.

system of stars to which the sun belongs, we will see how the motion of the local standard of rest itself is determined

Moving Cluster Parallax

As already noted, very precise trigonometric parallaxes can be obtained for a relatively small number of nearby stars. This number can almost be doubled by an indirect technique known as the *moving cluster method.* It depends upon the existence of what are called *galactic clusters,* groups of stars in which the individual member stars move more or less together through space. Three of these galactic clusters, the Hyades, the Scorpius-Centaurus, and the Ursa Major clusters, collectively contain nearly 500 stars and are favorably located for study by this method.

An everyday example can illustrate the general principle involved. Imagine that you are traveling down a highway and you observe a tree by the roadside in the rear-view mirror. As illustrated in Figure 11–8, the tree diminishes in apparent size as the car proceeds along the highway. If the tree appeared as large as your outstretched thumb at one moment and one minute later appeared only half as large as your thumb, you would know that the car must have doubled its distance from the tree in that minute. Assume the car is traveling at sixty kilometers per hour. Then in that one-minute interval the car would have traveled one kilometer. But if the distance doubled, and you traveled one kilometer in doubling the distance, the distance to the tree was exactly two kilometers at the moment it appeared half as big as your thumb.

A plot of some of the stars in the Hyades cluster is shown in Figure 11–9. Each star has an arrow which represents the size and direction of its proper motion. If the arrows were extended, they would essentially intersect in a point called the **convergent point** of the cluster. This point defines the direction in space in which the cluster is moving. Since the cluster stars are moving together, their velocities must be more or less parallel; and the convergent point represents the point at infinity where the parallel paths appear to converge. As time passes, the apparent size of the cluster will continuously diminish, eventually becoming vanishingly small as the cluster approaches its convergent point on the sky.

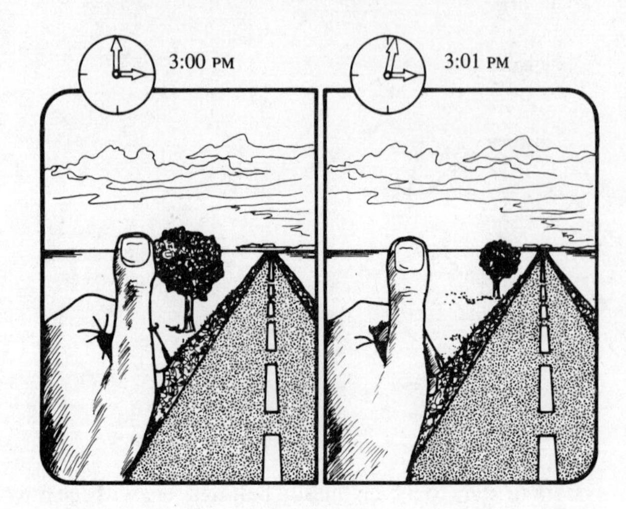

Figure 11–8 Measuring distance by the change in apparent size.

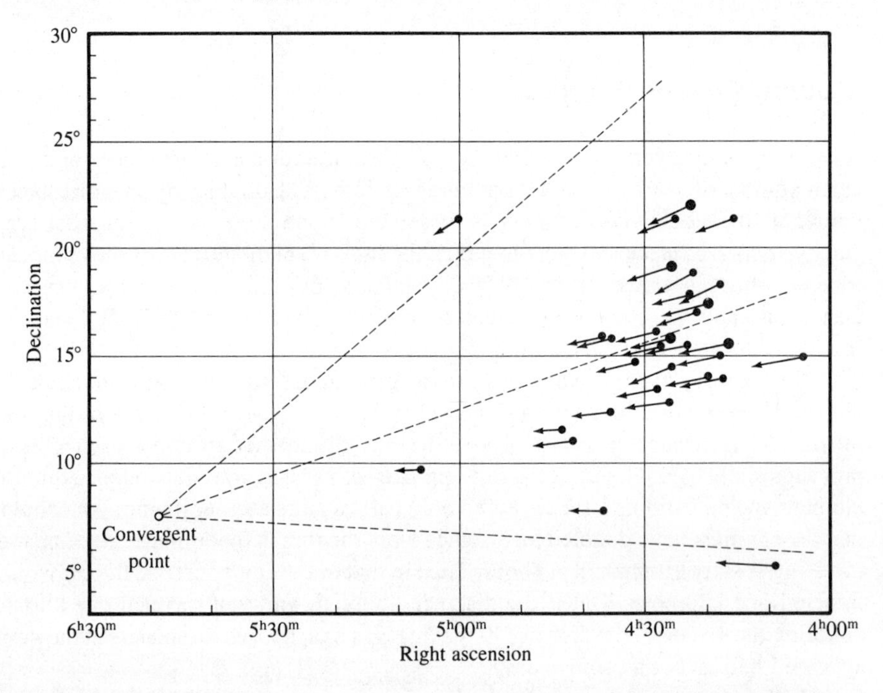

Figure 11–9 Proper motions in the Hyades cluster.

The distance to the cluster is so large that the cluster will change very slowly in angular size. However, from the known proper motions of its member stars, we could predict the size of the cluster at some future time. Now, if the radial velocities of a number of member stars can be measured, we can find how fast the cluster is receding from us. We could then do for the cluster what we did to find the distance to the tree. The distance to the cluster would be the fractional change in the apparent size of the cluster estimated from the proper motions during a specific time interval divided into the distance it would recede from us in that same time interval.

For example, if the stars in a cluster had radial velocities of 30 kilometers per second, the cluster would be receding from the sun 0.03 parsec every 1500 years. If this same cluster were observed to grow smaller by 0.1 percent in its angular size in that same interval, the cluster would be at a distance of 0.03/0.001 = 30 parsecs.

In practice, we use an equivalent method which gives greater accuracy. Since the convergent point lies in the direction the cluster would have as it would recede to infinity, the line from the sun to the convergent point must be parallel to the line of motion of the cluster. Therefore, the angle between the line of sight to a cluster star and the direction of the convergent point, labeled θ in Figure 11–10, is equal to the angle between the line of sight to that star and its space velocity, neglecting any small deviations the star may have from the cluster's general motion. But we can measure the velocity of the star in the line of sight, that is, its radial velocity. The tangential velocity of the star can then be found from the right triangle at the star formed by its space, tangential, and radial velocities by using simple geometry or trigonometry. But, we have already found the proper motion of the star to determine the convergent point. We can then use the relationship between tangential velocity, proper motion, and distance to calculate how far away the star is. Thus,

$$d = \frac{T}{4.74\mu} \text{ parsecs}$$

where T = the tangential velocity

μ = the proper motion

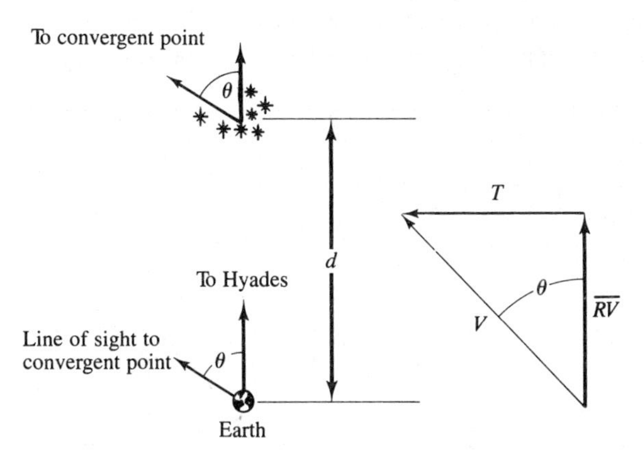

Figure 11–10 Determining the distance to a moving cluster.

This same process is carried out for the stars in the cluster bright enough to obtain good spectra for radial velocities. Since the size of the cluster is small compared to its distance from the sun, the cluster distance is taken as the average of the individual distance determinations. The averaging removes the effects of small deviations in the motions of the individual stars. Any other star in the cluster, even if it is not bright enough to give a good spectrum for radial velocities measurements, is assigned this same distance. The distance to the Hyades turns out to be 40 parsecs; alternatively, the cluster has a parallax of 0.025 second. While this parallax is two and one-half times greater than the usual limit for trigonometric parallaxes, the moving cluster method gives a more precise value of the distance to the Hyades.

Stellar Magnitudes

Besides just wanting to know how the stars are arranged in space, we use distance determinations for another reason. When we observe the stars in the sky, some appear brighter and some fainter. But we do not know which star radiates the most energy—that is, has the greatest intrinsic brightness—until we correct for its distance. Put another way, the apparent brightness of a star depends upon both its intrinsic brightness and our distance from it. A star which is really quite faint can appear brighter in the sky than a much more energetic star, if it happens to be close enough to the sun.

To designate the apparent brightness of a star, astronomers use a system invented by the early Greek astronomer Hipparchus. He called the brightest stars in the sky *first magnitude stars,* the next brightest *second magnitude,* and so on until those which he could just see were called *sixth magnitude*. We know that we receive 100 times as much light from the first magnitude stars as we do from sixth magnitude stars. Hipparchus had no accurate devices for measuring the light; he only knew that he could conveniently divide the naked-eye stars into the six magnitude groups. But the human eye responds in such a way that each magnitude step corresponds to the same fixed ratio in brightness. Since there are five steps in going from first to sixth magnitude, each step corresponds to a ratio of the fifth root of 100, or about 2.512. Thus, a third magnitude is 2.512 times brighter than a fourth magnitude star, and a first magnitude star is 2.512 times brighter than a second magnitude star. Table 11–1 shows how the brightness ratio changes over more than one magnitude. The minus sign precedes the magnitude difference because Hipparchus chose to label the brighter appearing stars with lower numbers.

Table 11–1

Brightness ratios corresponding to magnitude differences

Magnitude difference	Brightness ratio
−1	$(2.512)^1 = 2.512$
−2	$(2.512)^2 = 6.31$
−3	$(2.512)^3 = 15.85$
−4	$(2.512)^4 = 39.81$
−5	$(2.512)^5 = 100$

To make this way of designating the brightness of a star work, certain standard stars have been chosen and assigned magnitudes. To find the magnitude of any other star, we compare it with one of the standard stars. The difference in their magnitudes is computed from the ratio of their brightness. Thus, if the standard star had a magnitude of 4 and was 100 times as bright as the star we were interested in, our star would have a magnitude of 4 + 5, or 9. We would say it is a ninth magnitude star. Obviously, since telescopes were invented we have been able to detect stars much fainter than Hipparchus could. The faintest star which can be detected with the 200-inch telescope has a magnitude of nearly 25. When we give the magnitude of a star based upon its brightness as we see it in the sky, we call it the star's **apparent magnitude.**

If all the stars were at the same distance, a comparison of their apparent magnitudes would give us directly a comparison of their energy outputs. But they are not all at the same distance. Consequently, we define a new quantity: **absolute magnitude**. This is the magnitude a star would appear to have if it were placed at a distance of exactly 10 parsecs. While we cannot observe the absolute magnitude of a star directly, we can calculate its value if we know both the apparent magnitude and the distance to the star. Obviously, stars at greater distances than 10 parsecs will have fainter apparent magnitudes than their absolute magnitudes, while the reverse will be true for stars closer to us than 10 parsecs.

Ultimately, what we are interested in is the **luminosity** of a star, that is, the rate at which it radiates energy into space. Since we can measure the energy output of the sun with good accuracy, we can, by comparing the absolute magnitude of the sun with that of another star, find the luminosity of that star in terms of the solar luminosity. To see how this works, let us compare the sun, the object with the brightest apparent magnitude in the sky, with Sirius, the star with the next brightest apparent magnitude. The sun has an apparent magnitude of −26.7, while Sirius has an apparent magnitude of −1.5. The magnitude difference between these two objects is −25.2, which means that the sun appears about 12 billion times brighter than Sirius. The distance to Sirius is 2.7 parsecs, but the sun is only five millionths of a parsec away. Moving both of these bodies out to the standard distance of 10 parsecs to find their absolute distances, we would find their brightnesses would correspond to +4.8 for the sun and +1.4 for Sirius. The difference in absolute magnitude, now in favor of Sirius, is −3.4. Thus, Sirius actually radiates about 23 times as much energy as the sun. We would say that the luminosity of Sirius is nearly 23 solar luminosities.

It is only after we have measured the distances to the stars that we can appreciate the tremendous differences in the stars. Polaris, the pole star, has an absolute magnitude of −4.6, which means it is 5900 times as luminous as the sun. Its rather ordinary appearance in the sky results from the fact that it is 240 parsecs away. The closest star to the sun is Proxima Centauri, which has an absolute magnitude of +15.5. This means Polaris is 100 million times more luminous than Proxima. Even though Proxima Centauri is a little more than one parsec away, about 180 times closer than Polaris, it radiates so weakly that it cannot be seen without a rather good-sized telescope.

The search for parallax was started to try to find a proof for the heliocentric model of the solar system. Long before the first parallax had been measured, the ideas of Copernicus had already been accepted. The motions of the stars were, in one sense, a chance discovery, but they were not considered of overriding importance. Parallax and motions are now considered very important because they are at the very foundation of our knowledge of the true nature of stars and of the system of stars to which we belong.

KEY TERMS

parallax
trigonometric parallax
parsec
proper motion
tangential velocity
space velocity
peculiar velocity
solar apex
solar antapex
local standard of rest
convergent point
apparent magnitude
absolute magnitude
luminosity

Review Questions

1. What is parallax?
2. A parsec is a unit of
   (1) angle
   (2) size of a star
   (3) distance
   (4) motion
3. If a star is one parsec away, what will its parallax be?
4. If a certain star is found to have a parallax of 0.01 second of arc, how far away is it?
5. Why have trigonometric parallaxes been measured for only the nearest stars?
6. What is proper motion?
7. Suppose two stars are traveling away from us at 30 kilometers per second, but one star is twice as far away as the other. How will their Doppler shifts compare?
8. What is the difference between proper motion and tangential velocity?
9. If two stars have the same parallax but one has twice as much proper motion, how do their tangential velocities compare?
10. What is the difference between a star's peculiar velocity and its space velocity?
11. Suppose you're sitting in a train headed west at 40 kilometers per hour. Another train passes you, appearing to go westward 10 kilometers per hour faster than your train. How fast is the other train going?
12. If we look at many stars near the solar apex in the sky, on average we would expect these stars to show
    (1) Doppler shift to shorter wavelengths
    (2) Doppler shift to longer wavelengths
    (3) relatively large proper motion toward the apex
    (4) relatively large proper motion away from the apex
    (5) relatively large parallax
13. If we look at many stars 90 degrees away from the solar apex, on average we would expect these stars to show
    (1) Doppler shift to shorter wavelengths
    (2) Doppler shift to longer wavelengths
    (3) relatively large proper motion toward the apex
    (4) relatively large proper motion away from the apex
    (5) relatively large parallax.
14. What does the expression "local standard of rest" mean?
15. What is meant by the terms *apparent* and *absolute magnitude*?
16. Which appears brighter in the sky, a star of third or fourth magnitude? (Note: Unless specified otherwise, the word *magnitude* is taken to mean apparent magnitude.)

17. If two stars have the same absolute magnitude, why might one appear much brighter in the sky than the other?

Discussion Questions

1. If one star shows three times as much parallax as another, how far away is the first star compared to the second?
2. If a star changes position in the sky due to both proper motion and parallax, how could you decide how much of the change is due to just proper motion?
3. Why can proper motions often be measured for stars that are too far away to have their parallaxes measured?
4. If two observatories on nearly opposite sides of the earth both photograph the moon at the same instant, their two photographs show the moon appearing in quite different positions compared to the stars. Why? If these same two observatories then photograph the planet Venus at the same time, the position of the planet in the two photos will appear to change only slightly, much less than the change observed for the moon. What does this tell you about Venus compared to the moon?
5. In January, 1950, the position of a certain star of unknown distance was measured. In January, 1970, the same star's position was measured again. In this 20-year period, it had moved due south by 4 seconds of arc. What is its proper motion? Can you find the star's tangential velocity from this information?
6. What quantities must be measured if we are to find a star's space velocity?
7. What is the space velocity of the sun?
8. Our solar system is headed roughly toward the star Vega at 20 kilometers per second relative to the local standard of rest. You discover another star 90 degrees away from Vega in the sky which has a proper motion directed toward Vega but which shows no Doppler shift in its spectrum. What can you say about the space velocity and the peculiar velocity of this star?
9. If the star in the previous problem had its proper motion reversed so it was headed toward the antapex, what could you say about its space velocity and its peculiar velocity?
10. What quantities must be measured to obtain distances using the moving cluster method?
11. The star cluster Hyades, shown in Figure 11–9 with proper motions of its stars indicated by the arrows, is receding from the solar system, appearing to shrink at a rate of one percent in 10,000 years. Its radial velocity is about 38 kilometers per second, which amounts to 0.4 parsec in 10,000 years. How far away is the Hyades cluster?
12. Why will the moving cluster method work for stars too far away to have their trigonometric parallaxes accurately measured? Why can't the moving cluster method be used to find the distances to all clusters of stars?
13. How can you find the luminosity of a star?

12

Stellar Spectra and Classification

Classification is a human obsession. Dogs are classified. People are classified. Plants are classified. Diseases are classified. Stars are classified. Even yellow pages are classified.

The first comprehensive attempt at classifying stars was made in the mid-nineteenth century by the Italian Jesuit Angelo Secchi. He classified about 4000 stars into four basic types, according to their colors and spectral features. By 1920, Harvard College Observatory astronomers, with systematic thoroughness, had classified more than 100,000 stars according to their spectra. Instead of four types, their newer classification system had ten. However, as pointed out by the brilliant American astrophysicist Henry Norris Russell, more than 99 percent of the classified stars fell into one or another of the six major groups, arbitrarily called B, A, F, G, K, and M. He wrote, "That there should be so few types is noteworthy, but much more remarkable is the fact that they form a continuous series. Every degree of gradation between the typical spectra denoted by B and A may be found in different stars and the same is true to the end of the series . . . . "

Russell was of the opinion that the differences in stellar spectra were due to variations in some single physical property of the stellar atmosphere, such as the temperature. The seemingly obvious explanation for the difference in spectra of stars was that stars of different types have different chemical compositions. Indeed, different chemical elements have different spectral features, but laboratory experiments of the Englishman Sir Norman Lockyer showed that samples containing the same chemical elements showed different spectral characteristics, depending on the temperature of the sample. Unfortunately, Lockyer was quite unpopular among many American astronomers, and his ideas were often shunned. Most went along with the naïve idea that stars of different types have different chemical composition.

During the years 1920 to 1922, an Indian astrophysicist, M.N. Saha, applied the theory of atoms to astrophysics. Saha explained how the spectrum of a chemical

Spectrograph attached to a telescope

element depended on the extent to which its atoms are excited and ionized. By further explaining how the excitation and ionization of atoms depend on the temperature, Saha linked the spectrum of a chemical element to its temperature, thereby providing explanation for Lockyer's laboratory results.

Saha submitted a paper describing his theory to the **Astrophysical Journal,** a periodical founded by George Ellery Hale which has developed into the leading journal of astronomy. The editor of the **Astrophysical Journal** rejected Saha's paper. The paper was finally published in the **Philosophical Magazine,** a British journal; it ranks as one of the classic treatises of natural science.

The next editor of the **Astrophysical Journal** discovered Saha's manuscript in a box of rejected papers of several astronomers who went on to become internationally known. The previous editor was not perfect, however, for the box also contained several manuscripts that were absolutely worthless.

Saha explained the spectral sequence of stars on the basis of his theory. His work showed conclusively that the differences in the spectra of stars are largely due to the differences in their atmospheric temperatures.

At present, the spectral classification of stars is an integral part of the theory of the structure and evolution of stars. Our understanding of the nature of the stars is due not only to the scientists with insight into physical process—people like Lockyer, Russell, and Saha—but also to those who laboriously, systematically, and above all, carefully classified hundreds of thousands of stars.

Spectroscopic studies of starlight began shortly after the first stellar parallaxes had been measured. While parallaxes led to a knowledge of the luminosities of stars, spectroscopy led to an understanding of the physical processes in the stellar atmospheres. In fact, the application of physics to the understanding of the stars—that is, astrophysics—began with stellar spectroscopy.

Newton had demonstrated in 1665 that sunlight was composed of light of many colors. But it was not until the 1800s that the solar spectrum was studied in sufficient detail to see absorption features that were eventually identified with different chemical elements present in the sun. In the mid-1800s, similar absorption features were discovered in stellar spectra. The observations were made visually and were restricted to low-dispersion spectra of relatively bright stars. The introduction of photography permitted higher dispersion spectra and fainter stars to be observed. The stellar spectrum photographs also had the advantage of permitting more detailed measurements on permanent records. The early work in stellar spectroscopy was predominantly collecting and classifying spectra. With advances in understanding the atom and the formation of absorption and emission spectra, we finally are able to explain stellar spectra and to deduce the conditions in stellar atmospheres, even though we cannot place measuring instruments directly in the atmospheres.

Spectral Classification

The major pioneering work in spectral classification was the visual work of an Italian priest, Angelo Secchi. He divided most of the stellar spectra into three main types, based upon the presence of different absorption features and the color of the spectra. He used a fourth type for some relatively rare, but distinctive, spectra.

Secchi's work was followed by the work of Annie J. Cannon and E.C. Pickering at Harvard College Observatory. By placing a thin prism over a telescope objective, they simultaneously obtained the spectra of all stars in the telescope's field of view. Pickering, an early advocate of astronomical photography, recorded the objective prism spectra on photographic plates. The spectra were still low-dispersion, but he could now compare the image of one spectrum directly with that of another. Thus, the task of classifying the stars became both more reliable and more exact. All together, the Harvard workers classified more than 250,000 stars. They published their results as *The Henry Draper Catalogue* of stellar spectra. Their classification scheme is known as the Henry Draper, or HD classification. A modern color photograph of a field of objective prism spectra can be seen in Plate 10.

Originally, they arranged the spectra into groups of similar spectra, labeling each group by a letter of the alphabet from A through Q and leaving out the letter J. It soon became obvious that all of the groups were not needed. For example, what they thought was a distinct class, H, turned out to be spectra on defective plates. They actually were the same as class K, so H was dropped as a designation. Ultimately, they settled on seven major groupings and three auxiliary groupings. These groupings are now called the **spectral types** or classes. The process of locating a star in its proper spectral type is known as spectral classification.

The major spectral classes in the Draper system are O, B, A, F, G, K, and M. The three auxiliary classes are R, N, and S. An easy way to remember the classes is the sentence "Oh Be A Fine Girl Kiss Me, Right Now Sweet." The seven major classes are in a sequence of decreasing temperature from O to M. The R, N, and S types parallel the K and M types in temperature, but their spectra appear different because of slight differences in the chemical composition of their atmospheres.

Modern medium-dispersion spectra from each of the major spectral classes are shown in Figure 12–1. The numbers following the class letters locate the spectrum by tenths of a class. For example, K5 is a spectrum midway between K0 and M0. The differences in the spectra across the classes are quite striking. They do not arise from differences in the chemical compositions of the stars, as some early workers thought, but reflect changes in temperature.

To demonstrate the temperature run with spectral types, note the Balmer lines of hydrogen, labeled by H and followed by a lowercase Greek letter. For a hydrogen atom to absorb a Balmer photon, its electron must first be excited to the second permitted orbit. Therefore, if a sample of gas is to absorb the Balmer spectrum, it must be hot enough for enough hydrogen atoms to be excited to the second orbit by collision. When the temperature is low, as in the M stars, there are not many hydrogen atoms properly excited, and the Balmer lines are weak. With increasing temperature, the number of excited hydrogen atoms increases. The Balmer lines increase in strength through the K, G, and F stars, reaching maximum strength in the A stars. At the even higher temperatures of the B and O stars, collisional ionization of hydrogen becomes dominant and the number of neutral hydrogen atoms decreases. The strength of the Balmer lines therefore diminishes in the B and O stars. When we make proper allowance for the number of excited hydrogen atoms, the total content of hydrogen in stellar atmospheres turns out to be much the same in all spectral types. The great variation in the strength of the Balmer lines is seen to be almost totally dependent upon temperature and not on the chemical composition of the stars.

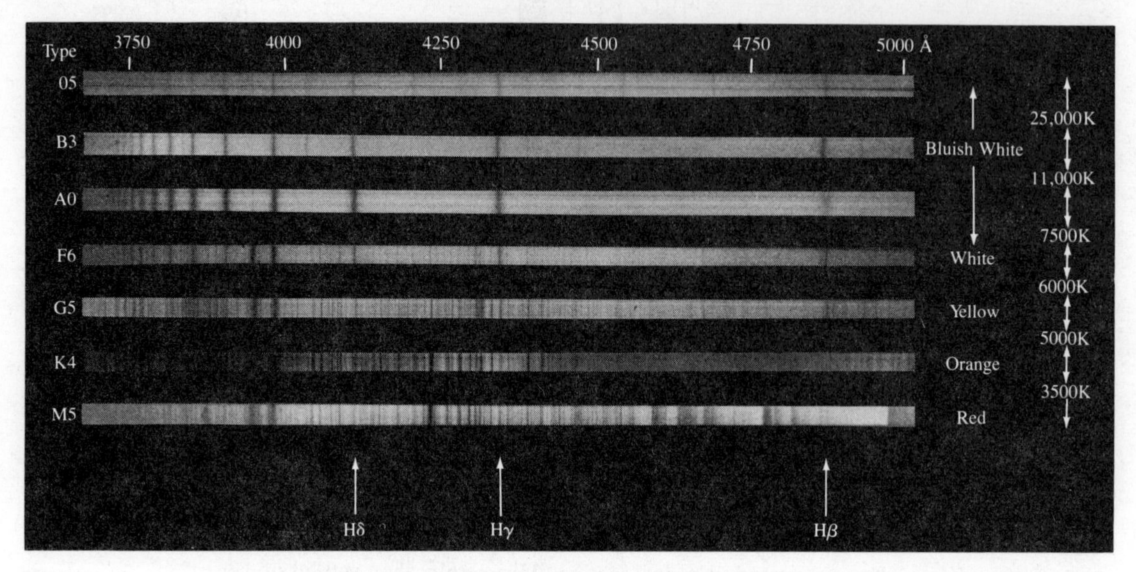

Figure 12–1 Representative spectra in the major spectral types.

The major features of the spectral types can be summarized as follows:

M stars: The most striking features are strong molecular absorption, mostly titanium oxide (TiO), and a multitude of neutral metal lines. The stars are red in color, with temperatures ranging from over 2000 to 3500K.

K stars: Neutral metal lines predominate, TiO bands disappear, but some molecular absorption due to CH persists. These stars appear red to orange to the eye and have temperatures between 3500 and 5000K.

G stars: The predominant lines are due to ionized calcium, with a mixture of other neutral and ionized metal lines. Hydrogen lines, while weak, become easily visible. G stars are yellow in color and are 5000 to 6000K in temperature.

F stars: Ionized metal lines are most significant, and hydrogen lines become conspicuous. The temperatures of these stars lie between 6000 and 7500K, and their color is yellow-white.

A stars: The Balmer lines of hydrogen are the strongest features, while weak neutral and singly ionized metal lines still persist. The color of the A stars is white at a temperature of 7500 to 11,000K.

B stars: The hydrogen lines weaken, while neutral helium lines appear. Doubly ionized silicon lines are seen. Their color is white or bluish-white. Their temperatures range from 11,000 to 25,000K.

O stars: Hydrogen lines are weak, with ionized and neutral helium lines present. O stars are hotter than 25,000K and are bluish-white in color.

After proper allowance for the temperature variation has been made, the amount of each element required to form the observed absorption features is nearly the same in all spectral types, at least for stars in the solar neighborhood. In general, it is found that most stellar atmospheres contain around 73 percent hydrogen, 25 percent helium, and less than 2 percent all other elements by mass. This result is the same as found for the sun. We can study the solar spectrum in great detail because of the sun's great brightness, and sixty-seven of the naturally occurring chemical elements have been identified in the sun. Probably all ninety-two elements are present in the sun, but some are in such small amounts that their lines are too weak to be seen.

The relatively rare red stars classified as R, N, and S are chemically different from the normal K and M stars which parallel them in temperature. The R and N stars are sometimes called the *carbon stars* because they show strong molecular absorption by such compounds as diatomic carbon (C_2) and cyanogen (CN) instead of the TiO absorption shown in the M stars. The spectra of the R and N stars, evident from Figure 12–2, are easily recognized as different from the more common K and M spectra. However, the remarkable change in appearance results from a slight change in overall chemical balance. These stars are still mostly made up of hydrogen and helium. Calling them carbon stars is somewhat unfortunate because they are very far from being composed of all carbon, as the name might imply. The major molecular absorption bands in the S stars arise from zirconium oxide and lanthanum oxide.

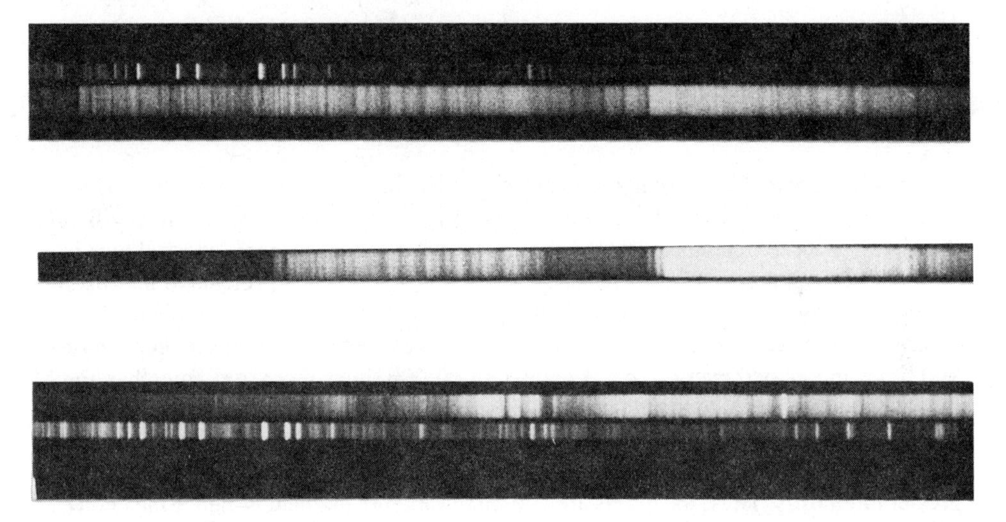

Figure 12–2 Representative R, N, and S spectra.

The H-R Diagram

A major step in our understanding of stars occurred when Henry Norris Russell, in the United States, and Einjar Hertzsprung, in Holland, combined the knowledge of stellar luminosities gained from parallaxes with that of stellar temperatures gained from spectral classification. The Hertzsprung-Russell plot, or **H-R diagram**, is a graph in which the absolute magnitudes of the stars are plotted against their spectral types. This process is equivalent to plotting the luminosities, or energy radiation rates, of the stars against their surface temperatures.

A representation of a modern H-R diagram, with many more stars than were available to earlier investigators, is shown in Figure 12–3. The most obvious feature of the diagram is the grouping of stars. The diagonal band containing the greatest number of stars is called the **main sequence**. Hertzsprung named the stars in this region of the plot **dwarfs**. The stars in the more or less horizontal grouping toward the right on the diagram were named **giants**. The scattering of very bright stars above the giants are called **supergiants**.

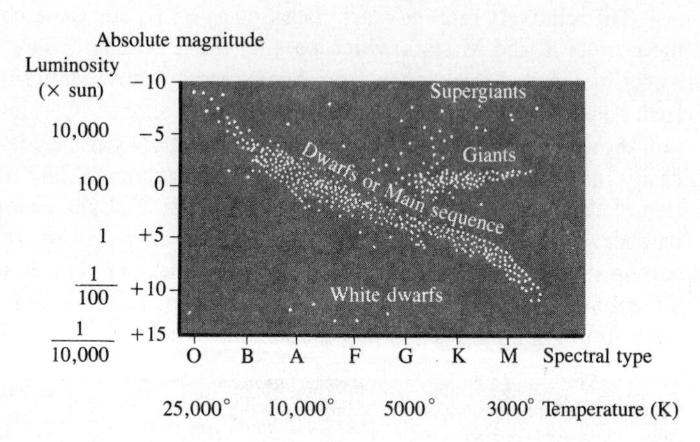

Figure 12–3 The Hertzsprung-Russell diagram.

The significance of these names can be found by interpreting the diagram. Imagine a vertical line on the diagram at spectral type M. Any stars located along this line will have the same surface temperature. But the stars in the main sequence at this spectral type have an absolute magnitude of +9. Those in the giant branch, however, have an absolute magnitude closer to −1. The giants are therefore ten magnitudes more luminous than the dwarfs for spectral class M. Since each five-magnitude step corresponds to an increase of 100 times in radiated energy, the M giants are 10,000 times more luminous than the M dwarfs. But, according to the black body radiation laws, the giants and dwarfs radiate the same amount of energy per unit area because they have the same temperature. Thus, the M giants must have 10,000 times the radiating surface of the M dwarfs to account for their greater luminosity. Thus, the giants are 100 times larger in diameter—hence the choice of names. Extending this argument to the supergiants indicates that they can be more than ten times larger than the giants.

In some respects the small group of stars in the lower left part of the diagram, labeled **white dwarfs,** is even more amazing. They are more than ten magnitudes fainter than the main-sequence stars of the same spectral type. Thus, even though they are very efficient radiators because of their high surface temperatures, they must have very small radiating surfaces to have such small luminosities. Calculations indicate that their radii are measured in thousands of kilometers rather than the hundreds of thousands of kilometers associated with ordinary stars like the sun. The white dwarfs are therefore planetary in size. They were named white dwarfs because they are very small and the first ones discovered were white in color. In any case, do not confuse the white dwarfs with ordinary dwarfs—they are very different.

Diameters of stars can be estimated when their luminosities and surface temperatures are known. Table 12–1 lists some representative values of stellar diameters and indicates the great range in sizes of stars.

Table 12–1
Calculated stellar diameters

Type of star	Diameter (in solar units)
Main Sequence	
O5	20
M5	0.25
Giants	
G0	10
M0	80
Supergiants	
G0	100
M0	500
White dwarfs	0.02

The H-R diagram gives a good example of why astronomers must be careful in generalizing. From the H-R diagram it would appear that giants and supergiants are relatively common, while white dwarfs are quite rare. If we redraw the H-R diagram but restrict it to stars within 10 parsecs of the sun, we obtain Figure 12–4. Notice that there are no giants or supergiants in the diagram. White dwarfs are still well represented. The most numerous stars in the main sequence are the cooler stars, and no stars hotter than spectral type A are seen. The supergiants, giants, and very hot O and B stars are definitely overrepresented in the general H-R diagram. This is because they are so luminous that, even at great distances, they are still among the brightest stars in the sky.

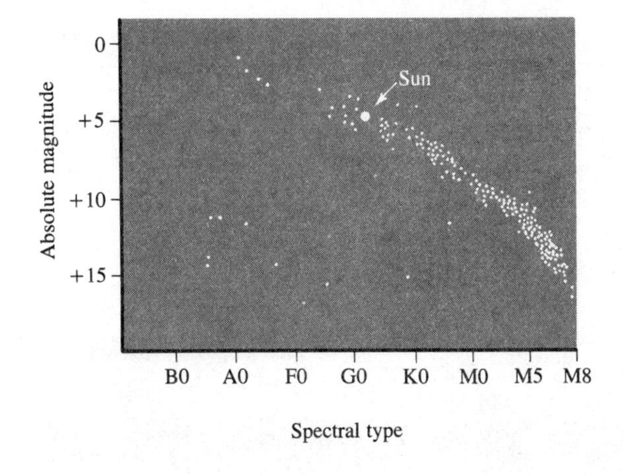

Figure 12–4 H-R diagram for stars within 10 parsecs.

When we allow for the distances at which various stars can be seen, we get a better idea of the number of the different kinds of stars in the solar neighborhood. On the average, 90 percent of the stars are ordinary dwarfs, nearly 10 percent are white dwarfs, and only about one percent are giants or supergiants.

Luminosity Classes

It is possible to recognize dwarfs or giants by a detailed examination of their spectra. Figure 12–5 shows an enlargement of a part of the spectrum of both a giant and a dwarf star of spectral type A; the H-delta line of the Balmer series has been singled out for examination. This line is much broader in the dwarf star than in the giant star. Since both stars have the same spectral type, they must have about the same temperature; hence, the difference in their lines is not due to temperature. The stars are hot enough to have a high degree of ionization in their atmospheres. Whenever hydrogen is present in a partially ionized gas under relatively high pressure, the Balmer lines are broadened. We therefore conclude that the dwarf star has a higher atmospheric pressure than the giant star. This conclusion implies that the surface gravity of the dwarf is greater, binding its atmosphere more tightly to it and causing the higher pressure. The surface gravity of a star depends upon its mass and size. But, as we will see in the next chapter, the range in stellar masses is too small to account for the observed variations in line widths. Instead, the variations must be mostly due to differences in size. The higher surface gravity of the dwarf therefore indicates a smaller diameter, and hence a lower luminosity, than that of the giant.

In different parts of the spectral sequence, other criteria, such as ionized line strengths relative to neutral line strengths, can be used to indicate differences in atmospheric pressure. Pressure, being related to the size of a star, allows these criteria to be used

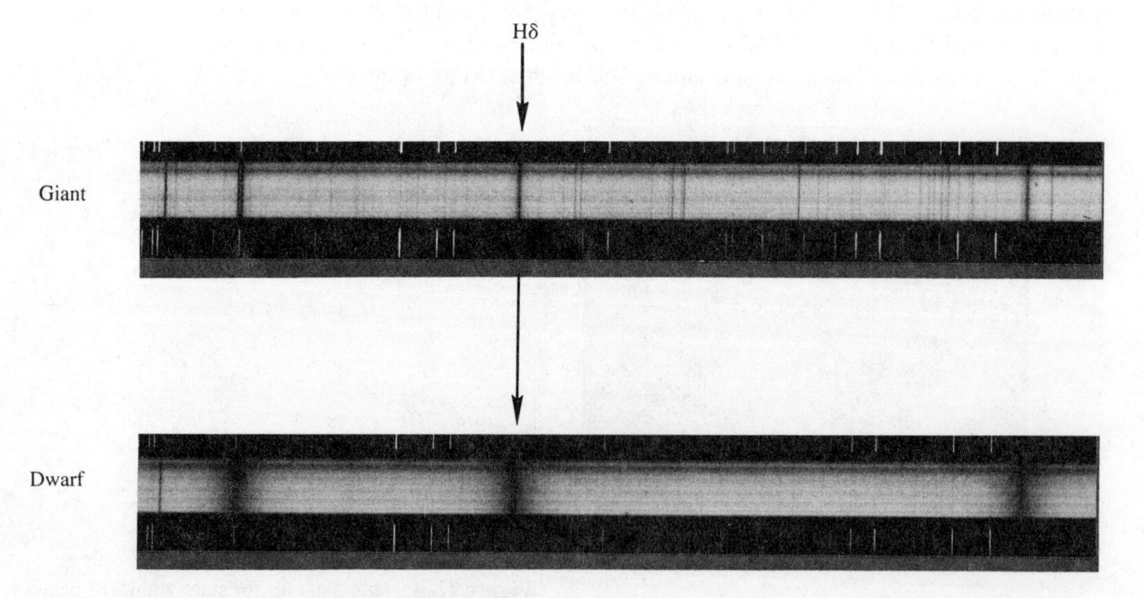

Figure 12–5 Pressure effects in the H-delta line.

to classify stars according to their luminosity. W. W. Morgan and P. C. Keenan have devised an orderly process of luminosity classification with rules for applying it to stellar spectra. The **luminosity class** of a star is designated by a Roman numeral written after its spectral type. Class Ia are the brightest supergiants, Class Ib the less luminous supergiants, Class II the bright giants, Class III the giants, Class IV the subgiants, and Class V the main-sequence, or dwarf, stars. The complete designation of the solar spectrum is G2V in the M-K (Morgan-Keenan) system.

Color-Magnitude Diagrams

The color of a star is related to its temperature and hence to its spectral type. This fact is particularly helpful when the apparent brightness of a star is too low to obtain a good spectrum for classification. If we use a blue-sensitive plate to photograph a star, we can obtain its apparent *photographic* magnitude. On the other hand, by using an appropriate yellow filter, we obtain the star's apparent *photovisual* magnitude, which is very close to its visual magnitude. When we take the difference between the photographic and photovisual magnitudes, we obtain the **color index** of the star. This index is a measure of how bright the star is in yellow light compared to its brightness in blue light. The color index is zero for an A0 star, can exceed two magnitudes for an M star, and is as small as -0.6 of a magnitude for an O-type star.

A more accurate tool for obtaining the color of a star is a photoelectric cell used with suitable color filters. The photoelectric technique, being more sensitive than photography, is also more useful in measuring the colors of fainter stars.

Color indices are particularly useful when studying the groups of stars known as *stellar clusters*. There are basically two types of stellar clusters. **Open** or **galactic clusters** are loose groups of stars containing several hundred to a thousand stars. The open cluster known as the Pleiades is shown in Plate 11A. **Globular clusters,** like the one shown in Plate 11B, are more tightly grouped. They typically contain around 100,000 stars. In both types, member stars normally occupy a small volume of space compared with the distance of the cluster from the earth. Consequently, when we measure the apparent magnitudes of the stars in a cluster, we obtain the relative luminosities of the cluster stars because they are all at essentially the same distance.

After the apparent magnitudes and color indices of the stars in a cluster have been obtained, we can form a plot, comparable to the H-R diagram, called a **color-magnitude diagram.** The plot for the Pleiades galactic cluster is shown in Figure 12–6. The stars in this cluster form a well-defined main sequence, but there are no giant stars evident. Notice, in particular, how the main sequence turns upward at the hot end.

The color-magnitude diagrams of galactic clusters show a number of interesting differences. The diagram shown in Figure 12–7 is for the Hyades cluster. This cluster shows a few giant stars, along with a well-defined main sequence. Also, the main sequence terminates near a color index of zero, while that of the Pleiades extends to -0.1. This is equivalent to saying that the brightest main-sequence stars of the Hyades are cooler than those of the Pleiades. In general, the main sequence in galactic clusters is well defined, but there is a large variation in the hot ends of the main sequences. Both the number and colors of giants in this type of cluster show great variations from one cluster to another.

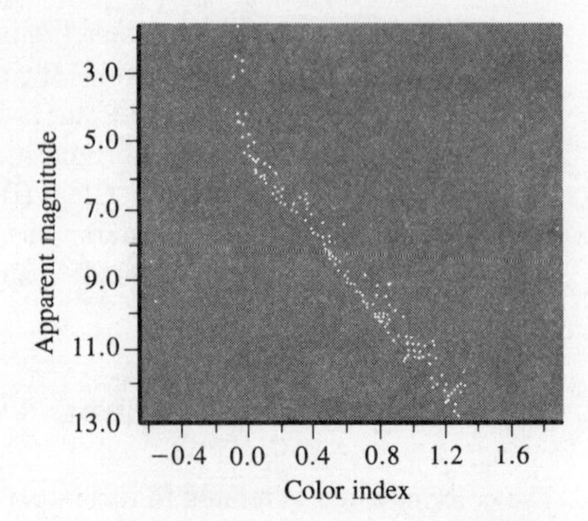

Figure 12–6 Color-magnitude diagram of the Pleiades open cluster.

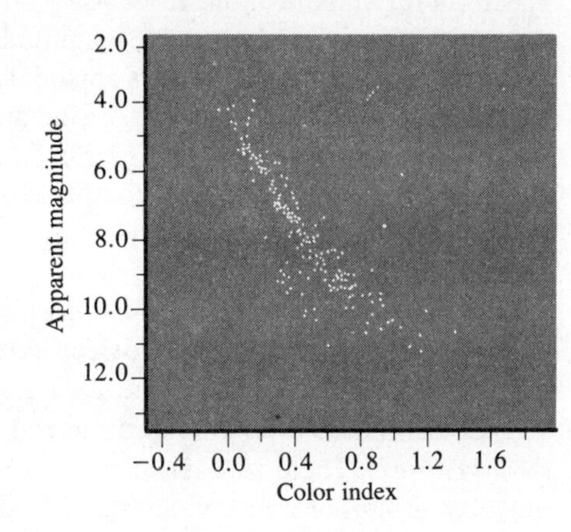

Figure 12–7 Color-magnitude diagram of the Hyades open cluster.

The color-magnitude diagrams of globular clusters are characterized by that of the cluster designated M3, as shown in Figure 12–8. The cool end of the main sequence is not plotted because the stars in that region of the diagram are too faint to be measured at the distance of M3. The main features of the diagram are the termination of the main sequence near color index 0.4, around spectral type F, and the presence of bright red giants. The giant branch of the diagram forms an S-shaped band running diagonally upward to the right from the upper end of the main sequence. There is also a horizontal band of stars running from right to left across the diagram. The gap in this band is called the *cluster variable gap*. This gap is where the stars which vary in both brightness and color would be located on the diagram. By convention, they are not plotted on the color-magnitude diagram. All globular clusters show more or less the same features as M3.

The color-magnitude diagrams of clusters raise a number of questions. Why do the diagrams for the galactic and globular clusters differ so much? What causes the upper ends of the main sequences to terminate at different color indices? These questions will be answered when we discuss stellar structure and evolution. The color-magnitude plots will, to a great extent, be the observational evidence against which to test our theories of how stars evolve.

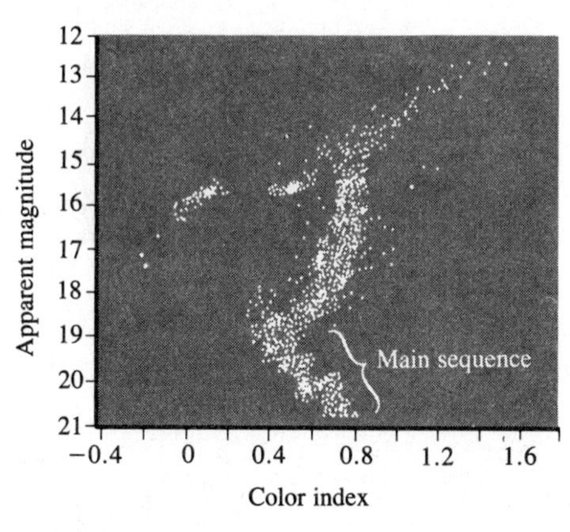

Figure 12–8 Color-magnitude diagram of the M3 globular cluster.

Color-magnitude diagrams can also be used to find the distances to clusters. If we assume that the main sequence of a cluster is similar to the main sequence of stars near the sun, we can convert the apparent magnitudes of the plot to absolute magnitudes. To do this we replot the H-R diagram of the nearby stars using color index instead of spectral type. We then place the color-magnitude diagram of a cluster, plotted to the same scale on tracing paper, over the plot for the solar neighborhood. The cluster plot is shifted up or down, keeping the color index scales on both plots in agreement, until we get the best fit of the cluster main sequence with that of the H-R diagram. Doing so establishes the absolute magnitudes of the cluster stars. The apparent magnitude minus the absolute magnitude is called the *distance modulus* of the cluster and can be used to calculate the cluster distance in parsecs.

Line Profiles

Even more information can be obtained by studying the lines of a high-dispersion spectrum in detail. The plot of light intensity in a line with wavelength is called a **line profile.**

The profiles of the same lines seen in two different stellar spectra are shown in Figure 12–9. Photographs of the corresponding parts of the stellar spectra used to obtain the profiles are shown above them. The lines of the spectrum on the right are much broader than the ones on the left, even though the spectra came from stars of the same spectral type. This broadening is not due to a pressure effect in the stellar atmospheres because it is much too large to be explained this way. Instead, we are seeing the result of a rapid rotation by the star giving the right-hand spectrum.

To explain the shape of the profile, imagine a rapidly rotating star viewed perpendicular to its axis of rotation. Divide the surface of the star into narrow strips parallel to its rotation axis. The light received from each strip will depend upon the radiating area within each strip. Close to the limb of the star, the strips will be short and therefore will not have much area, so they will not contribute much light to the spectrum. The light from the central strip will make the largest contribution because it will have the greatest length and, hence, the greatest area.

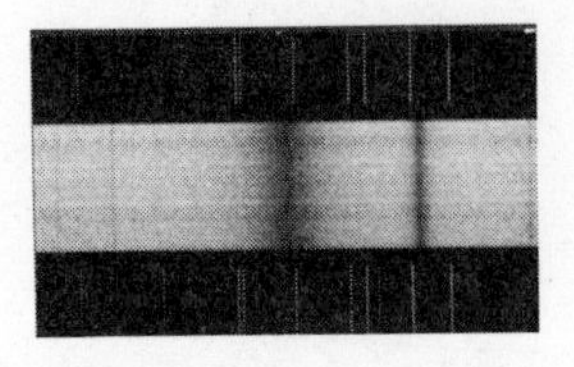

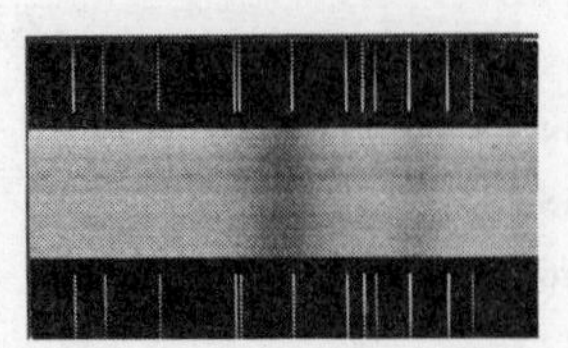

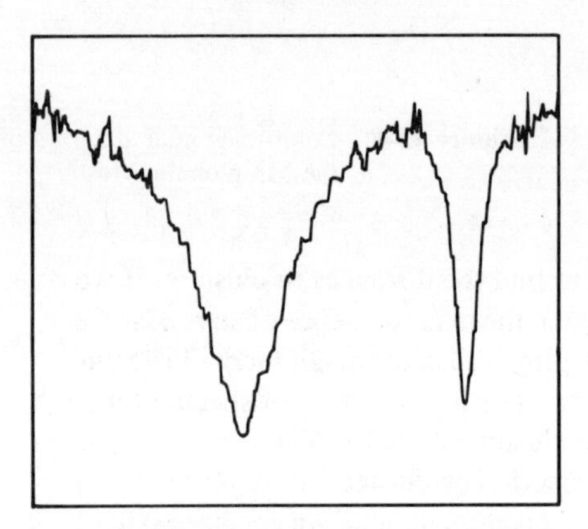

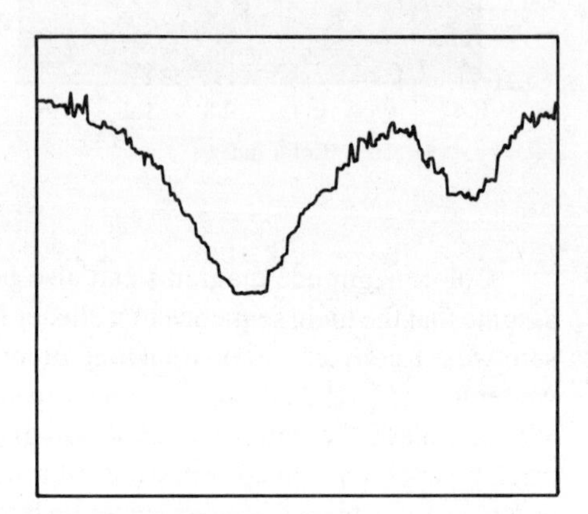

Figure 12–9 Line profiles demonstrating stellar rotation.

Since the star is rotating, one limb will be moving toward us and the other away from us. The light will be Doppler shifted to the red for one limb and to the blue for the other. Moving inward, each strip will have a smaller rotational Doppler shift, until at the center strip there will be no shift because the surface of the star is moving across our line of sight at that point. The observed line profile will be the sum of the contributions of each strip. The final line profile, therefore, is shallow and broadened because of the changing rotational velocity in our line of sight across the surface of the star.

If the star had little or no rotation or if it were viewed pole-on, the contributions from each strip would fall near the same wavelength and a sharp deep line would result. In Figure 12–9, the line profile on the left is therefore from a slowly rotating star (or more likely a star observed pole-on), while the one on the right is a rapidly rotating star, probably viewed nearly perpendicular to its axis of rotation. A number of B-type stars have been observed with equatorial speeds of 200 to 300 kilometers per second, approaching the limiting speed at which they would become distorted enough to lose part of their atmosphere.

The line profile shown in Figure 12–10 is thought to be from a star which rotates fast enough to lose some of its material. The lines appear to be broad absorption lines upon which are centered emission lines. Sometimes, very narrow absorption lines are seen in the middle of the emission features. Stars having these lines are usually B stars and are designated *Be* stars. The *e* denotes emission features in their spectra.

The line profile can be interpreted in terms of the model in the lower part of Figure 12–10. The broad absorption line arises from the rapidly rotating star. The emission feature is due to light scattered by a ring of gas formed by material which has escaped from the star. Since the matter must slow down in rotation after leaving the star, the emission

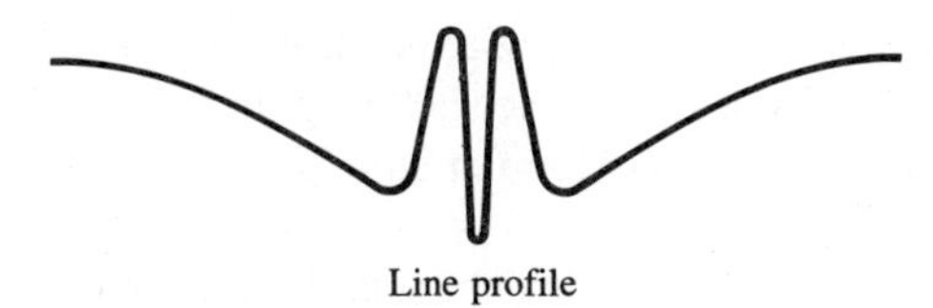

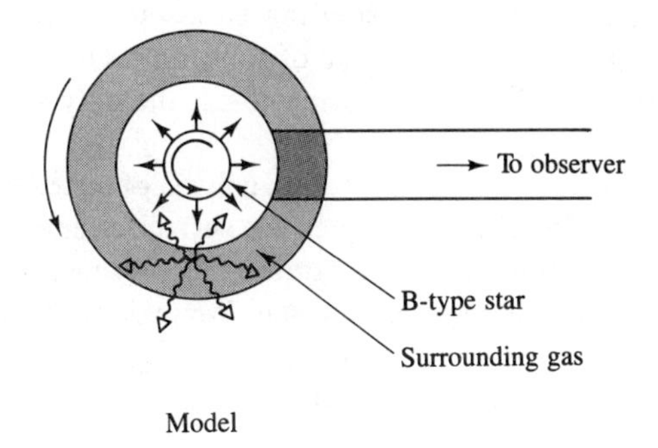

Figure 12–10 The line profile of a Be star.

line is narrower than the absorption line. Finally, the part of the ring directly in front of the star causes the narrow absorption, which is in the center of the emission line, because there is no Doppler shift as the ring is moving across the line of sight at that point.

Another example of an unusual spectrum is that of the star P Cygni, shown in Figure 12–11. In this star, the lines of helium and hydrogen appear in emission with a strong absorption feature to the short-wavelength side of the emission. To explain these lines, astronomers postulate an expanding shell of gas thrown off by the star. The emission lines

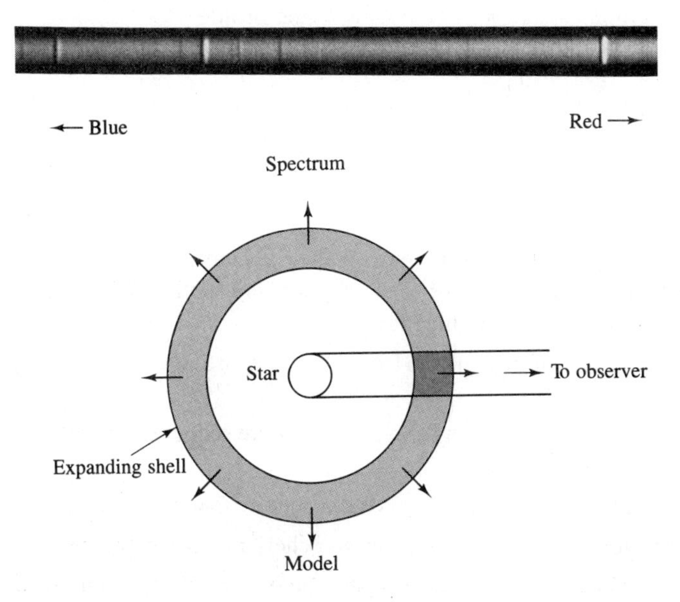

Figure 12–11 Explanation of lines in P Cygni stars.

come from the shell and are broadened by its rapid expansion. The absorption arises in the part of the shell directly in line with the star. Since this part of the shell is in the region of greatest speed toward the observer, the absorption line is Doppler shifted to the short-wavelength side of the emission feature.

Other stars besides the two kinds we have discussed also show emission features probably caused by material surrounding the star. The material may either have come from the star or may be matter falling into the star.

There are about ninety A stars in which Zeeman line splitting is observed. This indicates that these stars have strong magnetic fields. The amount of splitting, as well as the total light from these stars, varies with time. The sun is therefore not the only star which has a variable magnetic field.

The amount of information which can be found from a careful analysis of a stellar spectrum is quite amazing. Some of the facts which can be learned are surface temperature, atmospheric pressure, chemical composition, stellar rotation, the presence of magnetic fields, and whether or not shells of matter surround the star. We deduce these facts about a star even though it never appears as more than a point of light in our telescope. Without the use of spectroscopy, we would remain ignorant about many of the characteristics of the stars.

KEY TERMS

spectral type
H-R diagram
main sequence
dwarfs
giants
supergiants
white dwarfs
luminosity class
color index
open or galactic cluster
globular cluster
color-magnitude diagram
line profile

Review Questions

1. Secchi classified stellar spectra into four classes, while the Draper system uses ten main classes and even subdivides some of these classes into tenths. Why can the Draper system make a much finer classification?
2. What are the advantages of classifying spectra from photographs rather than visually, as Secchi did?
3. List the seven main spectral classes which form a temperature sequence, going from hotter to cooler stars.
4. Why do hydrogen lines get weaker as you observe stars cooler than spectral class A?
5. Why do hydrogen lines get weaker as you observe stars hotter than spectral class A?
6. Which two elements dominate the chemical composition of stars like the sun? What is the approximate fraction by mass of each of these two elements?

7. Spectral lines of neutral calcium are strong in cool stars, and ionized calcium shows prominently in warmer stars. In the hottest stars, no spectral lines of calcium in any form show up. Why not?
8. Why are the R, N, and S stars thought to have a somewhat different chemical composition than most stars?
9. If two stars have the same surface temperature, but one star has four times the luminosity of the other, how do their sizes compare?
10. Two particular stars have the same absolute magnitude, but their spectral types are G and A. Which star is larger?
11. What quantities are plotted in the H-R diagram? Which one corresponds to luminosity and which one to surface temperature?
12. Where on an H-R diagram would you find the following?
    (1) White dwarfs
    (2) Supergiants
    (3) Dwarfs
    (4) Giants
    (5) Main-sequence stars
13. If we list the absolute magnitudes of all the stars visible to the naked eye, most turn out to be more luminous than the sun. Yet most stars are actually less luminous than the sun. How do you account for this discrepancy?
14. The sun's spectral type is often characterized as a G2V spectrum. What do these three symbols mean?
15. How are the colors of stars measured?
16. What differences are there between open and globular clusters?
17. More giants in a star cluster usually are found in clusters
    (1) whose main sequence extends to very hot stars.
    (2) whose main sequence lacks very hot stars.
    (3) which are close to the Milky Way in the sky.
18. What is meant by distance modulus? How far away is a star whose distance modulus is zero?
19. What quantities must be measured to find the distance modulus of a cluster, using an H-R diagram of nearby stars for comparison?
20. What two factors determine how much broadening we will observe in a star's spectral lines due to the star's rotation?
21. How can magnetic fields be measured on some stars?

Discussion Questions

1. What characteristics of the light from an M star show that it is cooler than a G star?
2. If collisions in cool stars are too weak to excite electrons in hydrogen atoms to the

second orbit, why doesn't absorption of photons excite large numbers of electrons to the second orbit instead?

3. Where on an H-R diagram would the largest stars appear? The smallest stars? Describe the region in the diagram where stars of various temperatures but of the same size as the sun would be found.
4. A particular F star with a parallax of 0.02 second of arc has the same apparent magnitude as a G star with a parallax of 0.01 second of arc. Which star is larger?
5. What effects can cause the absorption lines in a star to appear broad?
6. Color index is always given as the blue minus the yellow magnitude, with an A0 star having a color index of zero. If a star has a color index of +0.6, is it hotter or cooler than an A0 star?
7. A basic method of obtaining distances to objects far from the solar system is comparing the apparent and absolute magnitudes of the objects. How can we find the absolute magnitudes of stars in a cluster so we can use this method?
8. Emission lines in a stellar spectrum usually result from clouds of gas around the star. Why does gas in the star's atmosphere cause absorption lines, while gas around the star can cause emission lines?
9. Why is the absorption line from the surrounding gas centered on the emission line in a Be star, while it is found on the short-wavelength side of the emission line in a P Cygni spectrum?

Plate 9 Aurora borealis, the northern lights

Plate 10 Objective prism spectra.

13

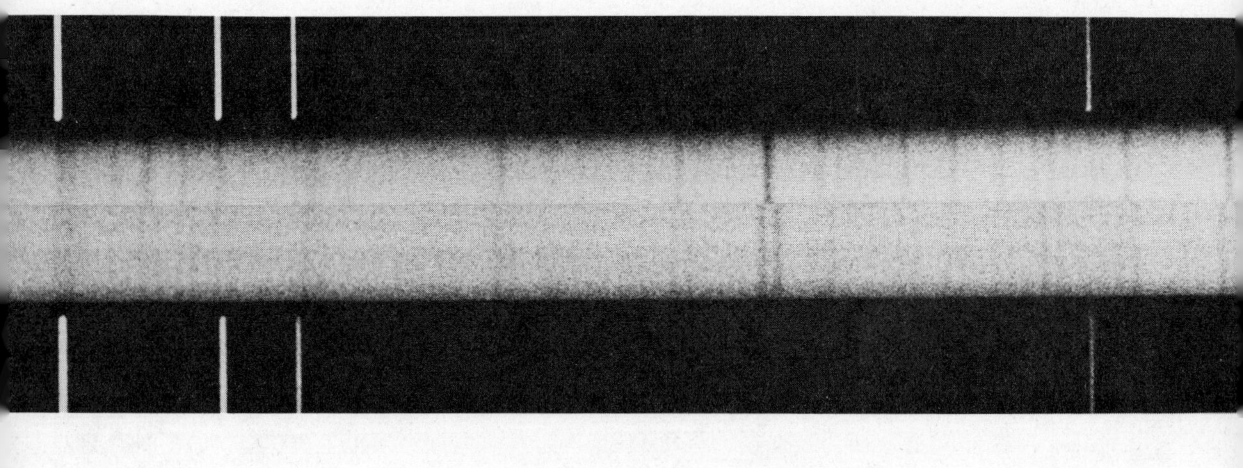

Binary Stars

There is a star called Algol. The name derives from Al Ghūl, the name given to this star by the ancient Arabs. Literally, Al Ghūl means "changing spirit." To the Hebrews, Algol was Rosh ha Sātān, "Satan's head." To the ancient Chinese, it was Tseih She, meaning "piled up corpses."

Known popularly as the "demon star," Algol time and again in different cultures connotes something grotesque and macabre. Why? Could it be that ancient people of many different lands noticed that at times Algol's brightness rapidly drops to one-third its normal value?

No one knows for sure, because the first existing record of Algol's occasional change in brightness was not made until 1670 by the Italian astronomer Montanari. Twenty-five years later, Maraldi also noticed Algol's strange behavior, but the astronomical community did not consider Algol seriously until 1783. In May of that year, a letter from the president of the Royal Society to William Herschel told of the discovery of the periodicity of Algol's changing brightness. This discovery had been made during the late months of 1782 by an eighteen-year-old, John Goodricke. He found that every two days and twenty-one hours, Algol completed a cycle of change in brightness.

To some, John Goodricke may be thought of as a tragic figure. He was unable to either hear or speak. His life was short, snuffed out at twenty-two. But, on the other hand, he had a fine mind and the ultimate playground in which to exercise it—the universe. Besides the periodicity of Algol, he discovered the variability of two other stars: Beta Lyrae and Delta Cephei. For his accomplishments, he was awarded a medal by the Royal Society and was elected a Fellow of the Society only two weeks before his death.

Goodricke published his data on Algol in the "Philosophical Transactions of the Royal Society." In the paper's conclusion, he offered the explanation that the variation in Algol's brightness resulted from an eclipse by a body revolving around it.

Spectrum of the binary star Mizar

Even though he had reviewed Goodricke's paper and approved it for publication, Herschel was not convinced of the validity of the conclusion. In fact, he had observed Algol repeatedly and had never noticed a second object in its vicinity. Not that Herschel would have been aghast at two stars being very close together. Less than fifty years after Galileo had first looked at the sky with a telescope, Riccioli had observed that Mizar, located in the middle of the handle of the Big Dipper, was actually a pair of stars. And, by Herschel's time, many other closely separated pairs of stars had been discovered.

Herschel initially assumed that double stars are two stars with different distances, seen lined up in projection. However, after cataloguing more than 800 double stars, he realized that it was extremely improbable that so many individual stars would be seen in projection with other stars. Ultimately, he decided that these "double stars" are physically connected by gravity and revolve about each other.

The proof of this hypothesis, at least for one pair of stars, was provided by the double star Castor in the constellation Gemini. Herschel noticed that the fainter star of Castor had slightly changed its direction from the other star. He concluded that Castor is a pair of physically connected stars.

Herschel's son John continued his father's work and published a catalogue listing star systems of two, three, or even more stars. The work of the Herschels supported Goodricke's belief that Algol was a pair of orbiting stars whose orientation is such that we observe eclipses. Firm confirmation of Algol's duplicity was made in 1889 by Vogel, who discovered two separate spectra for the star.

The behavior of gravitationally connected stars—or binary stars, as they are called—is quite well known at present. In fact, the motions of these stars as they orbit about each other can lead to determinations of their masses and sizes. These are most important pieces of information in understanding the nature of stars.

Parallax measurement and spectral classification permit us to find the luminosities and temperatures of the stars. These two quantities, with a knowledge of the black body radiation laws, allow us to estimate a star's size rather well. But there is still one important measurement of a star that we have not yet discussed—the mass of a star. We obtain our knowledge of stellar masses by applying Newton's law of gravitation to pairs of stars that are bound together by mutual gravitation.

There are a number of stars visible to the eye which appear to be two stars close to each other in the sky. When telescopes were invented, many of the stars which appear single to the eye were seen to be double. They were first thought to be two stars that happened to lie in the same direction, thus appearing close in the sky, although they are actually quite distant from each other in space. A double star of this kind is called an **optical double,** and as it turns out, they are rather infrequent.

Galileo had suggested that double stars might be good objects to use for the measurement of parallax. He reasoned that when the two stars in a double were greatly different in brightness, the brighter star should be closer to the sun. He also reasoned that it should undergo an unusual parallactic movement relative to the fainter star. Over a hundred years later, when telescopes had been greatly improved, Sir William Herschel began a search for double stars, hoping to find one to which he could apply Galileo's idea. In over twenty-five years of observing, he did not find a double star that could be used to measure parallax; instead, he found a whole new class of objects that he called **binary stars** or **physical doubles.** These are stars bound together by gravity. He had noticed six double-star systems in which the components showed a large apparent motion relative to each other. He knew that he had not observed parallax because he could not make all of the systems conform to the same orbital motion for the earth. He had to conclude that each pair of stars actually consisted of components which were moving about each other. They were, as he put it, "real doubles." Today, unless otherwise stated, we use the term "double star" interchangeably with "binary star," to imply a physical system rather than an optical one.

Visual Binaries

The stars that Herschel studied are now called **visual binaries.** These are systems in which the duplicity can be seen or photographed in a telescope. The star known as Krueger 60 is an excellent example. Figure 13–1 shows photographs of this star taken over a number of years in which the orbital motion of the visual binary can be easily seen. The third star in the photograph is an optical companion to the binary. The separation of the binary components is around 3 seconds, while the optical companion is close to 80 seconds distant. Due to the proper motions of the stars, the binary is slowly drifting away from the optical companion at 0.8 second per year. The visual binary components, however, will continue to move about each other indefinitely, making one revolution every 44.5 years.

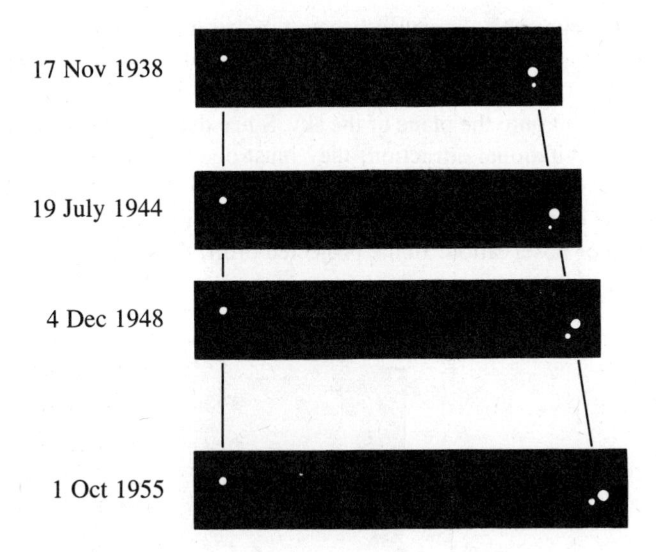

Figure 13–1 Krueger 60, a visual binary star.

The measurement of a visual binary consists of determining the **separation** of the fainter star, or **companion,** from the brighter one, or **primary,** in seconds of arc. The **position angle**—that is, the orientation of the line between the primary and the companion in the sky—is also measured at the same time. After enough measurements have been made, we can diagram the system. As shown in Figure 13–2, the position of the primary is located at the intersection of the north-south and east-west lines on the sky. Lines are laid out at the appropriate position angles corresponding to the time of each observation. A distance proportional to the measured separation is laid out along each line to locate the companion star relative to the primary. A smooth curve is then drawn through these points. The curve, which always turns out to be an ellipse, is the **apparent orbit** of the companion relative to the primary. Sometimes the apparent orbit is a circle or a straight line, but these are just special cases of an ellipse.

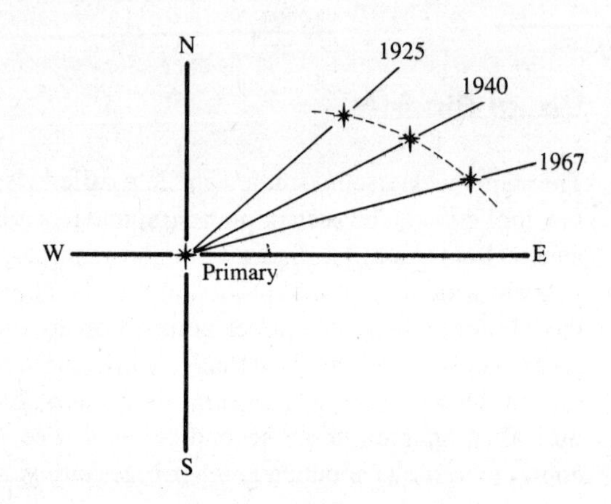

Figure 13–2 Plotting the apparent orbit of the companion about the primary star.

We normally do not see the **true orbit,** however, because the orbital plane is usually inclined to the plane of the sky. This is equivalent to saying that we usually do not view the system perpendicular to its orbital plane. As demonstrated in Figure 13–3, what we do observe is the projection of the orbit onto the plane of the sky. Since the stars are moving about each other due to their gravitational attraction, they must obey Kepler's laws of orbital motion. The true relative orbit is an ellipse with the primary star at one focus. No matter from what direction it is observed, an ellipse appears as an ellipse, although its shape depends on the direction of observation. In the projected orbit, the primary star is

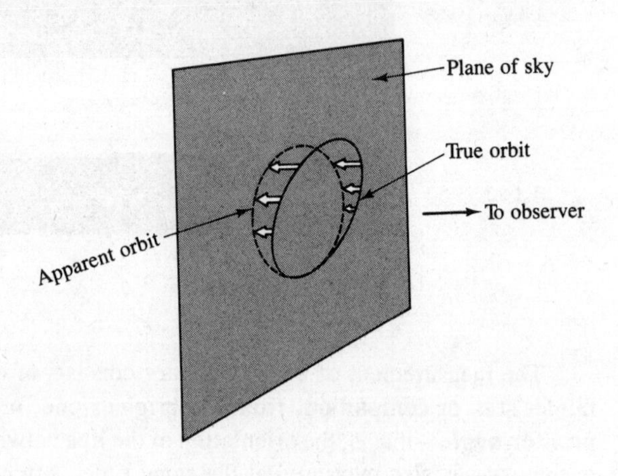

Figure 13–3 Projection of the true orbit onto the plane of the sky.

not at a focus unless the inclination is zero, in which case the true and apparent orbits are one and the same.

To demonstrate how the true and apparent orbits are related, refer to Figure 13–4. Two true orbits are shown, one with a large eccentricity and the other a circle with zero eccentricity. If these particular objects were to be inclined at 60° to the plane of the sky (about a vertical line in the diagram), the eccentric orbit would project into a circle and the circle would project into an eccentric ellipse. Notice that in the circular apparent orbit, the primary is displaced far from the center. In the eccentric apparent orbit, the primary is at the center of the apparent orbit, not at a focus. Whenever the primary is found at the center of the apparent orbit, the true orbit must be a circle. Starting with the position of the primary in the apparent orbit, it is possible to find the eccentricity, the inclination, and the semi-major axis of the true orbit.

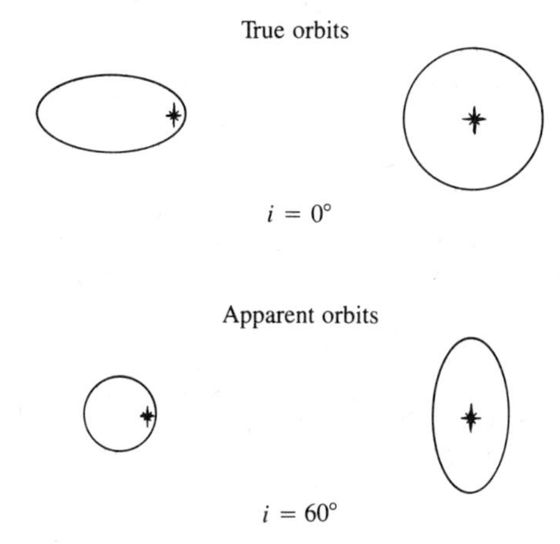

Figure 13–4 The relation of apparent to true orbits.

The period of the system can be found by timing how long the binary takes to complete one revolution. Unfortunately, most visual binaries have very long periods. For example, the apparent orbit of the double star Castor is shown in Figure 13–5. It has yet to complete one revolution since it was first measured in 1719. To find the period of such a

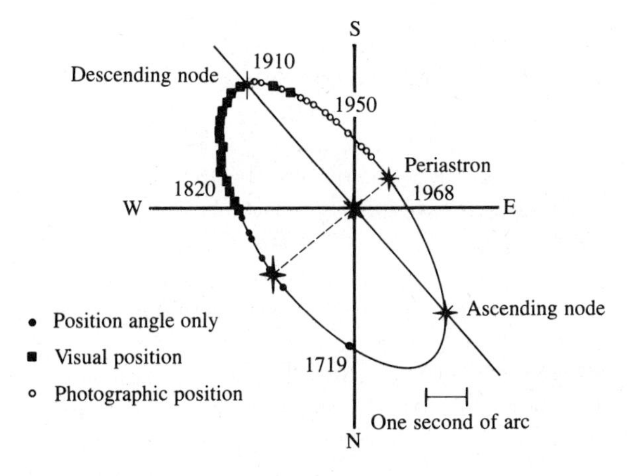

Figure 13–5 The apparent orbit of Castor.

system, we use Kepler's law of equal areas, which also holds for the apparent orbit if areas are measured relative to the primary star. We measure the area between the two lines from the primary star at two different times. Dividing the area by the difference in the times gives the rate at which area is swept out by the line joining the two stars in the apparent ellipse. The period is then found by dividing the area of the ellipse by this rate.

Returning to Castor and Figure 13–5 as an example, we see that the primary is not at a focus of the apparent orbit; hence the true orbit is inclined to the plane of the sky. In the true orbit, the major axis passes through the primary (at a focus) and through the center of the true orbit. In the apparent orbit, this major axis still goes through the center of the apparent orbit and the primary but not necessarily through a focus, as shown by the dotted line. When this orbit is analyzed, we find that the period of Castor is 420 years, the true semi-major axis is 6.30 seconds of arc, its eccentricity is 0.37, and the inclination is 65°.

The orbital solution only gives the angular size of the semi-major axis. If we can also find the parallax of the star and hence its distance in parsecs, the size of the semi-major axis in astronomical units is given by

$$a = a''d$$

where a = semi-major axis in AU

a'' = semi-major axis in seconds of arc

and d = distance to the system in parsecs

Once the linear size of a visual binary orbit is known, we can apply Newton's form of Kepler's harmonic law to find the sum of the masses of the components. We can write the harmonic law in the specific form required for visual binaries as

$$M_1 + M_2 = \frac{(a''d)^3}{P^2}$$

As an example, consider the system Krueger 60. It has a parallax of 0.25 second, or a distance of 4 parsecs. Its semi-major axis is 2$''\!.$3. Consequently, the linear size of the orbit is 9.2 AU. Its period is 44.5 years; and from the harmonic law, the mass of the system is 0.4 solar masses. Krueger 60 is one of the least massive systems known.

To determine the masses of the individual stars, we must find the individual orbits of the components about the center of mass of the system. This requires us to measure the motions relative to the background stars, which is much more difficult than measuring the relative orbit. Figure 13–6 shows the apparent motions of a primary and its companion. The straight line is the path of the center of mass on the sky, while the two curves are the paths of the two components. In the lower right, both the relative and the individual orbits are shown, after the motion of the center of mass has been removed. From Newton's laws of motion, we know that the less massive star will have the larger motion. The ratio of the masses of the components will be inversely proportional to the ratio of the sizes of their individual orbits. For Krueger 60, again, the sizes of the two orbits are in the ratio three to one. Therefore, the primary has a mass of 0.3 of a solar mass, while its companion has only 0.1 solar mass.

Some apparently single stars are found to have wavy paths, like one of those seen in Figure 13–6. This can only occur when the visible star has an unseen companion. Such systems are called **astrometric binaries** because they are discovered by measuring the position of stars. The bright star Sirius was found to be an astrometric binary by Bessel in

1844. It was not until 1864, when larger telescopes were available, that its faint and previously unseen companion was located. The companion turns out to have a mass around that of the sun, but it is so faint that its radiating surface is very small. Its radius is even less than the earth's. The companion of Sirius is a white dwarf—one of the first to be discovered.

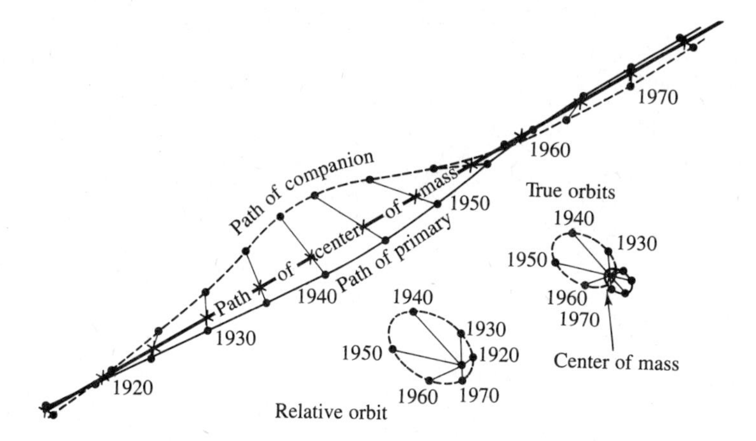

Figure 13–6 Paths of the primary and companion across the sky.

Spectroscopic Binaries

When Pickering obtained a spectrum of the primary star of the visual double Mizar, he was surprised to find its spectral lines were double. As he continued to obtain spectra on subsequent nights, the separation of the double lines changed. Every so often the lines became single. Pickering gave the correct interpretation of the spectra: The changing lines arise from a varying Doppler shift due to the orbital motion of two close, unresolved stars. A binary that shows its duplicity by its spectrum is called a **spectroscopic binary.**

The doubling of the lines is explained in Figure 13–7. In the upper diagram, we see the system at the moment when the primary is moving toward us and the companion away from us. The lines from the primary are therefore Doppler shifted to the blue and those from the companion to the red. Each spectral line appears double. In the lower diagram, both stars are moving across our line of sight. There is no Doppler shift, and the lines from both stars coincide; therefore, we see only single lines. Even when the system is so distant that it cannot be resolved, the Doppler shifts remain and the star reveals its duplicity.

By taking a series of spectrograms and measuring the line displacements, we find how the radial velocities of the components change with time. We can plot the observations, as shown in Figure 13–8, to obtain what we call the **velocity curves** of the binary. The amplitude of each curve, K, is one-half the largest velocity change in one complete cycle for that curve. By Newton's force law, the less massive star must have the larger motion—that is, the greater velocity. The ratio of the masses is inversely proportional to the ratio of the velocity amplitudes. The length of time for the system to go through one complete velocity variation is, of course, the period of the binary. Since spectroscopic binaries are very close systems, their periods are usually given in days.

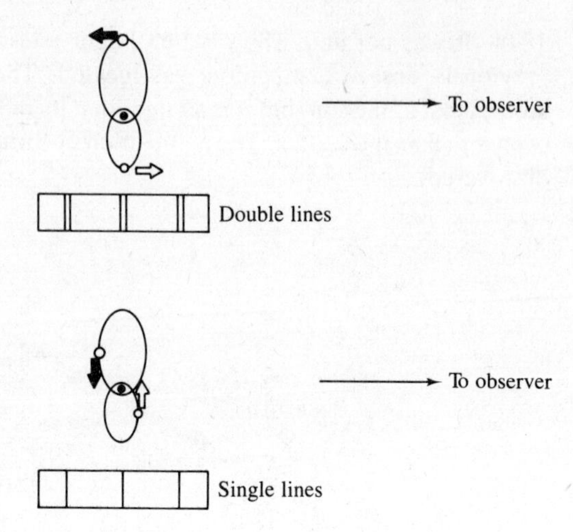

Figure 13–7 Doubling of spectral lines in a binary star.

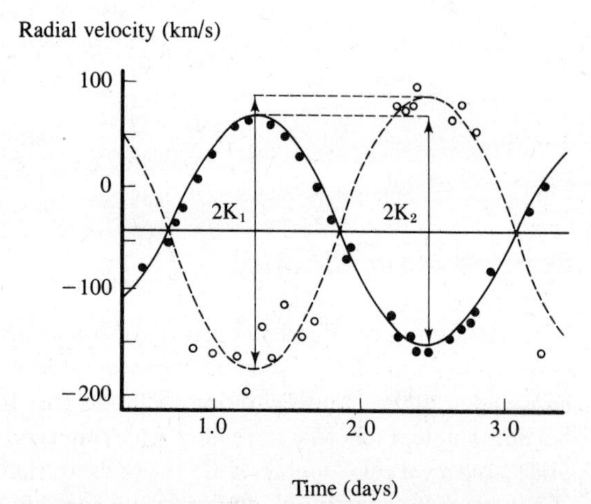

Figure 13–8 Velocity curves of a binary.

We can find the size of the orbit from the velocity curve. For simplicity, assume circular orbits which have nearly 90° inclination; that is, they are viewed nearly edge-on. The maximum velocity observed is consequently just the orbital velocity. Multiplying this velocity by the period gives the circumference of the orbit, from which we find the radius. If the orbit is elliptical, the velocity curve will have a different shape than if the orbit were circular. The precise shape will depend upon the eccentricity and how the major axis of the orbit is oriented to our line of sight. Finding the size of an elliptical orbit is more complicated than a circular orbit, but it is rather straightforward.

Whenever the orbit is not seen edge-on, the measured Doppler velocity will be less than the orbital velocity because the stars would never move directly toward or away from us. Thus, there is an inclination factor which depends on how the orbit is inclined. It will have a value between zero and one. It is one only when the inclination is 90°; it is zero when the orbit lies in the plane of the sky. Since we cannot determine the inclination of the orbit from the velocity curves, we cannot find the value of the inclination factor. This problem usually causes us to underestimate the mass of the system; because when we apply the harmonic law, we use too small a value for the semi-major axis of the orbit. Since we must cube the semi-major axis to use the harmonic law, we also cube the inclination factor. For example, if the factor is 1/3, we calculate a mass only 1/27 the true mass.

Eclipsing Binaries

Binary stars are also detected when their orbits are so close to 90° to the plane of the sky that the stars alternately eclipse each other. An **eclipsing binary** is studied by measuring the brightness of the system. The observations are plotted as a **light curve,** an example of which is shown in Figure 13–9. Normally, there are two dips in the curve for each revolution of the system. The deeper dip is called the *primary mimimum;* the weaker one, the *secondary minimum*. The interval between two successive primary minima is the period of the binary.

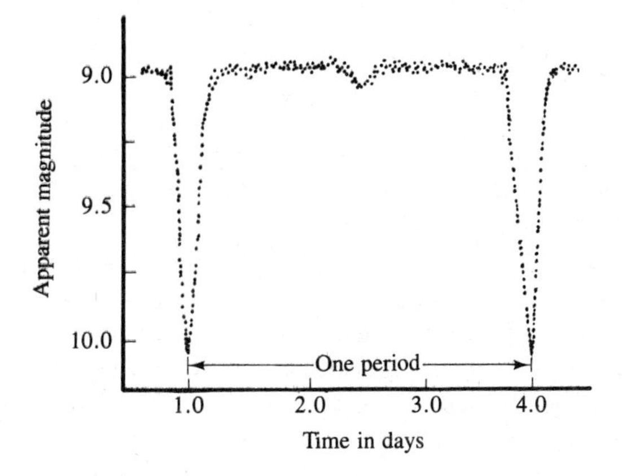

Figure 13–9 Light curve of the eclipsing binary, KO Aquilae.

A light curve may be interpreted with a diagram like that in Figure 13–10. We assume that a small star is moving about a large star in a relative orbit nearly in our line of sight. The numbered positions on the orbit correspond to the numbered points on the light curve. When the light of the small star is completely obscured, we say a total eclipse occurs. When the small star is in front of the big star, the eclipse is said to be annular. In total and annular eclipses, the minima have flat bottoms.

Primary minimum occurs whenever the hotter star is eclipsed. In our example we have assumed the small star to be the hotter one. Since the same radiating area is obscured whether the eclipse is total or annular, the most light will be removed when the more efficient radiator is eclipsed. Whether or not it is the larger one, the more efficient radiator is the hotter one. Notice that the primary eclipse need not occur when the more luminous star is eclipsed, because the cooler star, while less efficient, could have so large a radiating surface that it radiates more total energy than the smaller one.

If the inclination deviates enough from 90°, the stars only partially obscure each other and we have partial eclipses. In this case the minima tend to be pointed rather than

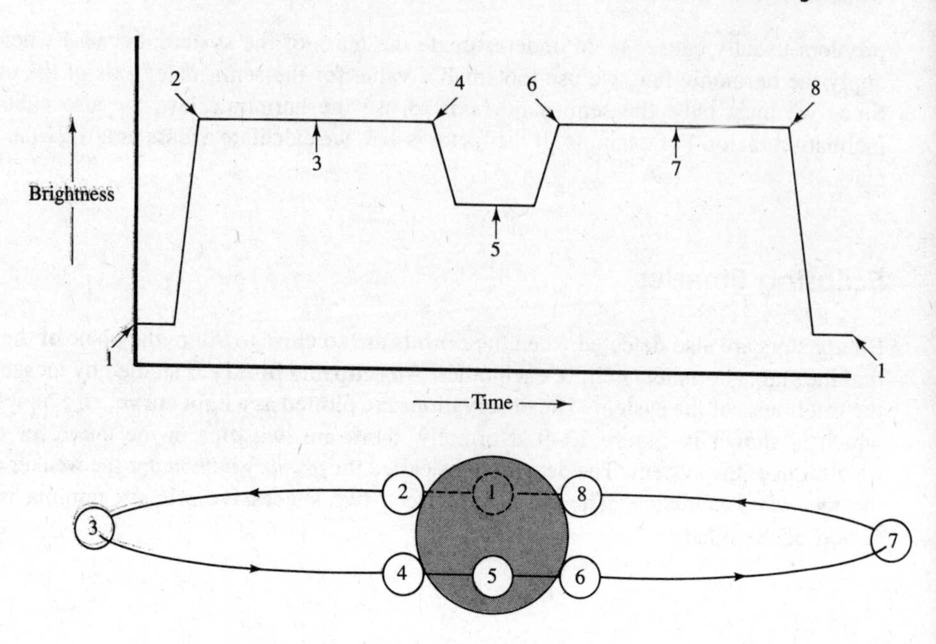

Figure 13–10 Light curve of a hypothetical eclipsing binary.

flat-bottomed. No matter whether the eclipses are partial or total, careful analysis of the light curve allows us to determine the inclination of the orbit.

In some eclipsing binaries, the stars are so close together that they severely distort each other by tidal interaction. In these systems, the light between minima continually varies because the radiating area of the distorted stars continually appears to change as the system revolves. Figure 13–11 illustrates the effect on a light curve due to tidal distortion. A light curve can also be complicated if the two stars differ significantly in temperature and are relatively close together. In this situation, the part of the cool star facing the hot star is heated, and consequently, it is brighter on that side. But this is the side that faces us when secondary minimum takes place. Therefore, the light curve increases in light from primary to secondary and decreases from secondary to primary. The influence of the reflection effect, as this is usually called, is shown in Figure 13–12.

One of the most important results of an analysis of a light curve is determining the diameters of the components relative to the size of the orbit. Again for simplicity, we assume circular orbits seen edge-on; then the orbital motion will be uniform, and a measurement of time will be a direct measure of the distance the smaller star travels relative to the larger star. In Figure 13–13, we see that the eclipse begins when the small star appears to just touch the large star, point 1 on the light curve. When the light has just reached minimum, point 2, the small star has moved a distance equal to its own diameter. When the small star has traveled a distance equal to the diameter of the large star, point 3 on the light curve has been reached. Consequently, the time from point 1 and 2 (t_s) and the time from point 1 to 3 (t_g) are proportional to the star diameters. By comparing these intervals with the period of the system, we can eventually find the diameters of the components relative to the major axis of the orbit. In some cases tidal distortion, reflection

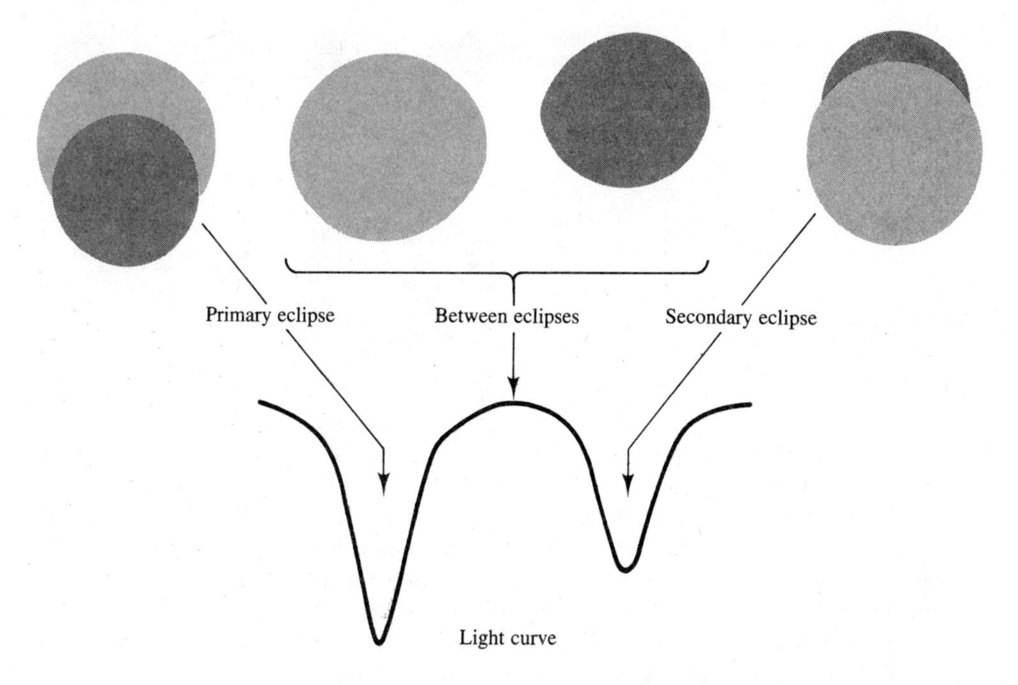

Figure 13–11 The influence of tidal distortion upon the light curve of a close binary star.

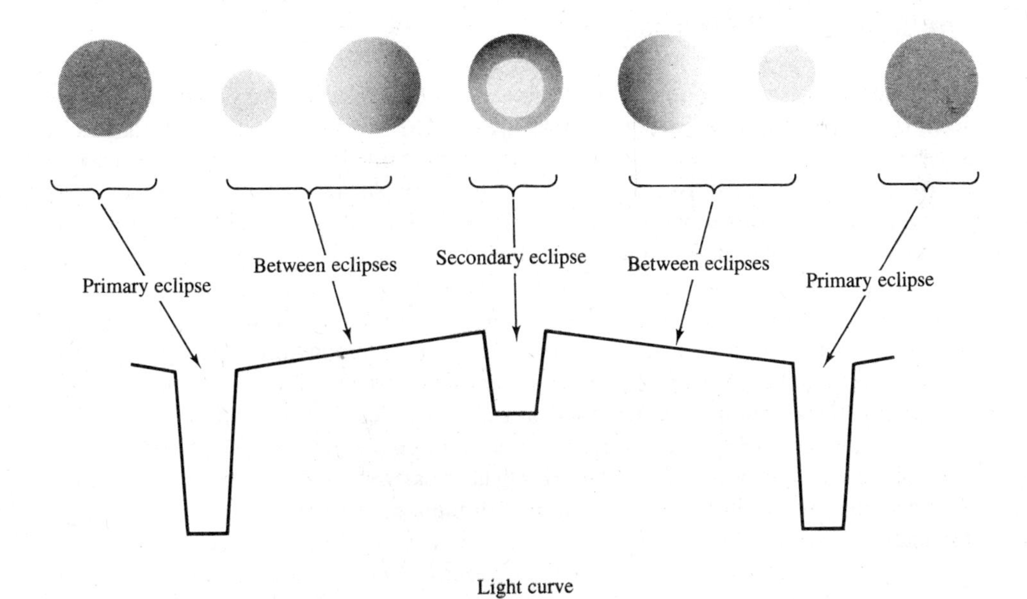

Figure 13–12 The influence of heating effects upon the light curve of a binary star with components significantly differently in temperature.

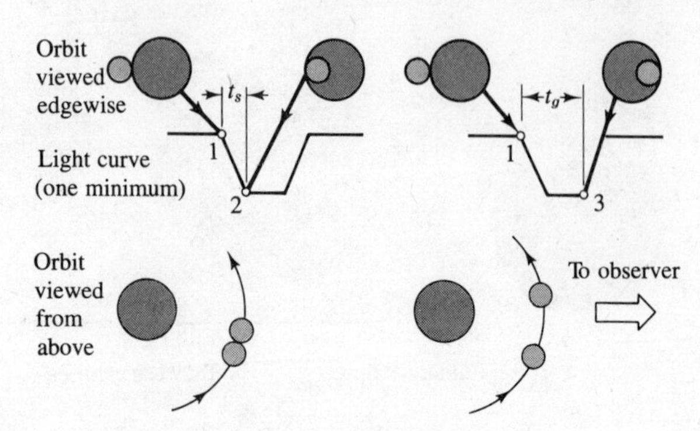

Figure 13–13 Relative diameters of the stars in an eclipsing binary system.

effects, or noncentral eclipses make the analysis more difficult, but nevertheless the relative sizes of the stars can still be measured.

If an eclipsing binary is bright enough, we can obtain spectra with enough dispersion to observe the system as a spectroscopic binary at the same time. From the solution of the light curve, we find the inclination of the orbit; but this allows us to correct for the inclination factor in the velocity-curve solution. We therefore find the true size of the orbit and the true masses of the components. In addition, since we now know the size of the orbit, we can convert the relative star sizes to linear sizes.

The Statistics of Binaries

Many star systems are multiple, not just binary. For example, the star Castor, as we have seen, is a visual binary with a period of 420 years. The two components are separated by about 86 AU. There is, however, a faint red companion moving slowly about the binary at a distance of 1000 AU. But when the spectra of what appear to be three stars are examined, each one turns out to be a spectroscopic binary. So one familiar bright star in the sky which appears single to the eye is actually made up of a physical system of six stars held together by gravitation. This is not the only multiple star; Mizar, the first visual double and spectroscopic double to be discovered, is a quadruple star.

Binary stars are quite numerous. Table 13–1 lists the known double stars by type. While a large number of visual doubles have been detected, only 2500 have shown enough orbital motion to be definitely identified as visual binaries. Furthermore, only 250 have shown sufficient movement to permit an orbital solution. Far fewer spectroscopic binaries have been found, because only systems which are relatively bright can be observed with a spectrograph. Somewhat more than one-half of them have velocity curves good enough for analysis. More than twice as many eclipsing as spectroscopic systems are known, and more than one-third of these have light curves good enough to attempt an orbital solution.

Why do we see such a large disparity in the numbers of different binary types? Visual doubles are easy to discover. If two stars are close together in the sky, they are a double star, although not necessarily a true visual binary. Statistical studies indicate, however, that the number of known doubles which are merely optical doubles must be small and that

Table 13–1
Known double stars

Type	Approximate number
Visual	65,000
Orbits	250
Spectroscopic	1,300
Orbits	800
Eclipsing	3,000
Orbits	1,200

most doubles are probably true binaries. Spectroscopic binaries, on the other hand, must be both bright enough to obtain good spectra and have short enough periods to give large Doppler shifts. Eclipsing binaries are easier to discover than spectroscopic binaries because we need less light to study them.

If we allow for the difficulties in observing the different types of star systems, the number which exist in space can be estimated. On the average, in a sample of 100 stars, 30 are single, 47 are double, and 23 are multiple. The total number of component stars in our sample of 100 is 205. Consequently, the chance of finding a single star by itself is less than finding two or more stars gravitationally bound together.

Mass-Luminosity Relationship

Collecting data on all binary systems for which we have both good masses and luminosities, we can make the plot shown in Figure 13–14. This plot is known as the **mass-luminosity relationship.** It shows that, in general, the more massive a star, the more energy it radiates per unit time. The physical basis for this is important in understanding

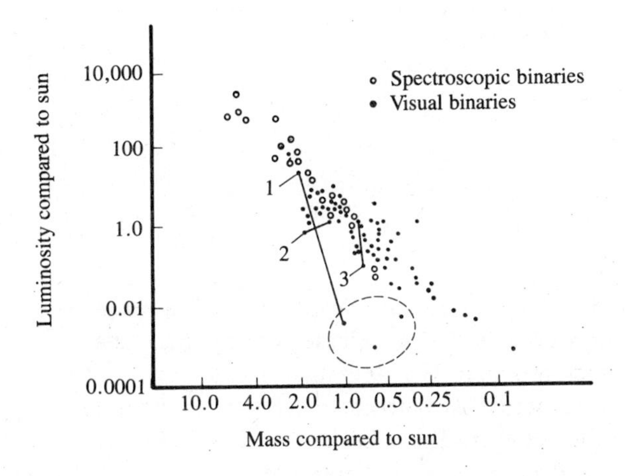

Figure 13–14 The mass-luminosity relationship.

stellar energy generation, but we will set it aside until we discuss that topic. The only stars which strongly violate this relationship are the white dwarfs, enclosed within the dotted circle. The main-sequence stars form a plot with rather small scatter, but the few giants found in binaries usually do not conform as well.

We have connected the components of three different systems in the figure to show how the components of a binary can disagree with the general trend of the mass-luminosity relationship. The system marked 1 is Sirius, which contains a white dwarf. The primary star is a main-sequence star. In system 2, the more massive component is underluminous; while in system 3, the less massive component is underluminous, at least according to the plot.

The components of a binary are thought to have formed at the same time. How can some systems contain two stars that behave so differently?

We think that these apparently peculiar systems develop because of **mass transfer** between the components. There is a point on the line joining the two stars where the gravitational attraction by the two stars is exactly equal and opposite. A particle of matter at that point would not belong to either star. Actually, there are imaginary surfaces, the so-called *critical surfaces,* about each star that intersect at this point and define the locations where a body could just move about both stars without being captured by either one. These surfaces therefore give the largest size that each component could have and still retain gravitational control of its matter.

As we will see when we discuss stellar evolution, a star changes its size as it consumes its energy source. In a binary star, the situation pictured in Figure 13–15 can occur. If one component expands until it fills its critical surface, some of its mass can be transferred to its companion. Some of its matter can surround the system in a huge cloud, and some matter can be lost in space. When this occurs, what was originally the massive component can transfer enough mass to the other component to make it the more massive component.

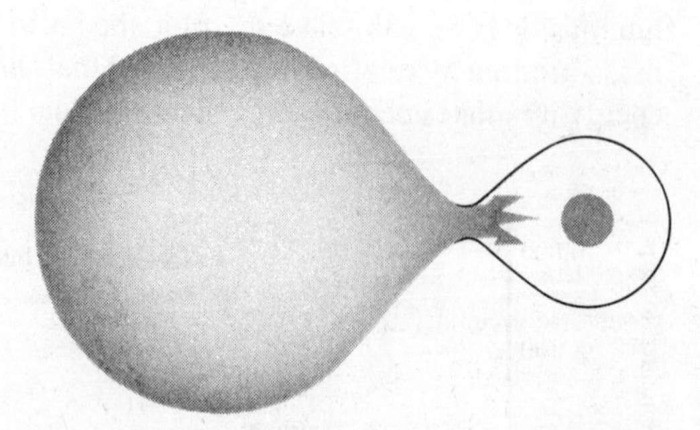

Figure 13–15 Mass transfer in a close binary.

The mass-transfer process seems to be the mechanism which causes a **nova.** A nova is an apparently ordinary star that experiences a sudden increase in brightness, sometimes becoming tens of thousands of times brighter. The nova then fades slowly back to obscurity. Figure 13–16 shows a nova before its outburst and just after it reached its maximum brightness at the time of its outburst. The name *nova* means "new" and was applied to this kind of star because it suddenly becomes visible where no star was noticed before. Some of these stars turn out to be close binaries. The evidence is becoming stronger that all of them are.

Figure 13–16 Nova Persei before and during an outburst.

One theory of the nova process proposes that the great brightening is always a result of mass transfer. The primary star in a nova binary is assumed to be a white dwarf, or very similar to one, and its companion is assumed to be a red star. When it expands beyond its critical surface, some of the matter of the red star is suddenly and rapidly transferred to the primary. The transferred mass collected on the dense primary causes a reaction which quickly releases a large amount of energy, causing the system to brighten. The system then slowly returns to normal brightness. It can flash forth again whenever the red component expands sufficiently.

Summary of Stellar Characteristics

The mass data from binaries and the sizes of stars obtained from both eclipsing stars and the radiation laws can be listed with spectral type. The average densities of stars can also be calculated from their masses and sizes. Table 13–2 compares some of the stellar categories we have already discussed. The range in masses is not terribly large, around 150. The range in size is much greater, about 25,000. The densities are quite surprising. For a supergiant, the average density of the star compares to that of a vacuum on the earth. The white dwarfs, at the other extreme, are so dense that a cubic meter of material would weigh one million tons. All of these stars, even the white dwarfs, are gaseous. The theory of stellar structure has much to explain. In the next chapter, we will try to do just that.

Table 13–2
Stellar masses, sizes, and densities

Star	Mass (solar units)	Radius (solar units)	Density (gm/cm^3)
Main sequence			
O5	32	20	$\frac{1}{100}$
M5	0.22	0.25	25
Giant			
M0	3.8	76	$\frac{1}{100{,}000}$
Supergiant			
M0	16	500	$\frac{5}{10{,}000{,}000}$
White dwarf			
A0	0.6	0.02	100,000

KEY TERMS

optical double
binary star or physical double
visual binary
separation
companion
primary
position angle
apparent orbit
true orbit
astrometric binary
spectroscopic binary
velocity curve
eclipsing binary
light curve
mass-luminosity relationship
mass transfer
nova

Review Questions

1. The discovery that binary stars exist, with two stars orbiting each other, was made
   (1) in prehistoric times.
   (2) by ancient Greek astronomers.
   (3) by Galileo with his new telescope.
   (4) soon after the American Revolutionary War.
   (5) in the twentieth century.

2. How are visual, spectroscopic, and eclipsing binaries distinguished?
3. What two quantities are directly measured in a visual binary system?
4. Which star is called the primary in a visual binary?
5. If we observe a visual binary where the companion appears to orbit the primary in a perfect circle, how could we decide if the orbit really is circular or if it is an ellipse being viewed from such an angle that it appears circular?
6. What quantities can be found from observations of the orbit of a visual binary without knowing its distance? What additional information can we obtain if we know its distance?
7. What must we measure to find the relative masses of the components of a visual binary?
8. In a spectroscopic binary, how would the spectral lines appear when one component is at its closest point to the earth and the other is at its farthest point?
9. How can the velocity curve of a spectroscopic binary be used to help find the size of the orbit? What additional information is necessary?
10. What quantities must be measured to find the masses of the components in a spectroscopic binary?
11. Under what circumstance can the orbital inclination of a spectroscopic binary be found?
12. How do the shapes of the light curves compare in two eclipsing binaries, one of which undergoes total eclipses while the other has only partial eclipses?
13. What are two effects which may cause an eclipsing binary to vary in brightness between eclipses?
14. Why are so many more visual doubles discovered than eclipsing or spectroscopic binaries?
15. *True or false:* Most stars are single stars.
16. What is the mass-luminosity relationship? Which shows a greater range, the masses or the luminosities of stars?
17. Which shows a greater range, the masses or the densities of stars?
18. Which stars follow the mass-luminosity relationship, and which do not?
19. What conditions are necessary if mass is to flow from one component of a binary to the other component?
20. What probably causes a nova outburst?
21. What observations indicate that a nova does not destroy the star giving rise to the explosion?

Discussion Questions

1. Several quantities, such as temperature, can be measured for a star. What measured quantities refer just to the surface of the star? Which quantities are measured for the entire star, interior plus atmosphere?

2. How must Kepler's laws be modified in order to apply to the apparent orbit of a companion about a primary?
3. If a binary star 10 parsecs away has a semi-major axis of 3 seconds of arc, what is the semi-major axis in AU?
4. How could you explain a stellar spectrum where the spectral lines always appear single but slowly oscillate back and forth in wavelength?
5. In a certain spectroscopic binary, the first component has a radial velocity that ranges between approaching at 40 kilometers per second and receding at 30 kilometers per second. The second component ranges between approaching at 30 kilometers per second and receding at 5 kilometers per second. What are their relative masses?
6. Outline the procedure used to find the masses of the components of a spectroscopic binary. What quantities must be measured to use this procedure?
7. Why does the deeper minimum in the light curve of a totally eclipsing binary occur when the hotter star is being eclipsed?
8. How can we measure the relative sizes of the stars in an eclipsing binary system? Assume that at mid-eclipse one star is exactly centered over the other.
9. In addition to visual, spectroscopic, and eclipsing binaries, how else might we conclude that a star is double?

14

The Structure and Evolution of Stars

The alchemists' dream was to change base metals into gold. Rumplestiltskin was an alchemist of sorts who didn't even need base metals to manufacture gold. As the story goes, he taught a young maiden how to gain favor with the king by spinning straw into the coveted metal. And Midas could turn anything at all into gold by merely touching it. But these are fairy tales; the old alchemists never did succeed in finding the so-called Philosopher's Stone, the legendary agent that had the power of transmuting one basic element into another.

To the alchemist, gold was more than just a precious metal. In accordance with the astrology of the Chaldeans, from which many of the precepts of alchemy descended, gold was embodied with the perfection and power of Shamash, the sun-god. In fact, each of the seven basic metals was associated by the alchemists with the seven heavenly bodies known in ancient times. As said by Chaucer, in his quaint and archaic English:

> The bodies seven, eek, lo heer anon. Sol gold is and Luna silver we declare; Mars yron, Mercurie is quyksilver; Saturnus leed and Jupiter is tyn, And Venus coper by my father kyn.

As relentless as the alchemists' quest for the Philosopher's Stone and the perfection and power of gold was the scientists' search for the answer to the riddle of the sun, the legendary source of gold's power. The basic question was "What makes the sun shine?" Scientists dreamed up a number of explanations to answer this question. For example, ordinary chemical reactions such as the burning of coal were suggested as the source of the sun's radiation. Gravitational contraction and heating were also considered. But neither of these processes could have provided the sun with the energy necessary to keep it shining for the billions of years required by fossil records on earth.

As it happened, the Philosopher's Stone and the answer to the problem of the sun's energy generation lie together in the realm of the tiniest known structures of

The Rosette Nebula

matter, the particles which compose the nuclei of atoms. Bombarding ordinary nuclei of heavy elements such as lead or uranium by neutrons leads to the formation of the nuclei of different heavy elements. In some cases, the newly formed nuclei are unstable and split apart into several particles, some of which are nuclei of much lighter elements. The splitting process is called fission. It is accompanied by the release of large amounts of energy.

Another transmutation process called fusion leads to even higher energy yields than are provided by fission. When nuclei of light elements such as hydrogen collide with enough force, they may fuse together, producing nuclei of heavier elements. The energy released in this process is enormous, as any witness of a hydrogen fusion bomb explosion can attest.

Starting with the proper ingredients and bringing into action the forces between nuclei, you could devise a sequence of reactions that would realize the alchemists' dream—the production of gold from the ordinary metals. And in considering the forces between nuclei of hydrogen atoms and other nuclei the scientist realizes his dream of finding the source of the sun's energy, for it is the fusion of hydrogen that supplies the solar energy. The Philosopher's Stone, ardently searched for by the alchemists, has been at work in the sun, changing one element into another since before the dawn of life on earth.

It is customary to begin a discussion of the source of energy in stars by considering what is required to supply the sun with enough energy to keep it shining. The sun radiates energy at a rate of about 4×10^{33} ergs per second. Thus, the energy from each square meter of the solar surface is equivalent to the full power output of more than 200 automobiles, each capable of generating 350 horsepower. The radius of the sun is nearly 700,000 kilometers, so there are many square meters of radiating surface (roughly 6 million million million square meters).

But how long has the sun been radiating at this tremendous rate? Geological fossils have been found with ages around 3 billion years. It appears that the sun must have generated energy at a rate not much different than its present rate for the living matter which formed the fossils to have existed. Studies by geologists indicate that the earth is at least 4.5 billion years old. The sun probably has been shining with an output close to 4×10^{33} ergs per second over this entire span of time.

What could supply this much energy for such a long time? It cannot be a chemical reaction like burning coal. If the sun were pure carbon and oxygen, it could only supply energy by the burning of the carbon in the oxygen for 2500 years—less time than recorded history, let alone the time required by the fossils. Furthermore, the sun is too hot, even at its surface, for carbon and oxygen to combine. Instead, the carbon dioxide molecule would be broken down into carbon and oxygen. Around the turn of the century, before the age of the earth was known, it seemed as though the sun could shine by slowly contracting, converting its gravitational potential energy to heat. But even if it had contracted from a size equal to the orbit of Jupiter, it would have released only enough energy for about 100,000 years or so, much too short to satisfy geologists. The only processes we know of which are capable of providing the required energy are nuclear reactions. In particular, the conversion of hydrogen to helium by a thermonuclear process is the source of energy which has kept the sun going for the great length of time required.

Stellar Energy

The self-gravitation of a body like the sun is continually trying to cause it to collapse. If there were no counterbalancing force, the sun would totally collapse in less than an hour. At any point within the sun, there must be an outward-directed pressure force just great enough to equal the weight of the overlying layers of the sun. At deeper and deeper layers the pressure must increase. Finally, at the center the pressure must be large enough to support the entire weight of the material of the sun.

The pressures required are so large that the interior of the sun can only be in a gaseous state at an extremely high temperature. Calculations indicate that near the center of the sun the temperature needed for the high pressure is about 15,000,000K. At such a high temperature, atoms are very highly ionized. Hydrogen is completely ionized; a heavier atom like iron, which normally has twenty-six electrons, will retain only one or two.

We cannot directly observe the chemical composition in the sun's interior. At the surface, it is made up of 73 percent hydrogen, 25 percent helium, and 2 percent of all other elements by mass. There is evidence that throughout most of the interior regions of the sun and other main sequence stars, the chemical composition is quite similar to that at the solar surface. Thus, most of the material in these stars exists as ionized hydrogen, which is more or less a gas of protons and free electrons.

Under normal circumstances, when a proton approaches another proton or a different atomic nucleus, the two particles repel each other and never quite collide; this is because all nuclei carry positive electrical charges. However, if the protons approach close enough, a stronger attractive force, called the **nuclear force**, overcomes the electrical repulsion, and the two protons will be bound together. This nuclear binding force is the strongest force known in nature, but it is effective only over very short distances, of the order of the size of the proton. This force is what holds together a helium nucleus, which consists of two protons and two neutrons. The **neutron** has a mass just about equal to that of the proton, but it has no electrical charge. The heavier elements have nuclei made up of protons and neutrons similarly bound by the nuclear binding force. Carbon, for example, has a nucleus consisting of six protons and six neutrons.

In the central region of the sun, where the temperature exceeds 10,000,000K, some protons move with enough energy to overcome the electrical repulsion and approach close enough to another nucleus for the nuclear force to become effective. Figure 14–1 is a diagram of what occurs when two protons collide with sufficient energy. They combine to form a deuterium, or heavy hydrogen, nucleus, which contains one proton and one neutron. In the combination process, two new particles are formed. One is a **positron**, or a positive electron, which carries away the excess electrical charge, since the deuterium nucleus has only a single positive charge. The other particle is a chargeless and essentially massless particle known as a **neutrino.** The positron is designated by $e+$ and the neutrino by ν.

The process of **hydrogen fusion**—that is, the combination of four protons to form a helium nucleus—can occur in one of two chains of nuclear reactions. In the chain most important in the sun and less massive stars, two protons collide to form a deuterium nucleus. The deuterium nucleus, in turn, captures another proton to form a light nucleus of helium. In the last step of the chain, two of these particles combine to form a normal helium nucleus, with two protons being released. The net result is the formation of one helium nucleus from four hydrogen nuclei. This chain is called the *proton-proton reaction* after the first step, which involves two protons.

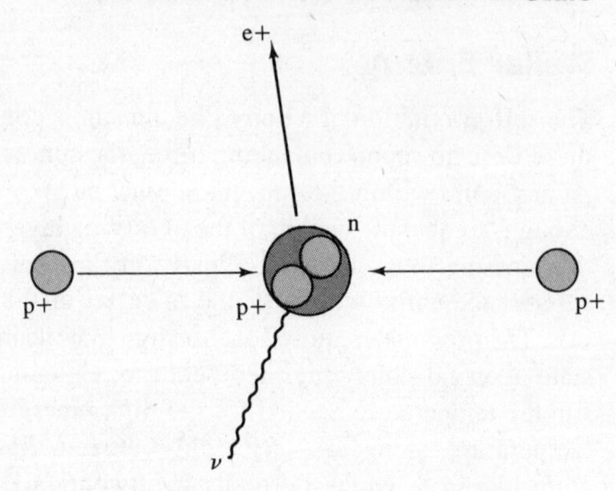

Figure 14–1 The formation of deuterium from two protons.

The other reaction chain, which begins with a carbon nucleus capturing a proton, is called the *carbon cycle*. After a number of intermediate steps in which three more hydrogen nuclei are captured, a carbon nucleus reappears along with a helium nucleus. Because the carbon nucleus contains six protons, its electrical repulsion of a proton is much greater than that of a single proton. Therefore, a proton must have more energy in order to penetrate the carbon nucleus than it must have to join with another proton. Consequently, the carbon cycle occurs for the most part in stars whose central temperatures are somewhat higher than the central temperature of the sun. In both processes, hydrogen is transmuted to helium. These processes, because they require high temperatures and involve nuclear reactions, are called *thermonuclear reactions*. The details of these reactions are given in Appendix VI.

In the formation of helium from hydrogen, there is a loss of mass. The four protons taken together have 4.033 mass units, but the helium nucleus has only 4.004 mass units. Thus, 0.029 mass unit, or about 0.7 percent of the original hydrogen mass, is lost. This mass does not just simply disappear. Instead, it is converted to energy according to Einstein's famous equation:

$$E = mc^2$$

where E = the energy equivalent of mass, in ergs
m = the mass in grams
c = velocity of light in centimeters per second

(A discussion of the relationship between energy and mass can be found in Appendix I.) The thermonuclear process is particularly energetic. Converting one gram of hydrogen to helium, for example, releases 6.3 million million million ergs of energy, enough energy to light a 100-watt light bulb for 200 years.

Since the mass of the sun is 2×10^{33} grams and about three-quarters of that is hydrogen, there are about 10^{52} ergs of energy available if all of the hydrogen could be converted to helium. Dividing this energy by the luminosity of the sun, 4×10^{33} ergs per second, gives the maximum length of time the sun could continue shining at its present rate by the transmutation of hydrogen. The maximum age turns out to be 2.5×10^{18} seconds, or nearly 100 billion years. But before the whole amount of hydrogen in the sun is converted to helium, some rather extreme changes will have taken place in the sun, and its age will be considerably shortened. In any case, a conversion of only 5 percent of the original hydrogen in the sun supplied the energy needed to keep it shining for the past 5 billion years.

In addition to the problem of how energy is produced in a star, there is the problem of how energy is carried from the interior of a star to its surface. There are three major possibilities for the transportation of energy: *conduction, convection,* or *radiation.*

A common example of conduction is the heating of an iron bar. While heat may be applied only to one end of the bar, the other end becomes hot by conduction. The heat energy is transferred by atoms in the bar becoming agitated by the addition of energy and, in turn, passing some of their agitation (and thus energy) to neighboring atoms. Also, some of the electrons present in the metal are loosely bound due to the tight packing of the iron nuclei in the solid iron, and they can rapidly carry energy along the length of the rod. There is no real flow of mass from one end of the rod to the other to accompany the heat flow. The transfer of energy is essentially a particle-to-particle transfer.

In a normal gas, and hence in most regions of the usual star, conduction is not very efficient. But in a white dwarf, conduction is the predominant process. The matter in a white dwarf is so highly compressed that many of the electrons present are free to move much as those in an iron bar. It is these freely moving electrons that give rise to the conduction and make the gas in a white dwarf act with some of the properties that we normally associate with a solid metal.

Convection, on the other hand, results in heat transfer by large numbers of atoms or molecules moving as groups from the hotter regions to the cooler regions of a gas or liquid. In turn, the displaced cooler particles move in groups to the hotter regions. Probably the most familiar example of convection is the rapid mixing seen in a container of water as it is heated, especially if some solids are present to mark the flow. A more astronomical example is the granulation of the solar photosphere. Convection, of course, can only occur in a gas or a liquid, where the individual atoms and molecules are free to move.

Even though some of the outermost regions of the sun are convective, energy in the sun, as well as in other main-sequence stars, is for the most part transported by the third process—radiation—and not by convection. Radiant energy is carried by a flow of photons and can occur with or without the presence of matter. The radiant energy produced in the interior of a main-sequence star by nuclear fusion is absorbed by the surrounding layers of material. This material re-radiates the energy which, in turn, is absorbed by the next surrounding layers. It is then re-radiated and re-absorbed until finally the energy leaves the stellar surface. Each time the energy is re-radiated, however, it is transformed into radiation characteristic of the temperature of the radiating material. So what starts out in the interior as a flow of very energetic photons of short wavelength, characteristic of the high temperature there, finally leaves the star's surface as a flow of less energetic, longer-wavelength photons, characteristic of the star's surface temperature.

Sir Arthur Eddington was one of the first to recognize the importance of radiation in transporting energy from the interior of a main-sequence star to its surface. He was able to show that a star of a particular mass and chemical composition in which energy is carried by radiation must have a particular luminosity. In other words, he showed that main-sequence stars should obey a mass-luminosity relationship, in agreement with what is observed. Thus, for a star of given mass (and composition) there is a particular luminosity; and furthermore, the greater the mass, the greater is the luminosity. This, by the way, is independent of the exact source of the stellar energy; it merely requires that the energy be transported through the star mainly by radiation.

Thc particular process of energy transport which is most important in a star depends upon the particular conditions within the star. All three processes may work in different parts of the same star, but one process will always predominate in a given region. The observed characteristics of a star will depend upon which method of transport is most effective and in what part of the star it assumes predominance. In addition, as a star uses its energy supply, the conditions within the star will change, thus favoring one or another of the energy transport methods. In such a case, the character of the star (giant or dwarf, for example) can also change.

The more luminous main-sequence stars consume their energy supplies in a much shorter time than the less luminous ones. As an example, consider an M8V and an O5V star. The M star has a mass of one-tenth the sun's and hence an energy supply only one-tenth that of the sun. The O star has a mass forty times the sun's and thus an energy supply forty times the sun's supply. But the M star has only one-thousandth of the solar luminosity. Its maximum lifetime is its energy supply divided by its luminosity, or 100 times that of the sun. The O star, on the other hand, is about 500,000 times as luminous as the sun. Consequently, the O star's maximum lifetime as a hydrogen-fueled star is somewhat less than one ten-thousandth that of the sun. We could also have predicted this outcome from the mass-luminosity relationship, which demonstrates that the luminosity of a star would increase much more rapidly than its mass as material is added to it. In any case, the more luminous a main-sequence star, the shorter lived it must be.

A mathematical **model star** can be found by calculating the temperature and pressure as you move inward along a radius of the star. For a particular assumed mass and chemical composition, one model with a definite radius, luminosity, and surface temperature, or spectral type, normally results. The calculated values of these three quantities are compared to the observed values for an actual star. If the calculated values disagree, the assumed mass and composition are changed and another set of calculations are completed until we obtain good agreement with the observations. When they agree, we say that we have found a model for the particular star.

When a series of models with the same chemical composition throughout their interiors as the sun but with a range in mass is calculated, the luminosities and spectral types obtained fit the observed main sequence of the H-R diagram for the solar neighborhood quite well. We therefore conclude that the different spectral types along the main sequence result primarily from different initial masses of the stars.

When a model of the sun is calculated with an assumed mass of 2×10^{33} grams and a uniform composition the same as the sun's surface layers, relatively good agreement between the model and the actual sun results. The computed radius is 6.2×10^{10} centimers, compared with the actual radius of 7.0×10^{10} centimeters. The computed luminosity is 2.8×10^{33} ergs per second, compared with the actual 3.8×10^{33} ergs per second. The small differences result from the fact that the sun, in the nearly 5 billion years it has been shining, has modified its chemical composition somewhat by transmuting hydrogen to helium in its deep interior. The calculated model, however, is thought to represent quite well the *zero-age* sun, that is, the sun when it first began the thermonuclear process.

A theoretical **zero-age main sequence** found by calculating models for stars of different masses is shown in Figure 14–2. The observed points for nearby stars all fall rather close to the theoretical curve. This finding indicates that main-sequence stars have quite similar compositions and generate their energy by the nuclear fusion process which changes hydrogen to helium. The small disagreements of the observed points with the theoretical curve result from small chemical composition differences from star to star.

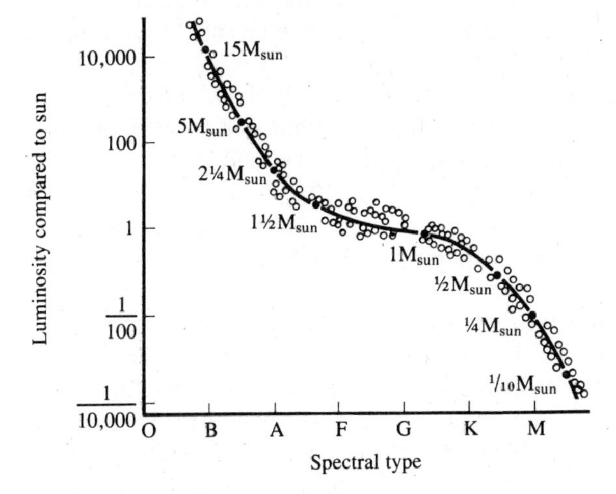

Figure 14–2 The zero-age main sequence.

Stellar Evolution

Obviously, as a star consumes its energy supply, its chemical composition must change, and hence its size, luminosity, and spectral type must change. As an example, consider the sun. The present location and the zero-age location of the sun on the theoretical zero-age main sequence are shown in Figure 14–3. Calculations indicate that the sun's present position relative to its initial position on the theoretical main sequence is consistent with

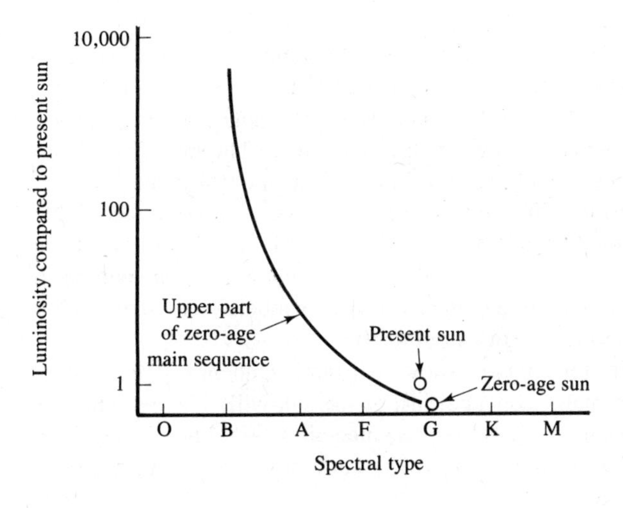

Figure 14–3 The present sun and the zero-age sun.

its converting hydrogen to helium for about the last 5 billion years. In the center of the sun, the hydrogen content has been reduced from 73 percent to about 38 percent during this time. Since the internal temperature of the sun is high enough to convert the hydrogen only within a distance of one-fifth of its radius from the center, it is only within this region that any enrichment of helium and depletion of hydrogen has occurred. This region is usually called the **core**. The present location of the sun on the diagram is slightly above its starting point as a result of the slowly changing chemical composition in its core. For a while the sun will continue to increase in luminosity slightly, but with very little change in spectral type.

By the end of the next 5 billion years, the sun's core will be almost depleted of hydrogen, and the core will be 98 percent helium. But the helium will be inert, because fusion reactions involving helium nuclei require temperatures around 100,000,000K or more, and the core will only have heated up to 20,000,000K by then. The inert helium core will be surrounded by a thin shell where the temperature is high enough for hydrogen fusion to continue. This **fusion shell** will be surrounded by the remainder of the sun, called the **envelope**, where no change in the original composition has occurred. At this time the change in the solar luminosity and structure will accelerate. The helium core—too cold to generate energy by fusion of helium—will begin to contract, causing its temperature to rise. This process in turn will heat the shell around the core, where hydrogen fusion still takes place, causing the thermonuclear conversion of hydrogen to helium to increase. The solar luminosity will increase accordingly. The surrounding envelope of material, that is, the rest of the sun, will begin to expand to compensate for the increased liberation in energy in the shell. As the luminosity of the sun increases, radiation alone cannot carry the energy to the surface. Convection takes on a larger and larger share of the load in transporting energy through the envelope. When convection transports a sizeable fraction of the energy in a star, the mass-luminosity law no longer applies. In contrast to a main-sequence star, the luminosity of a convective star depends not only on the mass but also on its size. In the case of the sun during the later stages of its life, the envelope will expand greatly to compensate for the increased liberation of energy in the hydrogen fusion shell. The helium core, although continually heating and increasing in density because of contraction towards its center, will not change much in size because it is continually fed new helium from the surrounding fusion shell. The envelope, as it expands, will cool considerably. The sun's spectral type will move toward the red stars of class M.

The path followed by a star in the H-R diagram as it consumes its fuel is called its **evolutionary track**. Figure 14–4 shows this track for the sun. For the first 10 billion years, the sun changes only slightly. But a rather sudden and dramatic change, requiring only 1 billion years, takes place when the inert helium core is fully formed. The sun will shoot up in luminosity by nearly 1000 times, while its outer layers expand 100 times and cool enough to make it a red giant. So in 6 billion years or so, the sun will no longer be the familiar sun we see in the sky now. As it expands to nearly 0.5 AU in radius, it will have swallowed up Mercury. The sun will span almost half the sky at noon, but there will be no one around to see it. By then the temperature of the earth will be much too high to support life because of the great increase in the sun's luminosity.

Calculations for stars of different masses show that main-sequence stars of other spectral types will also leave the main sequence, just as the sun will. The evolutionary tracks for stars of 15, 5, and 0.5 solar masses are presented along with the sun's track in Figure 14–5. The more massive stars do not have the remarkable increase in luminosity

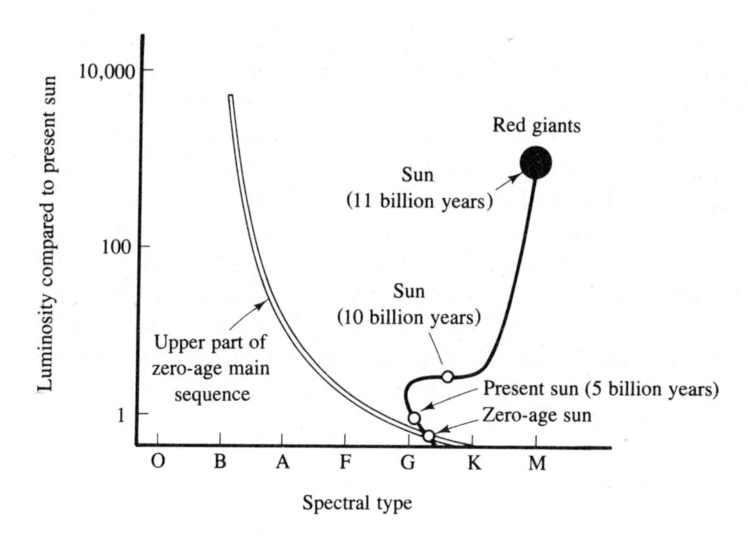

Figure 14–4 The evolutionary track of the sun.

that the sun will have; they move almost horizontally across the H-R diagram to become cool supergiants. Their behavior differs from that of the sun because as they evolve, the massive stars do not become predominantly convective. Radiation continues to be the predominant method by which energy is transported in these stars. Even so, the massive stars must still evolve much more rapidly than the sun because they consume their initial fuel supply so much more quickly.

At the other extreme, the stars less massive than the sun, like the one illustrated at 0.5 solar mass, will evolve rather lazily. Such a star will not even leave the main sequence until after it has been consuming its fuel for slightly more than 100 billion years. The hot B-type star with 15 solar masses, on the other hand, will only take around 10 million years to end its main-sequence stage and become a supergiant.

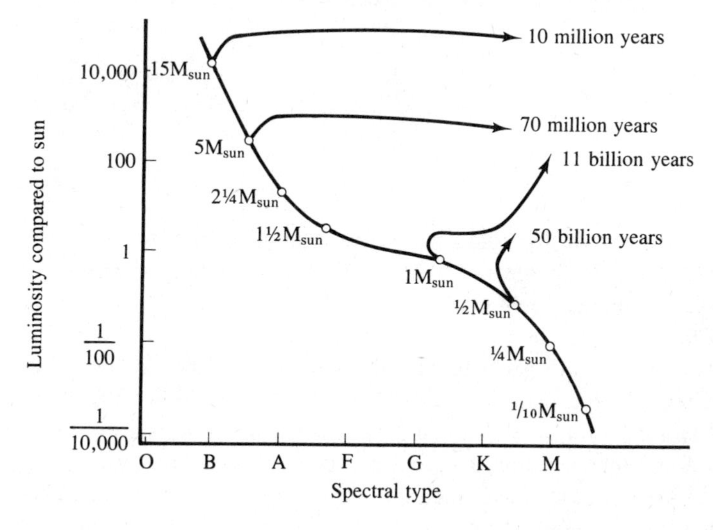

Figure 14–5 The evolutionary tracks for stars with different masses.

The H-R diagram can be likened to a snapshot which catches a group of stars in various stages of evolution at one instant. The supergiants and giants are former main-sequence stars that have already evolved when the hydrogen fuel in the cores was exhausted. The supergiants evolved from stars much more massive than the sun. Because a massive star spends a relatively short time on the main sequence, the fact that we presently see both supergiants and hot main-sequence stars implies that star formation is still going on, in essence replenishing the main sequence. The O-type stars in our snapshot must be very young, because their lifetimes on the main sequence only last about one million years. Although they are very young stars, they age the most rapidly of all.

The rapid departure of a star from the main sequence as its core is depleted of hydrogen can be used to determine the age of a group of stars such as a globular or open cluster. All members of a particular cluster are thought to have formed at about the same time. They therefore have been evolving for about the same period of time, but the more massive stars will have evolved the most.

In Figure 14–6 we have plotted the color-magnitude diagram for a globular cluster and indicated the location of the zero-age main sequence. The absence of main-sequence stars above point X is accounted for by the fact that the more massive stars have already evolved away from the main sequence. Point X is called the **cluster turn-off point**. The age of the cluster must be close to the main-sequence lifetime of a star just at the cluster turn-off point.

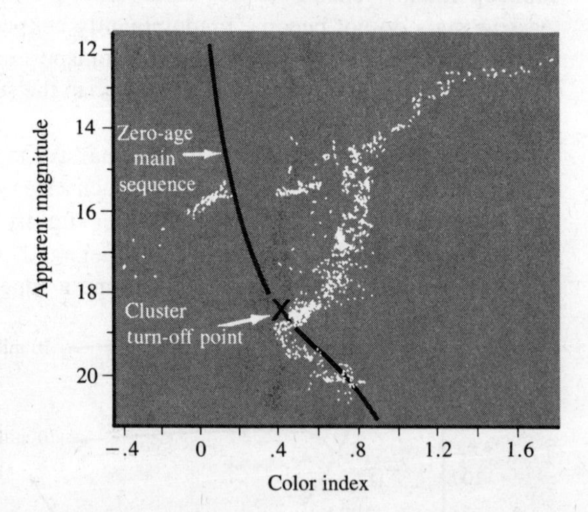

Figure 14–6 The color-magnitude diagram of the globular cluster M3.

A composite of a number of cluster color-magnitude diagrams fitted to the same zero-age main sequence is shown in Figure 14–7. All the diagrams came from observations of open clusters, except the one for M3, which is a globular cluster. The youngest cluster, by the turn-off point test, is obviously the one labeled NGC 2362. And, except for M67, all of the open clusters are younger than M3, the globular cluster. All globular clusters seem to be old like M3 because their turn-off points are well down the main sequence. The open cluster M67 is the oldest known open cluster. Most open clusters will eventually be dispersed and hence become unrecognizable as clusters because of the interactions with the large system of stars to which the sun belongs. M67 happens to be located at such a position that the interactions are not so severe, so it is still recognizable as

Plate 11A The Pleiades, an open cluster.

Plate 11B A globular cluster.

Plate 11C Reflection nebulae in the Pleiades.

Plate 12A Orion Nebula.

Plate 12B North American Nebula.

Plate 12C Horsehead Nebula.

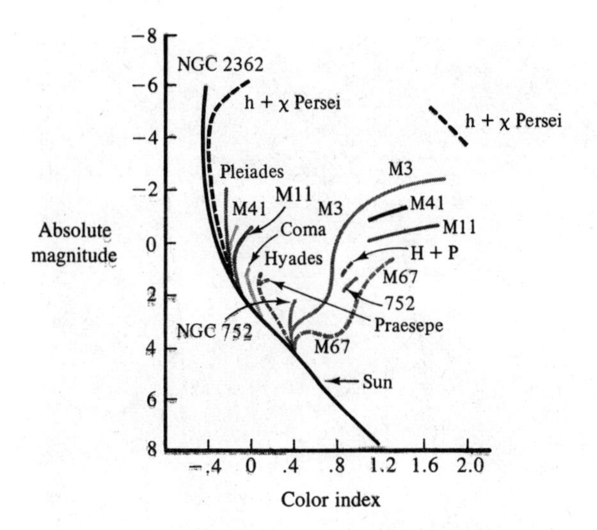

Figure 14–7 Combined color-magnitude diagrams of several star clusters.

a cluster even after a very long time. The globular clusters, because they are more densely packed and contain more stars, are gravitationally able to withstand the disturbances. Hence, they can remain identifiable as clusters more or less indefinitely. The oldest clusters identified by using the cluster turn-off method are around 10 billion years old and, except for M67, generally are globular clusters.

Star Formation

Stars do not magically appear as main-sequence objects; they must first be formed. The actual birth of a star is believed to occur in the dark, dusty, and gaseous regions of space between already formed stars. Figure 14–8 shows a photograph of one likely region in the constellation of Monoceros where star formation may even now be taking place. Dark extensions can be seen protruding into the bright surrounding nebulosity. In other regions, such as the one shown earlier in Figure 9–11, dark condensations or globules are visible.

The conditions necessary for the formation of a star are known, even though we do not yet know the details of how these conditions develop in the interstellar material. Consider a dark globule. If the inward force of its gravity exceeds the outward pressure force of the material in it, the globule will undergo a gravitational contraction, forming one or more stars. The density of matter in one of these condensations will initially be very small, around 100 million million times less than the density of water. The self-gravity is consequently extremely weak. Also, the temperature of the globule must be extremely low, near absolute zero, if the pressure is to be low enough to allow contraction. Some globules have been shown to have temperatures around 5° above absolute zero. Presumably, they are contracting into stars or even clusters of stars.

At the start of contraction, the process is so rapid that it would be better described as a collapse. After a short time, around 1000 years, the density of the material will increase enough to become opaque to radiation. Thus, instead of simply losing the radiant energy generated by the collapse, some of the radiation will be trapped within the contracting

Figure 14–8 Nebulosity in Monoceros.

material. The pressure will consequently increase, and the collapse will be transformed into a relatively slow contraction.

Calculations show that once enough material to make up a solar mass has contracted to the point where it can be called a new star, or more appropriately a **protostar**, it will decrease in luminosity as it contracts, but its surface temperature will remain essentially constant. The energy in the contracting star is transported principally by convection during this stage, allowing the luminosity to change drastically as the star shrinks in size. Figure 14–9 shows the path on the H-R diagram which a one-solar-mass protostar is expected to follow. In something like 9 million years, the protostar will have reached its minimum luminosity as a cool star. After this, the contraction continues; but now the surface layers begin to heat up, and the luminosity increases somewhat. The star begins to

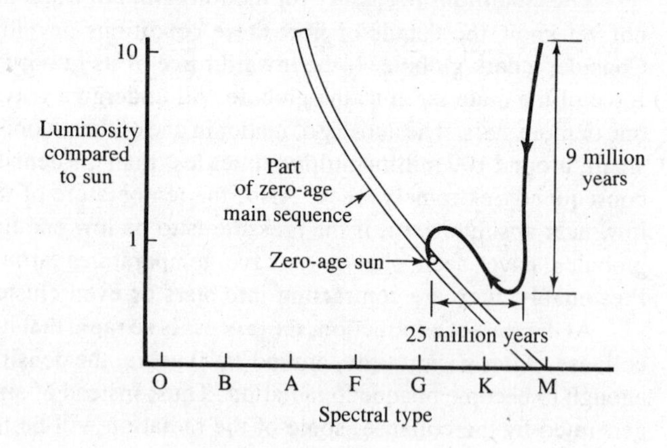

Figure 14–9 The pre–main-sequence evolution of the sun.

move more or less leftward on the H-R diagram. Roughly 25 million years after the leftward motion begins, the star takes a small dip in luminosity, just before it settles on the zero-age main sequence. By this time, the contraction has heated the central region to the point where the temperature is high enough to support the fusion of hydrogen into helium, and the contraction ceases. During the earliest parts of the evolution of a newly forming star, it is doubtful that the observed track would agree well with the theoretical track. This is because the dust particles in the outer regions of the in-falling material would most likely be opaque to the star's radiation and would render it essentially invisible. The dust particles, having been warmed by the star's radiation, would be quite bright in infrared radiation, however. The lifetime of the contraction stages of a one-solar-mass star is much shorter than its main-sequence lifetime. The same holds true for stars of other masses. The more massive stars, just as they exhaust their hydrogen faster than the sun during their main-sequence stages, will also complete their contraction more rapidly than the sun.

Our general ideas about star formation are apparently confirmed by the color-magnitude diagram of the open cluster NGC 2264, which lies near the nebulosity in Monoceros where the dark globules are seen. Notice in Figure 14–10 how the hotter stars seem to fit the zero-age main sequence rather well. The cooler spectral types, however, seem to lie above the main sequence. The cluster is still in the process of forming. Only the most massive stars have had time enough to contract to the point that hydrogen fusion has begun. The lower-mass stars, contracting more slowly, have not yet had their hydrogen fusion ignited. They still have more contracting to do before making it to the main sequence.

The contraction process can lead to a main-sequence star only over a limited range of mass. No star has yet been found with a mass definitely shown to exceed sixty solar masses. Nor has a star with less than a few hundredths of a solar mass been positively identified. Extremely massive stars, greater than sixty solar masses, would have to be so hot in their central regions that the pressure would be mostly due to radiation. The balance between radiation pressure and gravity would be so precarious that the star would be unstable, if it could even form at all. The protostars of very low mass, on the other hand, would never generate a sufficiently high central temperature to start the thermonuclear process and cause them to settle in as main-sequence stars.

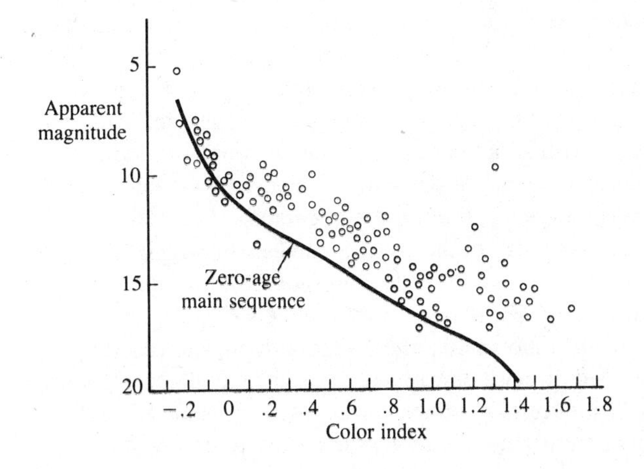

Figure 14–10 Color-magnitude diagram of the open cluster NGC 2264.

To summarize our ideas of stellar evolution, then, we have seen that the distribution of points in the H-R diagram representing the stars' observed luminosities and spectral types results from the initial masses of the stars and their subsequent changes as they consume their fuel supply. A star forms first as a condensation in the interstellar medium which contracts gravitationally. Eventually, the initial rapid collapse slows to a relatively slow contraction, as the internal pressure and gravity reach an essential balance. Ultimately, the central regions heat up to the point where thermonuclear reactions set in. With the conversion of hydrogen to helium, the star, now on the main sequence, begins the longest part of its lifetime. But as it converts its core to helium, the core contracts; and the outer envelope of the star expands, cools, and thus transforms the star into a red giant or a supergiant if it is massive enough.

What we have described are the more or less unspectacular events in stellar evolution. In the next chapter we will examine the more bizarre events in stellar evolution—mostly at the very end of a star's life, when it becomes erratic or unstable in its energy output.

The Neutrino Problem

Our theory of stellar evolution seems to fit rather nicely the H-R diagrams of evolved as well as newly forming star clusters. But recently we have found a possible difficulty. In the generation of energy by hydrogen fusion, the fundamental feature of our theory, large quantities of neutrinos should be released in the cores of stars and, in particular, in the sun. Neutrinos are chargeless and nearly massless and hardly interact with anything at all. The neutrinos released at the center of the sun, therefore, should stream outward through its surface essentially undisturbed. If a neutrino could be detected, we would in a sense be seeing directly to the center of the sun.

But unfortunately, the very properties that allow a neutrino to pass completely through the sun also make it extremely difficult to detect. There is one very weak interaction of neutrinos with the chlorine nucleus which has been discovered in the laboratory. If neutrinos of the correct energy, some of which should be emitted by the sun, are captured by chlorine nuclei, the chlorine is transformed into argon, with the emission of electrons. Based upon this reaction, Raymond Davis, Jr., has devised an experiment to detect solar neutrinos. He has placed a huge container of a commercial cleaning solvent, C_2Cl_4, 1.5 kilometers beneath the earth's surface in a deep gold mine in South Dakota to protect his equipment from other radiation. Occasionally, neutrinos react with the chlorine nuclei in the cleaning fluid. With sufficient care, the argon atoms formed can be collected and counted. The number of argon atoms detected is a measure of the number of solar neutrinos of appropriate energy passing through the apparatus.

So far Davis, after trying to account for all of the experimental errors which he thinks might arise, has concluded that, at most, the number of neutrinos from the sun must be below one-fifth the number which theory predicts we should receive.

Possibly something is wrong with some details of our theory of nuclear reactions in the sun, or there are as yet unexplained difficulties with the experiment to detect the solar neutrinos. If neither of these possibilities turns out to be correct, we may be faced with the fact that the sun is not currently generating its energy by the proton-proton chain as we have become quite thoroughly convinced. If that turns out to be true, we are really in trouble. Our ideas of stellar energy and stellar evolution may well require considerable revision. But most astronomers agree that before we panic and discard what is otherwise an apparently good description of how stars generate energy, we should wait for a more definitive review of both the theory and the experimental data of the neutrino process.

KEY TERMS

nuclear force
neutron
positron
neutrino
hydrogen fusion
model star
protostar
core
fusion shell
envelope
zero-age main sequence
evolutionary track
cluster turn-off point

Review Questions

1. Why doesn't the tremendous pressure at the center of the sun blow the sun apart?
2. What force holds two protons together when these particles fuse?
3. Why are very high temperatures required before hydrogen fuses into helium?
4. Why does the conversion of a tiny amount of mass produce large quantities of energy?
5. Why are the main-sequence lifetimes of the more massive stars, which actually contain more hydrogen, shorter than the lifetimes of stars of lesser mass?
6. What fundamental quantity determines the position of a star on the main sequence?
7. How can we discover the chemical composition in the interior of the sun?
8. Which of the following best characterizes main-sequence stars?
   (1) Have similar sizes
   (2) Are all converting hydrogen to helium in their cores
   (3) Are confined to the neighborhood of the sun
   (4) Are fairly young
   (5) Are fairly old
9. How have we estimated the amount of time for which the sun will still remain on the main sequence?
10. What is a zero-age main sequence? Why is it labeled "zero-age"?
11. Why will the sun eventually cease to be a main-sequence star?
12. When the sun leaves the main sequence, what will supply the energy to keep it shining? Why will it increase in luminosity?
13. Why doesn't the helium presently in the sun's core undergo thermonuclear reactions?

14. What kind of star will the sun become after it leaves the main sequence? How will the densities at its center and surface compare with those at the present time?
15. How can we demonstrate that some stars have formed very recently (within the last million years or so)?
16. What is the sun's main-sequence lifetime? How much of that time has already expired?
17. Suppose the stars in a cluster have a surface chemical composition similar to the sun's, but no stars hotter than the sun lie on the cluster's main sequence. How old is the cluster? Why does the chemical composition matter?
18. How do we conclude that globular clusters are generally older than galactic clusters?
19. What supplies the energy that allows a protostar to emit radiation before it reaches the main sequence?
20. What stops the contraction of a protostar as it reaches the main sequence?
21. How does the time it takes a massive protostar to reach the main sequence compare with the time it takes a low-mass protostar to reach it?
22. What sets the upper limit to the mass a star can have and still be on the main sequence? What sets the lower limit?
23. What is a neutrino? Why do most neutrinos escape the sun completely?
24. What unexpected result has been obtained in the measurement of solar neutrinos?

Discussion Questions

1. If we suddently added extra material to the sun's surface, making it start to contract, how would the internal temperature change? How would the temperature change affect the rate of hydrogen fusion? How would the changing fusion affect the sun's internal temperature and pressure? Would the sun end up smaller or larger than it was before the extra matter was added?
2. List the three major mechanisms for transporting energy and give some examples of each that occur in astronomy.
3. The sun's age on the main sequence has been used to support the nebular hypothesis for the origin of the solar system. How is the age of the sun determined, and how can the result be used to support the nebular hypothesis?
4. The fusion of hydrogen to helium is sometimes loosely referred to as "hydrogen burning" in a star. How is this different from what we mean by the burning of hydrogen on the earth?
5. When the sun leaves the main sequence and its surface temperature drops, why will the temperature on the earth increase?
6. Will the earth's orbit change significantly as the sun becomes a red giant?
7. Just before hydrogen fusion started in its core, the sun was contracting and moving to the left (also briefly downward) on the H-R diagram. Why will the sun not resume this contraction after hydrogen fusion ceases in its core?

8. In what ways will the evolution of a very massive star differ from a solar-mass star during and soon after their main-sequence phases?
9. Why must the dark globules of gas and dust seen in space have a very low temperature if they are to form into stars?
10. As a dark globule contracts in the process of star formation, it heats up. Why does this heat not allow the globule to dissipate?

15

Extraordinary Events in Stellar Evolution

John Herschel, the son of William Herschel and a brilliant astronomer in his own right, was once checking some calculations with a friend, Charles Babbage. The tedious nature of the work irritated Babbage to the point where, in frustration, he cried out, "I wish to God these calculations had been executed by steam." Herschel, without looking up from his numbers, calmly replied, "It is quite possible." Babbage did not let the matter drop. He actually built steam-powered machines capable of carrying out numerical computations. The machines could not work any faster than a skilled human, however, and did little to revolutionize the art of computation.

Over a century later, during World War II, the U.S. Army was involved in a project designed to speed up the calculation of projectile trajectories to compile artillery and bombing tables. A group of engineers and scientists at the Moore School of Electrical Engineering of the University of Pennsylvania collaborated with the army's Ballistic Research Laboratory to develop an electronic computer capable of making calculations over a thousand times faster than those carried out by hand. This machine, called ENIAC (the Electronic Numerical Integrator And Computer), was the first purely electronic computer that could perform a variety of functions.

The work on electronic computers continued after the end of World War II. In fact, ENIAC was not fully operational until after hostilities had ceased. A mathematician who was of crucial importance to the development of the atom bomb, John von Neumann, had become interested in electronic computers during the war and was thoroughly familiar with ENIAC.

Von Neumann was on a level with Gauss in regard to sheer intellect. Like Gauss, who had demonstrated a practical bent by applying mathematics to astronomy, inventing the telegraph, and doing sundry other things, von Neumann enjoyed the less pure aspects of mathematics as well as the abstract. Not only did he establish the mathematical basis of the electronic computer and the interactions of its numerous parts, he directed the Electronic Computer Project at the Institute for

The Orion Nebula

Advanced Study in Princeton, New Jersey. The aim of this project was to build an all-purpose, high-speed calculating machine.

The Institute for Advanced Study seemed to many to be a quite unlikely site for such a project. It had no experimental laboratory facilities. It had been founded with the purpose of providing theoretical physicists and mathematicians a place to work on purely abstract problems without the interruptions caused by classroom, laboratory, or administrative activities. This ivoriest of ivory towers was first directed by Albert Einstein. The trend toward attacking problems of the most fundamental nature regarding the universe was set early under his supervision. To say the least, it was a shock when electrical components, wires, and vacuum tubes were hauled through the hallowed doors of the Institute. Equally shocking was the influx of engineers who proceeded to carry out mundane activities such as soldering wires and attaching tube sockets to circuit boards.

In time, "von Neumann's machine" was finished. Although newer computers are sleeker, smaller, and much faster machines that utilize tiny transistors in place of cumbersome vacuum tubes, the basic concepts used in his design underlie the construction of practically all modern computers.

Two of the very first problems to which von Neumann's computer was applied were the study of the internal structure of stars that had been undertaken by Martin Schwarzschild of Princeton and the tabulation of mathematical functions for S. Chandrasekhar of the Yerkes Observatory. Since that time, electronic computers all over the world have been applied to the problem of the structure and evolution of the stars. The fantastic number of calculations that must be made to follow the life of a model star mathematically simply could not have been accomplished without electronic computers. As a result of their application, rare, strangely behaving stars which previously seemed to have no connection to normal stars now appear to fit into an overall picture of the births, lives, and deaths of stars.

In many respects, the most interesting periods in the lifetime of a star occur early, before it reaches the main sequence, or late, after it has completed its main-sequence phase. At these times the star may fluctuate in luminosity fast enough for us to measure the changes easily. The fluctuations are extremely rapid compared to the lifetime of a star, being measured in years, days, or even fractions of a second. Because a star spends most of its time evolving so slowly that its observable features appear quite constant over the lifetime of an astronomer, we usually call a star that can be easily seen to change in brightness a *variable star*.

Stars can vary in brightness because of some activity within the stars' interiors or atmospheres. These stars are called *intrinsic variables* and are the ones with which we will be concerned here. The eclipsing binaries, which we have already discussed, also vary in brightness; but their changes are normally due to the mutual eclipsing of the stars and not to an intrinsic property of either one.

Astronomers have discovered a great number of intrinsic variables. Many fall into categories which are frequently named after the first star of that category to be observed in detail. We shall, however, restrict our discussion here to only the major categories which seem to represent the most significant stages in the evolution of stars of different masses. There are two major kinds of variables. Those that vary in brightness rather modestly and with a reasonable degree of regularity are called **pulsating variables**. The other kind is more spectacular; they suddenly burst forth with a tremendous increase in luminosity. These are frequently called the **cataclysmic variables**.

We will begin first by describing the general properties of the variable stars and then show how they fit into the pattern of stellar evolution.

T Tauri Stars

The star T Tauri is characteristic of the category of variables called the **T Tauri stars**. T Tauri is imbedded in a nebula and has rapid and irregular changes in its luminosity. Many of the T Tauri stars are found in young clusters containing pre–main-sequence stars. These stars are brighter than they should be for their spectral type if they were main-sequence stars. They lie to the right and slightly above the main sequence on the H-R diagram (*see* Figure 15–6). They are in the region of the diagram which stars must cross to reach the main sequence as they form out of interstellar material. We associate the observed variability with gravitational contraction as the central temperature of these stars nears the threshold for hydrogen fusion.

One of their most striking features is illustrated in Figure 15–1. The stars are surrounded by a huge shell of material, five to ten times the size of the star itself. Much of the radiation from the star is absorbed by dust particles in the shell, which heats the particles. The dust particles, in turn, re-radiate the energy. But because they are lower in temperature, their radiant energy occurs mostly in the infrared part of the spectrum. Consequently, the continuous part of the spectrum of a T Tauri star is a composite of that from the shell and the star itself. The right-hand panel of Figure 15–1 demonstrates this. The net result is that the T Tauri stars emit far more infrared radiation than we expect from their spectral types as determined by examining the absorption lines formed in their atmospheres. Presumably, the shell is matter left behind during the contraction which formed the star. And, possibly, some of the shell may have been ejected from the star.

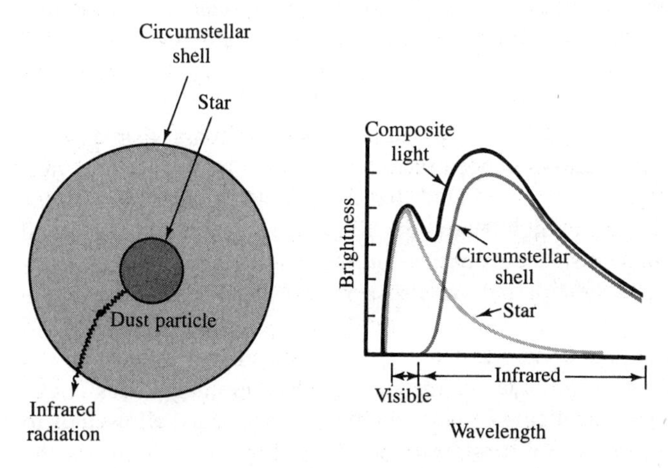

Figure 15–1 Explanation of the excess infrared radiation from T Tauri stars.

Cepheid Variables

Another group of variables, one in which the stars show very regular and repetitive light variations, are the **cepheid variables**. Figure 15–2 shows the light curve of Delta Cephei, the star after which this category was named. The cepheids include some stars which change in brightness by as much as six or seven times and some by as little as 10 percent. However, the more common variation is around two and one-half times, or roughly one magnitude. The period of light variation is different from star to star. A few cepheids have periods as short as several days, and there are some with periods over 100 days.

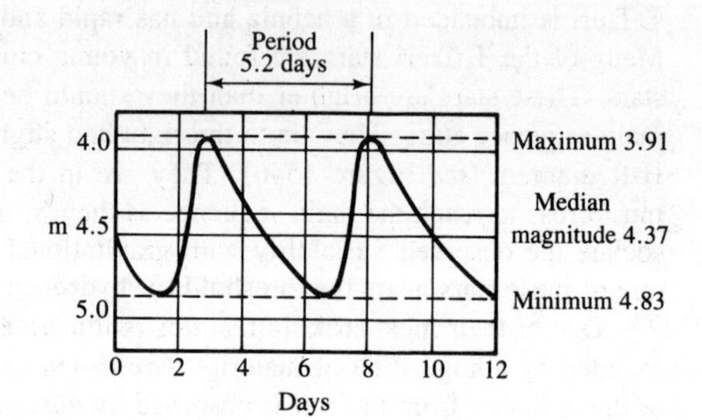

Figure 15–2 The light curve of δ Cephei.

The cepheids are yellow giants or supergiants around spectral types F to G. The variation in their light is accompanied by a variation in their radii. In the case of Delta Cephei, its radius changes by nearly 3 million kilometers in the course of one cycle. Since it is a giant star, however, this represents only a 7 or 8 percent variation in its radius. The cepheids are truly pulsating variables.

Being giants or supergiants, the cepheids are very luminous and can be seen at great distances. If they happen to be located in a distant group of stars which consequently appear faint in the sky, they are readily identified by their light variation. These two properties of the cepheids, along with a discovery concerning them in 1912 by Henrietta Leavitt, have made them quite important in estimating the size of the universe.

While studying the Magellanic Clouds, two large systems of stars visible in the southern hemisphere of the sky, Leavitt concentrated her attention on the cepheid variables in the Small Magellanic Cloud. She found that the brighter a cepheid is on the average, the longer is its period of light variation. Since the cepheids in the Small Cloud are all nearly at the same distance from the sun, the relation between their average apparent brightnesses and their periods is also a relation between their average luminosities and their periods.

Unfortunately, Leavitt did not know the distance to the Magellanic Clouds, so she could not convert the apparent magnitudes of the stars to absolute magnitudes. Later studies of cepheids, not in the Magellanic Clouds and much closer to us, allowed us to fix the luminosity scale in the **period-luminosity relation**. Figure 15–3 shows the period-luminosity relation in its present form. The absolute median magnitude, halfway between the brightest and faintest, is plotted against the period in days. We now

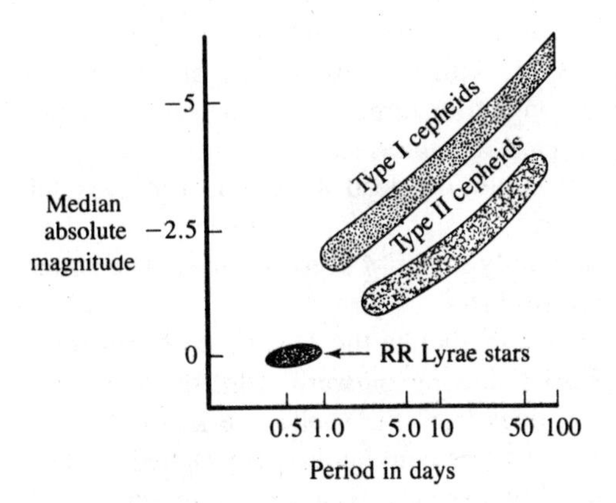

Figure 15–3 The period-luminosity relation for cepheid variables.

know that there are two kinds of cepheids. Those like the ones discovered in the Small Magellanic Cloud are called *classical,* or *Type I, cepheids.* The other group, called the *Type II cepheids,* are slightly less luminous than the Type I cepheids.

While the cepheids are interesting as a stage in stellar evolution, their major astronomical importance has been their use as distance indicators. Once the period of a cepheid has been measured, the period-luminosity relation gives its absolute magnitude. But in determining the period, we usually measure the star's apparent magnitude. We thus know the difference between the apparent and absolute magnitude of the star, which can be converted by a simple calculation to the star's distance in parsecs. Because the cepheids are so luminous, they can be seen at great distances. They are readily identified by their light variations. Consequently, they have been used to find distances much greater than are possible from trigonometric parallaxes. In fact, they were the first objects to allow us to find distances to other systems of stars not unlike the system of stars to which our sun belongs.

There is another type of pulsating variable quite similar to the cepheids in the nature of their light variation. Their periods are much shorter, however—always less than one day. They were first called short-period cepheids but are now more commonly called the **RR Lyrae stars** after RR Lyrae, the prototype star for this group. Since they are very common in globular clusters, they are also frequently called *cluster variables.*

The RR Lyrae stars are believed to pulsate much as the cepheids do. But, being smaller, denser stars, the periods of their pulsations are much shorter. They are not quite as luminous as the faintest Type II cepheids. Even though the periods are different from star to star, they all have nearly the same median luminosity; therefore, they are useful in finding the distances to globular clusters. If one is identified in a cluster and its apparent magnitude determined, we in essence know the distance to the cluster, since all RR Lyrae stars have about the same absolute magnitude. The RR Lyrae variables are hotter than the cepheids, running from spectral type A to F. A comparison of the RR Lyrae and cepheid positions in the H-R diagram is shown in Figure 15–6 on page 291.

Mira Variables

There are also highly evolved red giants which vary in brightness. They are not as regular in their light variation as the cepheids, and some of them are so unpredictable that they are called *irregular variables*. The ones that have some degree of regularity are called **Mira variables** after the prototype star, Omicron Ceti, which is also known as Mira, "the wonderful." The periods of the Mira variables range from three months to over two years, and they are frequently referred to as *long-period variables*.

The Mira-type stars are huge. Mira, for example, if placed at the sun, would extend beyond the orbit of Mars. With this large size, even though they are quite luminous, they have low surface temperatures and are found mainly in the M spectral types. The hotter ones are the brighter ones, as shown in Figure 15–6.

The Mira variables have large circumstellar shells like the T Tauri stars, but for a different reason. The shells consist of material flowing outward from the stars. By measuring the Doppler shifts in the emission lines from the shells, we know that the material is moving outward fast enough that it must be continuously resupplied by new matter from the stars, or the shells would quickly dissipate. The shells cool enough in expanding away from the parent stars that dust particles are able to condense out of the gas. These particles are continually warmed by the starlight but are still cool enough to radiate mostly in the infrared portion of the spectrum. Mira variables are in a post–main-sequence stage of evolution where they are undergoing a pronounced mass loss through the continued feeding of their circumstellar shells.

There is apparently a kind of recycling process at work. The star forms from interstellar matter composed of dust and gas. In contracting, the star heats up, vaporizing the dust and becoming completely gaseous. After spending its appropriate time as a main-sequence star, it evolves into a red giant. It then develops a circumstellar shell, ejecting some of its mass into space. This matter, in cooling, forms dust particles which return to the interstellar medium. Some astronomers argue that this process is so efficient that the dust particles we see in interstellar space now are all composed of material that has been processed and manufactured in cool red giants like the Mira variables. Others argue that most of the dust particles condensed directly out of the cold gas between the stars. The argument is not yet settled, but both groups agree that dust particles are destroyed when stars form. The stars are much too hot as main-sequence stars for the dust to survive in solid form.

Planetary Nebulae

Stellar material is also being returned to space in the **planetary nebulae**. They are so named because when they were first discovered, their telescopic appearance was similar to the appearance of Uranus and Neptune. More modern observations indicate that they always consist of an expanding nebula surrounding a faint, but very hot, star. The ultraviolet radiation from the star ionizes the gas in the nebula, causing it to glow, predominantly in emission lines characteristic of the constituents of the nebula. The nebula itself frequently is brighter in the visible portion of the spectrum than the star which illuminates it. Figure 15–4 is a photograph of the Helix Nebula, in which the central star is faintly seen compared to the brightly glowing nebula around it. At first, that is puzzling until you realize that the extremely hot central star radiates most of its energy in ultraviolet light, which, in turn, supplies the energy which causes the nebula to glow. Even though only a part of the ultraviolet light is converted to visible light, that fraction is still much greater than the visible light which we receive directly from the star.

Figure 15–4 The Helix Nebula—a planetary nebula.

On the average, a planetary nebula contains about one-quarter solar mass. But this material is spread very thinly over a more or less spherical region with a diameter around one light-year. The emission lines from the nebulae are Doppler shifted, indicating that the typical planetary nebula is expanding with a velocity between 15 and 30 kilometers per second. Since planetaries are usually found in regions of low dust and gas concentration, we interpret the nebulae as shells of matter ejected by their central stars. They represent a more abrupt ejection process than that which forms the Mira shells. This rather abrupt ejection leads to some rather beautiful objects such as the Ring Nebula (Plate 13A) and the Dumbbell Nebula (Plate 13B).

In general, after the luminosities and temperatures of the central stars have been measured, the application of Stefan's law to estimate their radii indicates that the radii of the stars are related to the sizes of the surrounding nebulae. On the average, we find that the larger the nebula is, the smaller the central star is. But the expansion velocities of all planetary nebulae are quite similar; thus the larger nebulae must be the older ones, because they have been expanding longer. For the central stars to become smaller as the planetaries become larger, the central stars must contract with time. The observed radii of these stars range from that of the sun down to as small as one percent the radius of the sun.

The locations of the central stars in the H-R diagram are shown in Figure 15-6. They are similar in temperature to the hottest stars in the main sequence but, of course, are less luminous because of their smaller sizes. They fall along a band which makes them appear to be contracting toward the white-dwarf region of the diagram.

The white dwarfs are truly remarkable. They are extremely dense; they can contain as much mass as the sun but can be as small as the earth in size. A cubic centimeter of material from a white dwarf would weigh a ton, or more, if placed on the earth. They are hot throughout, and their atoms are ionized. The free electrons are packed so tightly together that a white dwarf can no longer contract under its own gravity. The gaseous material, therefore, behaves as if it were a solid; that is, it is incapable of compression. Furthermore, the freely moving electrons make the gas a very efficient conductor of heat, giving the gaseous material the characteristics of a solid metal. We know of no source of new energy in a white dwarf, so it simply shines by its internal heat which is conducted to and radiated away from the surface. The rate at which it cools is rather slow, however, and the cooling slows down even more as the temperature falls. Consequently, it takes a very long time for a white dwarf to lose its energy. As a white dwarf cools, it will become redder and redder, eventually cooling to the point where it becomes invisible. It will then be a cold **black dwarf,** never again to radiate so it can be seen.

Because the white dwarfs cannot change much in diameter as they cool, their evolutionary tracks on the H-R diagram are straight lines like those shown in Figure 15–5. Their paths are, therefore, lines of constant radius. White dwarfs with different radii obviously must evolve along different lines. Theoretical calculations predict that more massive white dwarfs evolve along lines of smaller radius. The most massive white dwarf star possible would be the one corresponding to the line of zero radius. The mass of such an object would be about 1.25 solar masses, called the *Chandrasekhar limit.* A white dwarf with any larger mass is impossible and cannot exist.

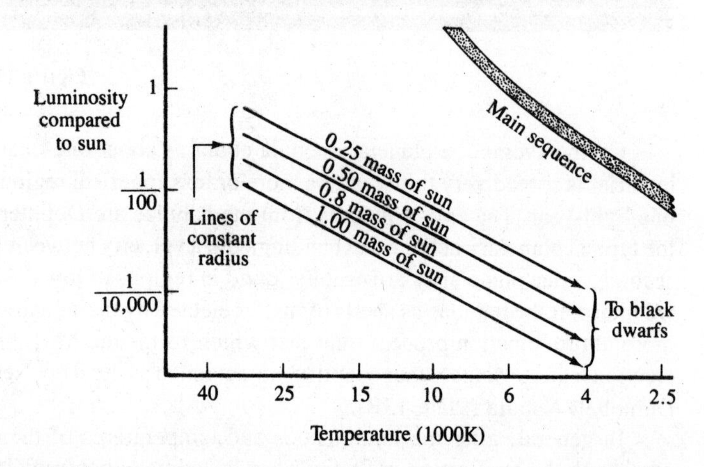

Figure 15–5 The evolutionary tracks of white dwarfs.

The locations of the different types of pulsating variables and the central stars of planetary nebulae and white dwarfs in the H-R diagram are shown in Figure 15-6. Is there any way to tie these seemingly very different objects together? Can an object in the lower left-hand corner of the diagram, a white dwarf, have any connection with a Mira variable in the upper right-hand corner? Our current theory of stellar evolution indicates that there is a connection, but the precise connection will depend greatly upon the mass of the star when it reaches the main sequence.

Evolution and Variability

The theory of stellar evolution predicts that stars of moderate mass, from about 0.5 to 2.25 solar masses, will pass through most of the phases of variability that we have described. The evolutionary path of a star of 1.25 solar mass is shown in Figure 15–7.

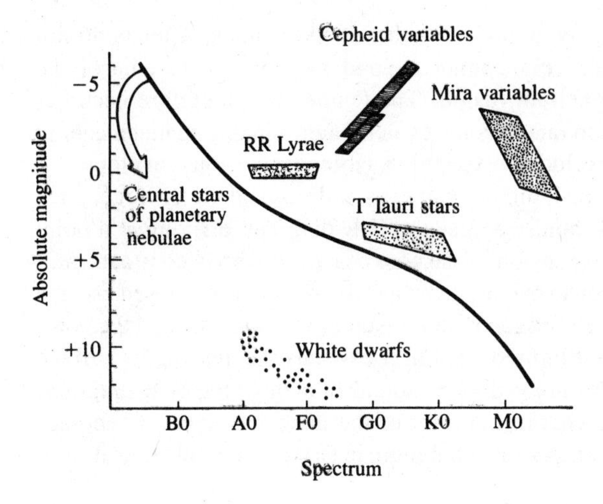

Figure 15–6 The location of post– and pre–main-sequence stages in the H-R diagram.

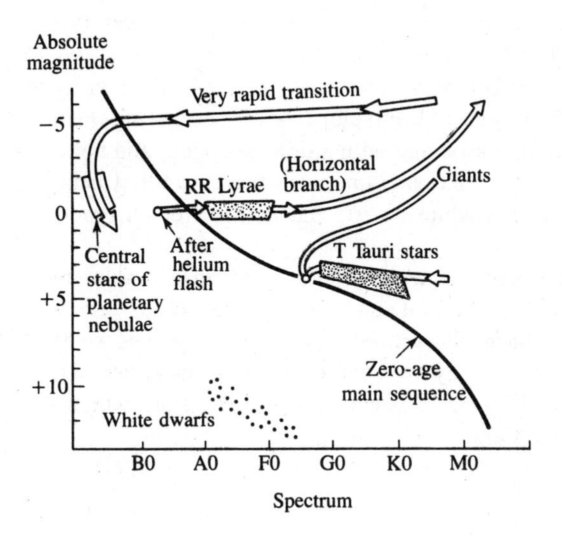

Figure 15–7 The evolutionary track of a 1.25 solar mass star.

Such a star forms by the gravitational contraction of interstellar matter. In the later stages of its contraction, it enters the region of the T Tauri variables. Within a million years after becoming a T Tauri variable, the central temperature of the star is hot enough for hydrogen fusion to begin. The star then settles on the main sequence, where it remains for about the next 3 billion years.

When the hydrogen in the central region becomes completely exhausted, the hydrogen fusion takes place in a shell surrounding what has now become a hot helium-rich core. The core contracts and the envelope of the star surrounding the hydrogen fusion shell expands, turning the star into a red giant. The helium core continues to heat up gravitationally until, abruptly, it is hot enough to fuse helium into carbon. The onset of the fusion takes place rapidly—in a matter of hours—and is called the **helium flash.** The flash occurs because when the helium core becomes hot enough, it is so compressed that it has the property of a white dwarf—namely, a gas that behaves like a metal. Metals do not expand much when heated, and the core cannot expand fast enough to control the temperature increase caused by the helium fusion. The temperature therefore becomes even higher, making the helium fusion rate become even greater. Finally, so much energy is released that the core virtually explodes, expanding violently into the envelope.

With the expansion of the core, the luminosity of the star drops and the hydrogen-rich envelope contracts. The star moves rapidly across the H-R diagram, arriving at a point near the upper main sequence. The expansion of the core ceases. It slowly contracts until it is again hot enough for helium fusion to occur, but a flash does not take place because the core is not sufficiently compressed. Hydrogen fusion resumes in a shell around the core. The star keeps an essentially constant luminosity, but it expands, decreasing its surface temperature and moving to the right along the horizontal branch of the H-R diagram. When the star reaches a point somewhat to the right of the main sequence, it becomes unstable, pulsating as an RR Lyrae star. A somewhat more massive star would pulsate as a cepheid variable instead.

After the helium in the very center of the star has been completely converted to carbon, a helium fusion shell develops inside the continuing hydrogen fusion shell. The inert carbon shell contracts, raising the temperature in the helium and hydrogen fusion shells. The luminosity shoots up, and the star's envelope expands greatly. It begins to pulsate as a Mira variable. Thus, the star begins to lose mass. The pulsations grow in amplitude until virtually the entire hydrogen-rich envelope is ejected as a planetary nebula. The hot carbon core, highly compressed and with a thin hydrogen- and helium-rich surface layer, remains as the central star that illuminates the expanding nebula. Ultimately, the central star contracts into a white dwarf; and its life, for all practical purposes, has ended.

Stars can also become white dwarfs in other ways. A star of less than 0.35 solar mass will contract directly to a white dwarf after its giant stage, because its helium core will never get hot enough to set off a helium flash. Stars more massive than the ones we have been discussing can become white dwarfs if they are components of binary stars. Their evolution is, in essence, short-circuited by the process of mass exchange which we discussed when we talked about binary stars.

The Problem of Different Initial Chemical Compositions

We have already noted in our discussion of stellar spectra that some stars have more heavy elements in their atmospheres than other stars. This difference in chemical composition is thought to arise in the interstellar matter from which the stars formed.

When we examined the color-magnitude diagrams of globular clusters, we saw that their main-sequence turn-off points corresponded to ages of about 10 billion years. Analyses of the spectra of objects in globular clusters, including one known planetary nebula, show that all of these objects have nearly the same chemical composition; but their composition differs from that of the interstellar medium, the sun, other planetary nebulae, and the stars found in young clusters. The abundance of the heavier elements in the old stars of the globular clusters is only 1 to 10 percent that found in the young stars and the interstellar matter from which they were formed not too long ago. The interstellar medium must have been enriched by heavier elements since the globular clusters were formed.

Could this enrichment have come from the material recycled to the interstellar material by the Mira variables and the planetary nebula process? No, because the material ejected in both of these mechanisms comes from the hydrogen-rich envelopes of the stars. The transmuted heavier elements end up trapped in white dwarfs. At best, during the development of a planetary nebula, only a very slightly higher amount of nitrogen is returned to the interstellar medium than was present at the formation of the original star. So these two possibilities fail.

Stars can also lose mass at times other than during the more advanced stages of evolution. The sun, which is a middle-aged star, is losing matter through its solar wind. Satellite observations of O-type main-sequence stars show that these stars have stellar winds that carry matter into the interstellar space. There are also groups of rather rare stars, such as the Wolf-Rayet stars, that eject mass. Not only is the matter given off by all of these objects insufficient to account for the replenishment, it also comes from their atmospheres and is not significantly different in composition from the interstellar medium.

Possibly the cataclysmic variables known as novae can supply the needed heavy elements. A nova results in the rapid ejection of material from a white dwarf member of a binary system. For example, the material ejected in a nova which occurred in the constellation of Perseus can be seen as a nebular shell in Figure 15–8. The ejected matter may contain a high proportion of heavy elements, but a typical nova only ejects one-thousandth or less solar mass at each outburst. The number of novae occurring each year is too small to change the chemical composition of the interstellar medium significantly.

Figure 15–8 The shell of matter ejected by Nova Persei.

There is one last mechanism to consider—the supernovae process. A supernovae is a spectacular outburst of radiation and matter. At the time a supernova is at its brightest, it can outshine the equivalent of billions of suns. It has been estimated that a supernova occurs in our galaxy of stars about once every thirty years. But because of the dust in the interstellar matter, many are obscured and are undetected. Every 200 years or so, one does occur close enough to the sun to be seen. When this happens, it may be bright enough to be seen with the unaided eye, even during the daytime. One such supernova occurred in 1054 and was recorded by Chinese astronomers. They called it a "guest star" because of its sudden appearance in the constellation Taurus. It was visible even in the daytime sky for about three weeks. During the short time of its great brilliance, it radiated about as much energy as the sun has radiated in the past 500 million years.

Today we see the remnant of this supernova as the Crab Nebula, shown in Plate 14A, at the location recorded by the Chinese astronomers about a thousand years ago. The nebula, however, is too faint to see without a telescope. It contains at least 0.1 solar mass and is still expanding away from the site of the explosion with a velocity over 1000 kilometers per second. We are sure that the nebula was ejected by the supernova, because if we imagine it to reverse its expansion velocity and count time backwards, it would take just enough time to appear to coalesce in 1054, the date of the supernovae's discovery.

Although we can identify sites of previous supernovae, such as the Crab Nebula and the Veil Nebula in Cygnus (Plate 14B), no positive identification of an object at the location of a supernova prior to its explosion has ever been made. Because we do not know what a pre-supernova looks like, what we think we know about the cause of a supernova is mostly conjecture. The most generally accepted theory is that a supernova occurs in the last stages of a star with many times the mass of the sun. The core will be composed of almost pure carbon by the time helium fusion has exhausted the original helilum core. In such a massive star, the core temperature can increase enough to cause fusion to proceed to heavier elements. Ultimately, the core is converted to iron. The iron cannot fuse to heavier elements in exactly the same manner as it was built up from lighter elements, because the mass of a nucleus heavier than iron is greater than its component parts. In other words, to form the elements heavier than iron, energy is required, in contrast to the release of energy when the lighter elements are formed. Therefore, the iron core cannot release nuclear energy and must continue to collapse gravitationally. The pressure in the core increases to the point that all of the electrons present in the core are forced to combine with protons to form neutrons. The core is consequently transformed to a neutron core. When it becomes so small that it has the characteristics of a solid body, it undergoes very little further contraction. The material in the envelope, however, continues its violent collapse. Gaseous matter from the envelope piles up on the surface of the neutron core and recoils. This recoil of the collapsing envelope results in a violent expansion—that is, an explosion. The stellar envelope is quickly heated to a high temperature. Before the expanding envelope can again cool, a great many nuclear reactions occur, forming nuclei of all the chemical elements in different amounts. Thus, a very different chemical composition is returned to the interstellar medium than was originally collected to form the star. Calculations indicate that supernovae contribute about 100 times less material to the interstellar medium than the Mira variables and planetary nebulae, but what they do contribute appears to enrich the heavy-element component of the interstellar matter. Figure 15–9 is a schematic diagram illustrating the recycling of material in interstellar space. It should be noted, however, that some supernovae appear to have arisen from older stars which were not as massive as those we

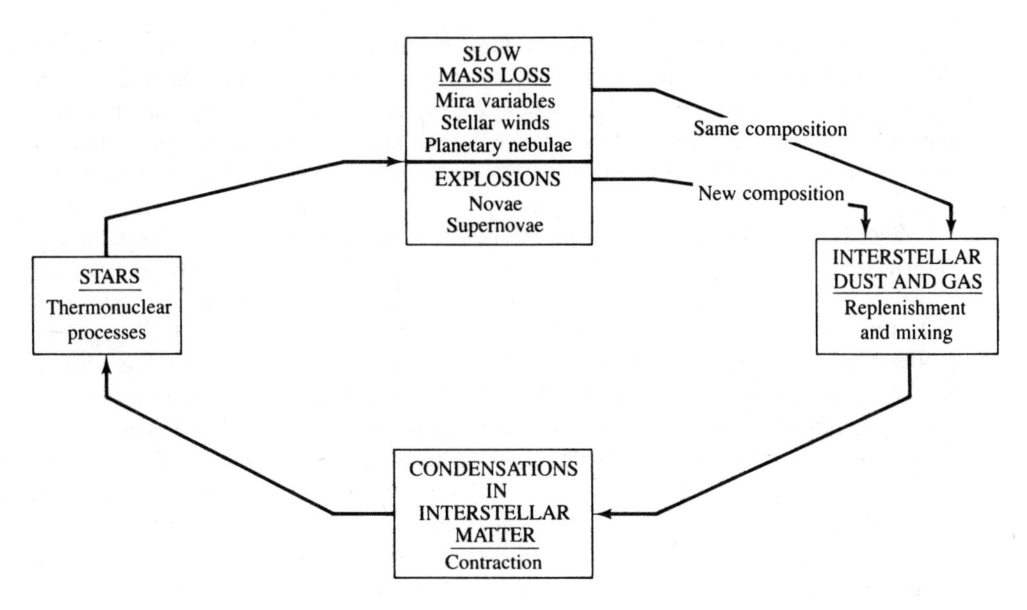

Figure 15–9 The cycling of interstellar matter.

have just considered. There is as yet no generally accepted explanation for these supernovae events.

Possibly the most striking feature of a supernova explosion is the remnant star left behind. The former neutron core is now a **neutron star.** The remnant neutron core is so dense that one cubic centimeter of its material would weigh one billion tons. A solar mass in the form of a neutron star would occupy a sphere only 20 kilometers in diameter.

The theory of neutron stars was developed in the 1930s, but they were considered to be entirely speculative until 1967. During that year Jocelyn Bell, a student at Cambridge University, unexpectedly discovered sources of radio radiation that emit short, rapid bursts of energy at extremely regular intervals measured in fractions of a second. The sources are called **pulsars** because they emit pulses of energy. The pulsars are interpreted as small objects, the expected size of neutron stars, which are undergoing rapid rotation with periods equal to the periods of their pulses.

A pulsar has been found near the center of the Crab Nebula. Its period is only 0.033 second. The period has been observed to be increasing gradually, indicating that the rate of rotation is decreasing. Calculations show that the rotational energy lost by this slowing down is just about the same as the energy radiated by the Crab Nebula. While the precise mechanism for the transfer of this energy is not yet known for certain, the mystery of how the Crab Nebula has continued to shine for nearly a thousand years appears to be solved. The pulsar in the Crab Nebula must be the neutron star remnant of the supernova of 1054.

Black Holes

The **black holes** are the last extreme objects to discuss. Theoretical calculations show that a neutron star could not contain more than about three solar masses. If a more massive neutron star tried to form, the object's self-gravity would overcome its structural strength and it would undergo a contraction that nothing could stop. The gravitational field about 10 kilometers away from the object would be so intense that not even light could escape from it. The name *black hole* comes from this property; the object could not radiate energy into space. If you tried to illuminate it to see it, the light would simply disappear into the black hole. It would be a black hole in space, completely invisible. It could be detected only by its gravitational effect upon nearby matter in the form of stars, dust, or gas.

Shortly after Einstein announced his general theory of relativity in 1916, it was shown that his theory predicted the possibility of black holes. The first observational evidence of their possible existence came with the discovery of the X-ray source Cygnus X-1 from satellite observations. The radiation from Cygnus X-1 can be interpreted properly only if the object is a binary system. Several astronomers think that it is a bright B-type star coupled gravitationally to a black hole, with a mass around eight times that of the sun. Figure 15–10 is a possible schematic diagram of the system. The B star exceeds its critical surface, and mass is transferred to the black hole. The material spirals into the black hole, accelerating to velocities close to the speed of light as it approaches the black hole. The matter moving at such speeds would be accelerated with great energy, causing it to become hot and to radiate strongly in the X-ray region of the spectrum by a complicated process of collisions between the accelerating particles. The observations are still too sketchy as yet to accept this interpretation as definitive. Possibly it is really just speculation; but, in any case, many astronomers are convinced that black holes must exist. If Cygnus X-1 does not turn out to be the first black hole to be discovered, they feel that surely others will eventually be found.

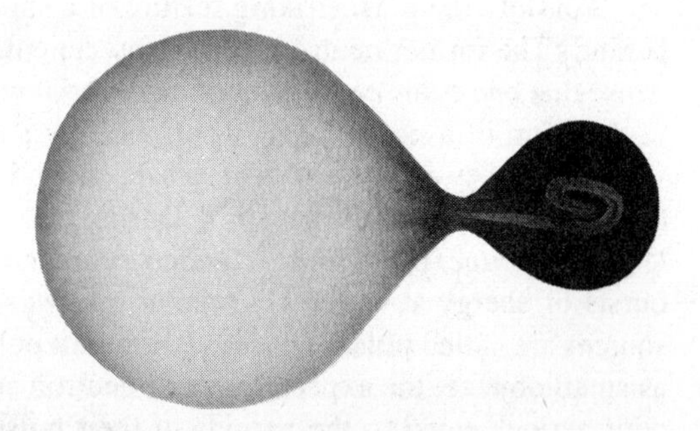

Figure 15–10 A possible model of Cygnus X-1.

Figure 15–11 illustrates the relative sizes of a giant star, a main-sequence star, a white dwarf, a neutron star, and a black hole. The size indicated for the black hole is at a distance from it where the escape velocity is just equal to the speed of light. Notice that there are extreme size differences between any two of these objects, except for the sizes of the neutron star and the black hole. If the material in a neutron star could be squeezed down just a bit further, it would collapse into a black hole.

Just as there are mass limitations on normal stars, there are also mass limitations on neutron stars, white dwarfs, and black holes which might evolve from normal stars. If the mass of the remnant core of a highly evolved star is less than about 1.25 times the mass of the sun, it will become a white dwarf. If the mass is between about 1.25 and 3 times that of the sun, the core will become a neutron star. Finally, according to theory, in a case where the mass is greater than about 3 solar masses, nothing can stop the collapse of the remnant core and it will become a black hole.

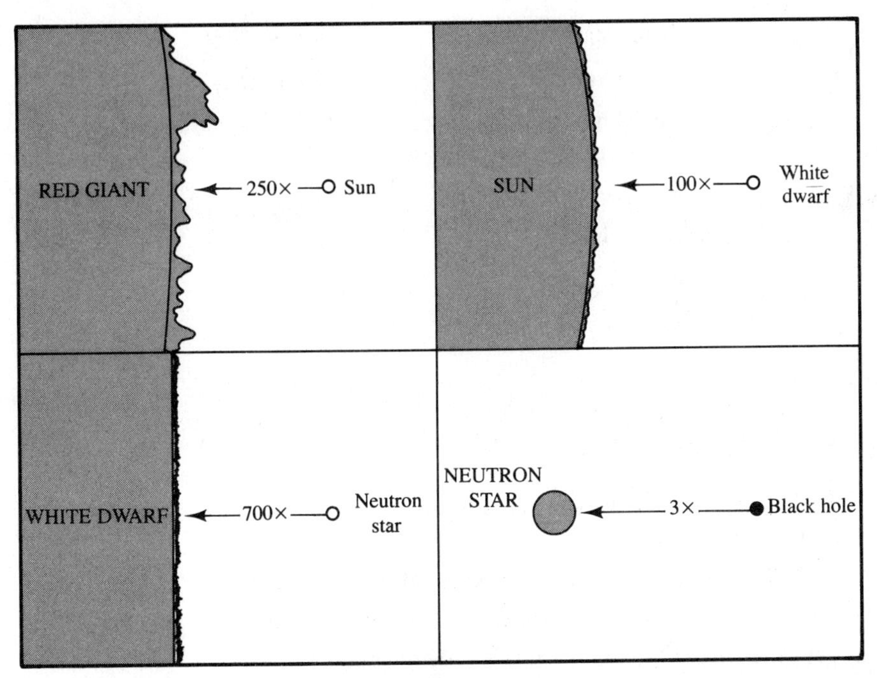

Figure 15–11 The relative sizes of giants, the sun, white dwarfs, neutron stars, and black holes.

In 1973 scientists, using artificial satellites to insure that the nuclear test-ban treaty was not being violated, reported that they had indeed detected signs of powerful explosions. They had detected photons of extremely high energy known as gamma rays. But the gamma rays did not originate on earth; in fact, they actually originated outside the solar system.

Continued observations in this region of the spectrum, known as *gamma ray astronomy,* have shown that the gamma-ray photons arrive in bursts varying in length from 0.01 to 10 seconds. The bursts are often multiple. Sometimes as many as five bursts will occur in quick succession, altogether spanning an interval of up to 100 seconds. Some of the pulses take less than 0.002 second to reach maximum strength. Such a rapid growth in intensity can only occur if the pulses are emitted from a small region, at most 600 kilometers in size.

Unfortunately, the resolution of the gamma-ray telescopes is not very good, so it has not yet been possible to locate the precise positions in the sky from which the bursts arrive. But, the observations do indicate that there must be many different sources scattered across the sky.

We do not yet understand just what causes the gamma-ray bursts. Most suggestions for the mechanism generating the extremely high-energy photons in a very small volume of space involve matter which falls into intense gravitational fields, such as those which would exist around neutron stars and black holes. This idea has received some observational support because one or two bursts may have come from the general direction of Cygnus X-1. Of course, this may be simply a case of pure coincidence; or, possibly, the gamma rays may eventually give us another important clue to just what is happening in this remarkable object.

KEY TERMS

T Tauri stars
cepheid variables
period-luminosity relation
RR Lyrae stars
Mira variables
planetary nebulae
black holes
helium flash
black dwarf
pulsating variables
cataclysmic variables
neutron stars
pulsars

Review Questions

1. How do T Tauri stars fit in with our theory of stellar evolution? What evidence supports this conclusion?
2. Why do T Tauri stars emit so much excess infrared radiation?
3. What changes accompany the variation in the light of a cepheid variable?
4. What does the period-luminosity relation demonstrate?
5. For what purpose is the period-luminosity relation so important? What quantities must be measured if we are to apply the relation for this purpose?
6. What are RR Lyrae stars? How are they different from cepheids?
7. What are Mira variables?
8. Where does the excess infrared radiation from Mira variables originate? Where did the material emitting this radiation come from?
9. What are planetary nebulae? How did they acquire the name "planetary"?
10. Where does the energy come from which causes a planetary nebula to glow?
11. In general, do the radii of the central stars of planetary nebulae get larger or smaller as the size of the nebula gets larger? What does that tell us about the central stars?
12. Where do planetary nebulae fit into our theory of stellar evolution?
13. How do the sizes of stars vary for different masses along the main sequence? For white dwarf stars?
14. Where does the energy come from to keep the surface of a white dwarf hot?
15. What is the greatest mass a white dwarf can have? Why is there an upper limit to the mass?
16. Of the following stages of evolution, pick out those that the sun has experienced or probably will experience. List them in the order in which they occur.

    black dwarf
    black hole
    cepheid variable
    main-sequence star
    neutron star
    planetary nebula
    red giant
    supernova
    T Tauri stage
    white dwarf

17. What is the helium flash?

18. Where do stars on the horizontal branch of the H-R diagram fit into our theory of stellar evolution?
19. What do we think is the chemical composition inside a Mira variable?
20. Which events in the lives of stars supply most of the hydrogen and helium returned to the interstellar medium? Which supply most of the heavy elements?
21. What differences in chemical composition do we see between the old and young stars? Why do these differences occur?
22. What is a supernova, and how is it different from a nova?
23. What is a pulsar?
24. What evidence is there that neutron stars are created by supernova explosions?
25. About how big is a neutron star?
26. What is a black hole?

Discussion Questions

1. How might we detect the expansion and contraction of cepheid variables?
2. How might we demonstrate that planetary nebulae are expanding?
3. Planetary nebulae would not be seen if their central stars were not so hot. Why is this high temperature required?
4. The apparent outer edge of the glowing gas in a planetary nebula is sometimes the actual edge of the gas. But in some cases we believe the gas continues to a greater distance—it just is not glowing. Why not?
5. Why do white dwarfs keep about the same radius as they cool off?
6. The lifetime of each stage in stellar evolution is related to the luminosity of that stage—the greater the luminosity, the quicker the energy which supplied the stage will be exhausted. How would the lifetime of a white dwarf compare with main-sequence and red giant stages of a star?
7. The following are the net results of nuclear reactions that can occur in stars:

   4 hydrogen to 1 helium nucleus
   12 hydrogen to 1 carbon nucleus
   56 hydrogen to 1 iron nucleus

   The maximum energy available by fusing hydrogen into heavier elements occurs when iron is formed. How efficient are each of the above examples? That is, what percentage of the initial mass is converted to energy? The masses of the nuclei are: hydrogen, 1.00813; helium, 4.00389; carbon, 12.0039; and iron, 55.9533.
8. If a star in the late stages of evolution has too much mass to become a white dwarf, what are its other possible end states?
9. The only supernova for which an outburst was clearly documented and which has left a pulsar remnant was the one which led to the Crab Nebula. The Crab pulsar is also the shortest-period pulsar known (out of about one hundred discovered so far). Why might we expect it to have the shortest period?
10. How might a black hole be detected?

16

Our Galaxy— The Milky Way

In the early part of this century, many astronomers outwardly scoffed at the possibility of the existence of matter lying between the stars and nebulae, but in truth they harbored feelings to the contrary. On this subject, the brilliant English astronomer Sir Arthur Stanley Eddington said of his colleagues: "They are like the guest who refused to sleep in a haunted room and who, when asked whether he believed in ghosts, replied: 'I do not believe in ghosts, but I am afraid of them.' "

In 1930 an American, R. J. Trumpler, provided direct evidence for the existence of a general stratum of interstellar matter. Trumpler constructed color-magnitude diagrams for many galactic star clusters. Since, in general, the clusters are at different distances, their main sequences plotted in apparent magnitudes do not coincide. The stars in the more distant clusters would appear fainter than similar stars in nearer clusters. By shifting the main sequences so that they did coincide, Trumpler was able to infer the relative distances of these clusters.

Trumpler had also measured the angular diameters of the clusters. Consequently, determining their relative distances allowed him to determine their relative sizes. An astounding result was obtained: the more distant the cluster, the greater its size. Certainly star clusters, even those chosen carefully by Trumpler to be similar, vary a bit in size, but why should the sun be in a preferred position with larger and larger clusters being progressively more distant?

Trumpler suggested an alternative explanation. Light from a distant cluster travels through interstellar material which absorbs some of it. On the average, the more distant the cluster, the larger is the fraction of the light absorbed. So the more distant a cluster, the greater is the difference between its brightness and the brightness it would have if its light had not been diminished by obscuring material. Thus, a cluster's distance and its size are systematically overestimated when absorption is not taken into account. Allowing for interstellar absorption, Trumpler was able to remove the discrepancy between the sizes of nearby and distant galactic clusters.

The Milky Way

Subsequent investigations resulted in the conclusion that the interstellar matter is a mixture of gas and dust particles. But no one was able to discern just how this material was distributed among the stars. As sometimes happens in science, a seemingly disconnected sequence of events leads to a coherent picture of an aspect of nature—in this case, how the interstellar material is distributed and, in fact, how the galaxy of stars to which the sun belongs is structured.

Karl Jansky, a scientist at the Bell Telephone Laboratory, conducted a series of experiments. He intended to find the source of static in the earth's atmosphere. He was able to separate the static into three distinct types: static from local thunderstorms, static from distant thunderstorms, and a steady hiss of unknown origin. Jansky concluded that the source of unknown origin was fixed in space somewhere beyond the solar system.

Surprisingly, the astronomical community was not excited by Jansky's discovery. Very few tried to explain the nature of the source of the radio waves he had detected. However, a radio engineer named Grote Reber did appreciate the importance of Jansky's discovery. During the middle of the 1930s, at his home in Wheaton, Illinois, Reber built a 30-foot diameter parabolic radio antenna, the first of that particular design which became the prototype for modern radio telescopes. After making measurements for some time, he went to the University of Chicago to take a course in astrophysics so that he could interpret his observational data. There the astronomers encouraged him. In 1940, Reber published a paper in the **Astrophysical Journal** in which he announced that the maximum intensity of radio-wavelength radiation came from the Milky Way. He also reported that radio radiation from bright stars was too weak to be detected. So the general radiation was not coming from stars, but, more likely, from material lying between the stars.

A Dutch astronomer, Jan Oort, was very interested in Reber's results. He suggested to a young colleague, H. C. van de Hulst, that he read Reber's papers and think about the possible existence of spectral lines in the radio region. Oort knew that if a line could be found, measurements of Doppler shifts in the radio spectra obtained from sources in different parts of the Milky Way could be found and used to map the motions and distribution of interstellar gas. Oort's discussions with van de Hulst took place during the Nazi occupation of the Netherlands, when most Dutch observatories were closed. Theoretical calculations could be made, however, and van de Hulst proceeded to study the structure of the hydrogen atom. He predicted that a hydrogen atom could undergo a transition between two energy states which would result in the emission of a photon with a wavelength of 21 centimeters, in the radio region of the spectrum.

In 1951 two physicists at Harvard, T. I. Ewen and E. M. Purcell, detected the 21-centimeter radiation predicted by van de Hulst. Their detection of the line did two things. It spurred the development of radio astronomy and allowed a detailed mapping of our Galaxy, which we see as the Milky Way. Radio waves are less likely to be absorbed than visible radiation, so parts of the Galaxy that are hidden from view by interstellar matter can be observed by using radio techniques.

The study of the interstellar medium is now as important as the study of stars, and radio techniques every bit as useful as optical techniques. To some of us, who have seen our knowledge of the interstellar medium and radio astronomy develop within our lifetimes, it is all quite amazing.

The system of stars to which the sun belongs is called **the Galaxy.** The name is derived from the Greek word *galaxias,* which means "milky way." The term **Milky Way** is properly reserved for the appearance of our system of stars on the celestial sphere, while the term *Galaxy* is reserved for its actual physical arrangement. It was not until after the invention of the telescope that we began to understand just what the Milky Way really is. The ancients thought that this faint band of light which circles the celestial sphere was a permanent cloud of glowing material in the sky. Galileo, with his crude telescope, was able to identify stars in the Milky Way that are quite invisible to the unaided eye. As larger and better telescopes became available, it became obvious that the Milky Way is made up of myriads of faint stars, the combined light of which causes its appearance on the sky. With this realization, the study of the Galaxy began—slowly at first, because as we have seen, it was not until quite late that the distances to the stars were comprehended.

The Shape of the Galaxy

What many would consider to be the first major step in finding the nature of the Galaxy took place in the eighteenth century with observations made by William Herschel. In Herschel's own words, "On applying the telescope to a portion of the *via lactea* [Latin for Milky Way], I found that it completely resolved the whole whitish appearance into small stars." Herschel systematically surveyed the sky by counting the number of stars which he could see with his telescope as he looked at several hundred regions selected to be more or less uniformly distributed about the celestial sphere. He found that his star counts were high when the telescope was pointed toward the Milky Way, but fell off dramatically when he observed 90° away from its plane. To explain his observations, Herschel proposed that the sun is at the center of a huge disc composed of stars with more or less uniform distribution throughout the disc. Figure 16–1 shows a cross section through such a disc. An observer at the center, looking towards points *c* and *d,* would see a large number of stars in his line of sight. Towards points *a* and *b,* however, the observer would see a much smaller number. Herschel's actual model was somewhat more complicated since he added extensions of stars in the directions where he obtained higher than average star counts along the Milky Way; but the basic disclike concept, with the sun at its center, seemed to account quite well for the observations.

This concept of the Galaxy held up until Harlow Shapley announced the results of his studies of globular clusters more than a century later. The globular clusters, being made up of tens or hundreds of thousands of stars, are quite bright and can be seen at great

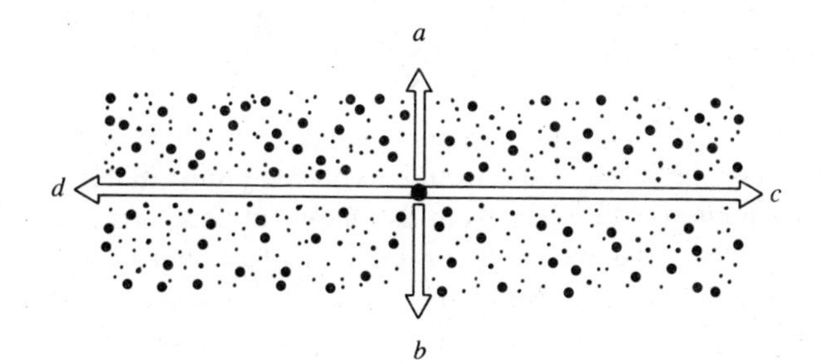

Figure 16–1 The disc model of the Galaxy.

distances. Many globular clusters contain RR Lyrae stars. Shapley used these stars to estimate the distances to the clusters in which they could be recognized. Noting that clusters which he found to be more distant had smaller angular distances, he calculated their real diameters from his measured distance and apparent diameters. Since there was not much variation in the calculated diameters from cluster to cluster, he assumed that all clusters had about the same diameter, which he took to be the average of those he had determined. He could now estimate distances for clusters in which he was unable to detect RR Lyrae variables by comparing their observed apparent diameters to the linear diameters of his average clusters.

Next Shapley noted, as shown in Figure 16–2, that the globular clusters are not distributed along the narrow band of the Milky Way. In fact, they appear to be centered on the constellation of Sagittarius, but can be found quite far from the plane of the Milky Way. By combining the directions to the clusters with his estimates of their distances, it was possible to make a three-dimensional plot of the clusters. Shapley found that the globular clusters have a spheroidal distribution in space, but the distribution is not centered about the sun. Instead, it is centered about a point in the direction of Sagittarius, which he estimated to be about 8000 parsecs from the sun. He took this point to be the center of the Galaxy.

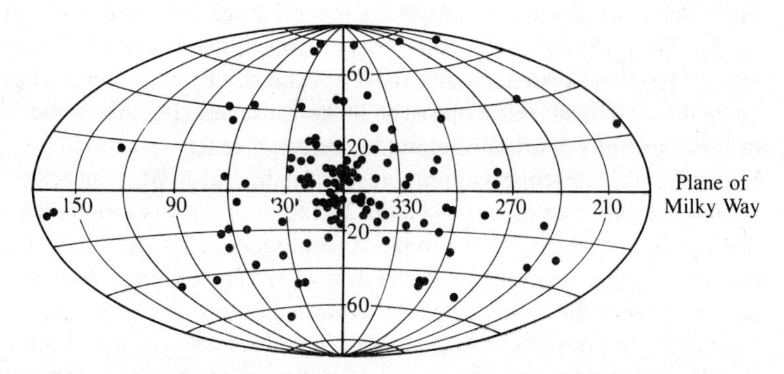

Figure 16–2 Distribution of globular clusters in the sky.

Shapley's interpretation has been confirmed by later observations. A modern cross-sectional diagram of the Galaxy is shown in Figure 16–3. Most of the stars form a flattened system, called the **galactic disc,** with the sun about two-thirds of the way from the center to the edge at a distance of 10,000 parsecs. The diameter of the disc is 30,000 parsecs, or about 100,000 light-years across. The central bulge in the disc is around 4000 parsecs across, while the disc itself is roughly 1000 parsecs thick away from the central region. The whole system is surrounded by a spherical halo, very sparsely populated by stars and more or less defined by the distribution of globular clusters. The halo has a diameter slightly larger than that of the disc.

The star density in the central bulge of the Galaxy is much greater than in the solar neighborhood. The average distance between stars would only be a few hundred astronomical units, instead of the hundreds of thousands of astronomical units near the sun. We might expect the bulge to be clearly visible in the constellation of Sagittarius, but it is not; radio and infrared observations must be used to detect it. The reason that the central bulge does not stand out clearly in the sky is because obscuring clouds of

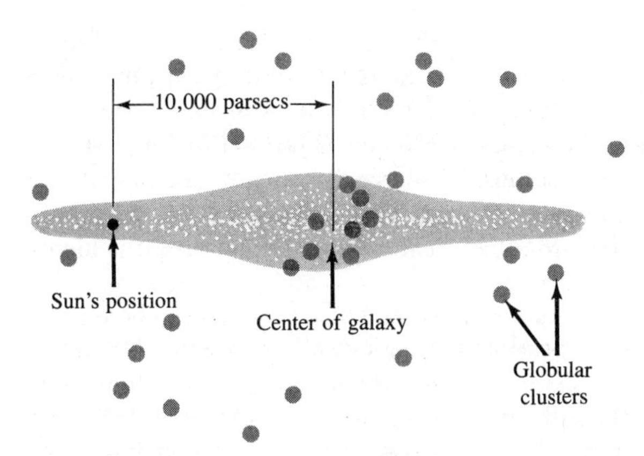

Figure 16–3 Cross section of the Galaxy, showing the globular cluster distribution.

interstellar matter block the intense light from the center of the Galaxy. The dust and gas of the interstellar matter is confined to the regions near the plane of the disc. Because of the obscuration, star counts like Herschel's give more or less the same number of stars in all directions around the Milky Way and obscure the eccentric location of the sun in the Galaxy. Herschel did, of course, notice the Great Rift which appears to divide the Milky Way into two bands from the constellations of Cygnus to Scorpius, nearly one-quarter of the way around it. In addition, he noticed other patches in the Milky Way where star counts were abnormally low. Herschel did not recognize the patches or the rift as evidence of interstellar matter; instead, he attributed them to an absence of stars or, as he put it, "holes in the heavens."

Interstellar Matter

The matter between the stars can be seen in a variety of ways. **Interstellar gas** can become evident when it surrounds extremely hot, luminous O and B stars. Plate 12A shows the Orion Nebula, which is a large cloud of interstellar matter in which several hot stars are imbedded. The intense ultraviolet light from the stars ionizes the gas, heating it to a temperature around 10,000K and causing it to glow. The nebula is spread over a large volume—its diameter is greater than one parsec. Although the gas is very tenuous, making it mostly transparent, the nebula contains more than 100 times as much mass as the sun.

The Orion Nebula has an emission line spectrum and hence is called an **emission nebula.** The bright lines in the spectrum are characteristic of the gas in the nebula and not of the stars which supplied the energy that ionized the gas. The emission lines seen in the Orion Nebula spectrum show that the gas has roughly the same chemical composition as the outer layers of the sun. The massive O stars supplying the ultraviolet radiation have very short main-sequence lives and could not have formed more than a few million years ago. In this short period of time they could not have moved very far from where they formed. Consequently, we think that the nebula is material left over when the stars were born and that the nebular region is one where star formation took place a relatively short time ago.

Emission nebulae are probably the most photogenic objects in the universe. In some cases, their names are derived from their shapes, such as the North American Nebula (Plate 12B) and the Horsehead Nebula (Plate 12C). Two other striking emission objects are the Trifid Nebula (Plate 13C) and the Eta Carina Nebula (Plate 13D). This last one, however, is not a typical emission nebula because its source of excitation is different than the usual nebula. While the exact source which supplies the energy by which the nebula shines is not yet known, it must differ from the usual O-type star because it is highly irregular in its energy output.

It has long been known that O and B stars are not distributed at random in the sky, but instead occur in groups called **stellar associations.** An association can contain from 5 to 50 stars in a volume of space between 30 and 200 parsecs in diameter. The stars of an association presumably formed at about the same time and are usually associated with nebulosities, like the Orion Nebula. There are two types of associations: the O-associations contain the luminous O and B stars, while the T-associations contain T Tauri stars. The latter are the stars that we identified with stars in the process of formation. About seventy associations have been catalogued. It is estimated that there may be as many as several thousand in our Galaxy, but most are hidden from view by obscuring interstellar matter.

The O-associations and the bright nebulosities found with them have been useful in discovering something about the structure of our Galaxy. The associations are always found close to the plane of the Milky Way. In other systems of stars like our Galaxy, the interstellar matter is seen to occur in **spiral arms** imbedded within the discs of these systems and spiraling outward from their central bulges. The distribution of interstellar material in these systems is easier to see than the distribution in our own system. The very interstellar matter in our Galaxy that we are trying to study obscures the galactic structure so that the spiral arms do not stand out in visible light. Once it was recognized that the hot O and B stars and bright nebulosities are always found in spiral arms in other galaxies, these objects, because of their great luminosities, were used to locate portions of spiral arms in our own system.

Using O and B stars and bright nebulosities as tracers, we have identified three short pieces of spiral arms. Figure 16–4 shows a crude map of spiral-arm segments. The sun lies in the segment known as the *Orion arm,* because much of its material appears to lie in the direction of the Orion constellation. The arm towards the galactic center, about 2000 parsecs away, is called the *Sagittarius arm*. In the opposite direction is the *Perseus arm,* about the same distance away as the Sagittarius arm.

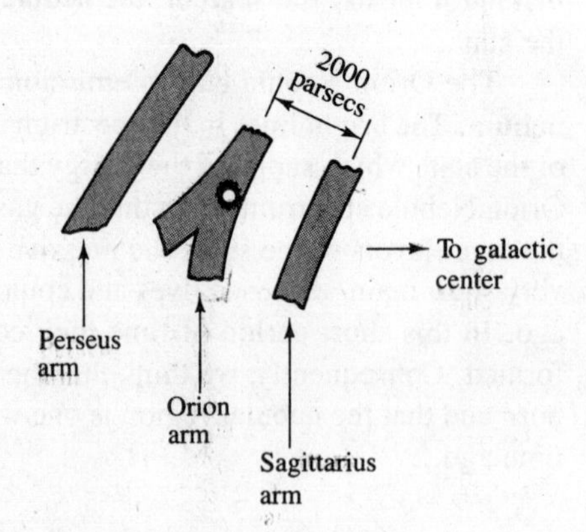

Figure 16–4 Spiral arms of the Galaxy in the solar vicinity.

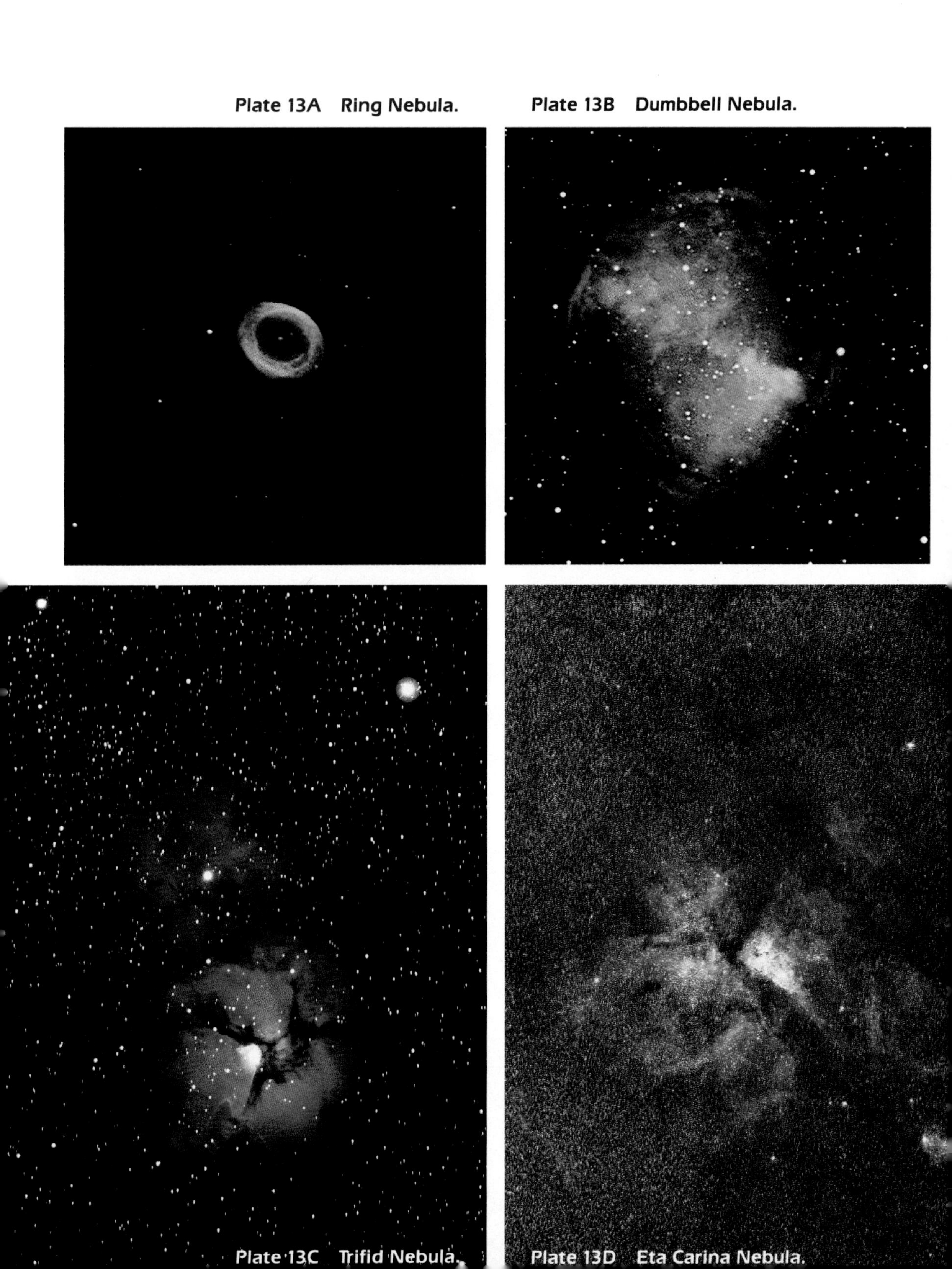

Plate 13A Ring Nebula.

Plate 13B Dumbbell Nebula.

Plate 13C Trifid Nebula.

Plate 13D Eta Carina Nebula.

Plate 14A Crab Nebula.

Plate 14B Veil Nebula.

Interstellar gas may also be detected in the spectra of some distant stars lying near the plane of the Milky Way. The spectra of these stars show narrow absorption lines like those in Figure 16–5. These lines are Doppler shifted relative to the corresponding lines due to absorption in the stellar atmospheres. They arise from absorption by clouds of interstellar gas moving relative to the stars and lying between us and the stars. Sometimes the lines are multiple, indicating that more than one cloud, each moving with a different velocity, lie in the line of sight to the stars where the multiple lines are evident. The correct interpretation of the lines, now called **interstellar lines,** came from a study of spectroscopic binary spectra in which they appear. The stellar lines in such spectra show continually changing Doppler shifts due to the orbital movements of the binary components. But the sharp interstellar lines have a constant Doppler shift and hence cannot arise in the stars' atmospheres. They are not due to circumstellar matter either, because they do not have the emission features which we would expect to see, particularly for hot B stars capable of ionizing any nearby circumstellar matter with their profuse ultraviolet radiation. The sharpness of the lines also confirms the interstellar interpretation. Such narrow lines can only arise from gas at extremely low pressure and temperature—just what we expect to find in the interstellar spaces far removed from bright stars.

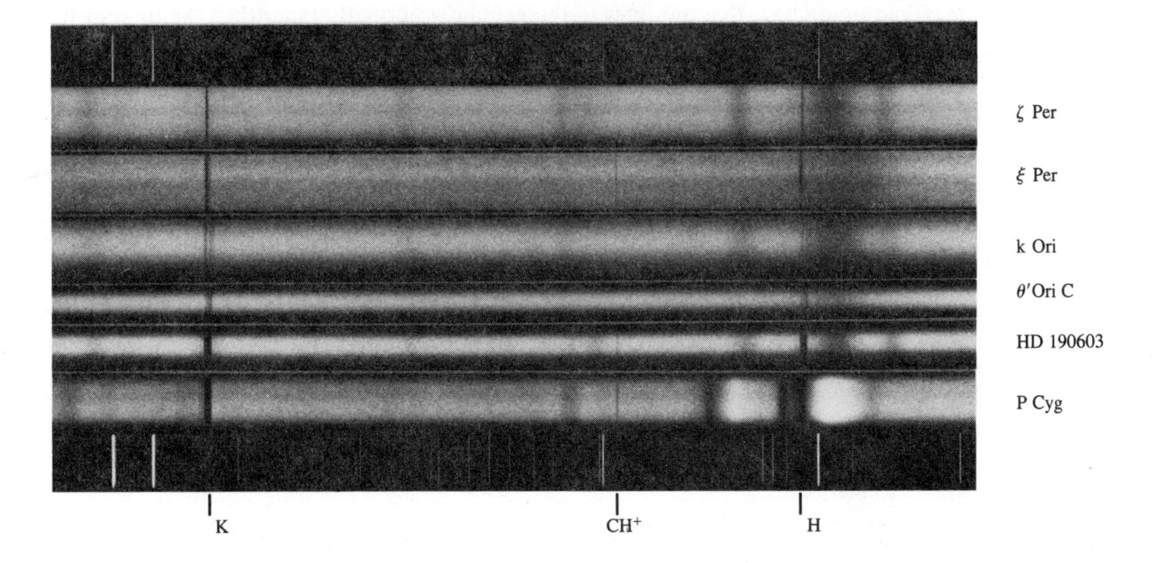

Figure 16–5 Interstellar absorption lines.

Besides the gas, there is also evidence that there are **interstellar dust** particles. The obscuration which we have already noted must be due to particles containing millions of atoms rather than simple atoms or molecules. A diffuse gas, like that in the interstellar medium, is transparent to visible radiation, but not to that which corresponds to the absorption lines of the constituents of the gas. Therefore, most of the light of a distant star can pass through the gas essentially undisturbed, and the apparent brightness of the star is not affected very much. At the extremely low densities of the interstellar medium, only particles are capable of causing the observed general decrease in light. These particles should be solid at the low interstellar temperatures.

The particles must be small, however, because stars seen in heavily obscured regions are always much redder in color than we expect from their spectral types. Particles comparable in size to the wavelength of visible light have the property of scattering short-wavelength light more efficiently than long-wavelength light. Larger particles, perhaps like a grain of sand or bigger, would affect all wavelengths of light equally. Thus, for example, when we see a yellow or red star which we know to be a B star from its spectrum and which should be bluish-white in color, we know that the light from the star has passed through a region of the interstellar medium containing what are commonly called *interstellar grains* or *dust particles*. In such a case we say that the star has been reddened. The degree of **interstellar reddening** is found by subtracting the expected color index of the star, based upon its spectral type, from its observed color index. Whenever color indices are used as an indication of stellar temperature, they must be corrected for any reddening to obtain true temperatures.

Some stars, like those in the Pleiades shown in Plate 11C, are immersed in a nebulosity. If the stars in a nebulosity are cooler than about spectral type B2, we do not see emission lines in the light of the nebula; instead, we see absorption lines which correspond to the lines of the star illuminating the nebula. Such a nebula is called a **reflection nebula** because we see it by starlight scattered from the dust grains which it contains. This is verified by the fact that the light of the nebula is normally bluer than the light of its illuminating star. When the star is positioned so we see it through the nebula, the light of the star is redder than it should be. This, of course, is what we would expect when the scattering occurs from particles the size of the interstellar dust grains. The star illuminating a reflection nebula must be quite bright or the nebulosity would be too faint or too small to see. Thus, unless the star is a supergiant, most reflection nebulae are seen in the vicinity of stars of spectral type A through B2. If the star is hotter than B2, the ultraviolet radiation will be intense enough that the gas present will shine brighter than the light reflected by any dust.

Dust particles absorb energy as well as scatter it. If the concentration of dust around a star is great enough, almost all of the visible light from the star will be absorbed, heating the dust particles. The particles will radiate more or less like black bodies, but because they are relatively far from the star, they will have temperatures much lower than the star itself. Most of the radiation from such nebulae will be emitted in the infrared, and the nebulae will be too faint to be seen in visible light. Such dust concentrations are called *infrared nebulae*.

Galactic Rotation

The stars and the interstellar medium are in continual motion. When we discussed the motions of stars in the solar neighborhood, we saw that they appear to move at random, relative to the local standard of rest. A representative space velocity of a star in the sun's vicinity would be around 25 kilometers per second. But the globular clusters, in many cases, have radial velocities as large as 250 kilometers per second—ten times that of the nearby stars.

An analysis of the values and distribution of globular cluster radial velocities indicates that the sun, along with its neighbors—or if you prefer, the local standard of rest—is moving relative to a spheroidal distribution of globular clusters toward the direction of Cygnus. But, as indicated in Figure 16–6, this is in the plane of the disc of the Galaxy and at right angles to the direction of its center. The entire disc of the Galaxy seems to be rotating, with the solar neighborhood in particular moving on a circular path about the galactic center at a velocity of 250 kilometers per second. The high radial velocities measured for some globular clusters are not due to their actual motions but are just the reflection of the motion of the sun about the galactic center.

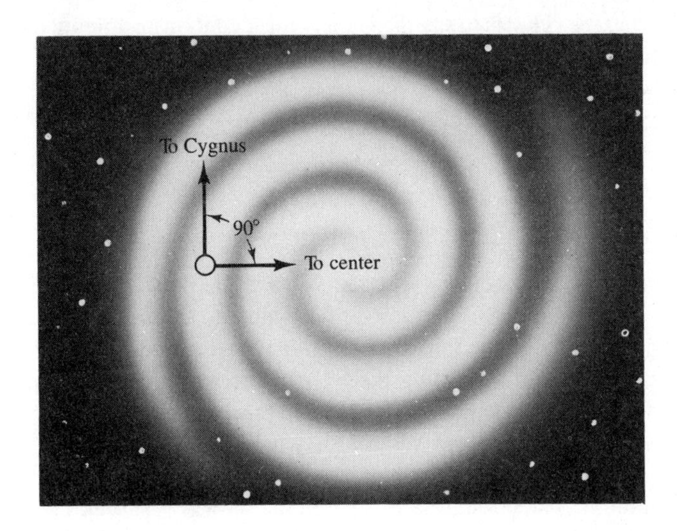

Figure 16–6 The galactic motion of the sun.

We have seen that the sun is about 10,000 parsecs from the center of the Galaxy. If we calculate the circumference of a circle with this radius, convert it to kilometers, and divide by the sun's velocity determined from the globular clusters, we obtain an estimate of how long it would take the sun to circle the Galaxy once. It turns out to be roughly 200 million years. The whole disc of the Galaxy does not rotate like a solid phonograph record, although we still use the term **galactic rotation** to refer to this motion. Instead, the stars farther from the center than the sun take longer for one complete trip, while those closer take less time. The Galaxy is said to undergo *differential rotation*. Put a different way, the larger the orbit of a star about the center of the Galaxy—at least for stars not too far removed from the sun—the longer is its period. This type of motion characterizes Keplerian motion. In fact, the change of periods with distance seems to fit that required in Keplerian motion quite well. The inner parts of the Galaxy do rotate much like a solid disc, however.

We can use Newton's form of Kepler's harmonic law to estimate the mass of our Galaxy. We assume that the mass of the Galaxy inside the sun's distance from the center controls the sun's motion gravitationally and behaves, for this purpose, as if it were concentrated at the galactic center. In other words, we treat the sun as a component of a binary system, with the rest of the Galaxy as the other component. Then we have that

$$M_{\text{Galaxy}} + M_{\text{sun}} = \frac{a^3}{P^2}$$

where M_{Galaxy} = mass of the Galaxy
M_{sun} = mass of the sun
a = the radius of the sun's orbit (10,000 parsecs, or about 2 billion AU)
P = the period of the sun's orbit (200 million yr)

Solving this equation for the sum of the mass of the Galaxy plus that of the sun gives an answer of 200 billion solar masses. That is to say the Galaxy must contain an amount of matter equivalent to about 200 billion stars to explain the motion of the sun about the galactic center.

Radio Studies of the Spiral Arms

With visible light, we can only see interstellar hydrogen when it has been heated by ultraviolet light from nearby stars. We do not see interstellar hydrogen absorption lines because the interstellar gas is too cool outside the heated regions for the hydrogen to be appreciably excited and hence capable of absorbing the Balmer spectrum. Most of the matter between the stars, which is hydrogen, is consequently inaccessible to us with visible light. Fortunately, neutral hydrogen can radiate in what is called the **21-cm radiation** in the radio part of the spectrum when the hydrogen is at the temperatures and pressures of the interstellar medium. The possibility of such radiation was first predicted in 1944 by the Dutch astronomer H. C. van de Hulst. It was not detected until 1951, however, when sufficiently sensitive equipment became available. Because the 21-cm radio waves are only slightly impeded by the matter which obscures our view of the Galaxy in visible light, we can use large radio telescopes to observe parts of the Galaxy otherwise hidden from us.

To utilize the 21-cm radiation for galactic studies, we use the Keplerian movement of the Galaxy, that is, its differential rotation. Figure 16–7 shows a schematic drawing of our Galaxy. Along the line of sight from the sun, the points labeled *A* and *B* will have different radial velocities relative to the sun due to the differential galactic rotation. Since the material farther out than the sun moves more slowly, the material at point *B* would show a larger apparent motion with respect to the sun than that at point *A*. If the neutral hydrogen is not uniformly distributed throughout the Galaxy, we would expect to receive 21-cm radiation only from the regions with the concentrations of hydrogen. Thus, in Figure 16–7, we would receive radiation from hydrogen near the sun and from hydrogen at points *A* and *B* if it is distributed as indicated in the diagram. But the radiation from *A* and *B* would be Doppler shifted because of the differential rotation. The 21-cm line from *B* would be shifted more than from *A*, and the radiation from the hydrogen near the sun would not be shifted at all. The strength of the displaced lines would indicate the amount of hydrogen present at each location. By using a Keplerian model of the Galaxy, we can relate the observed Doppler shift of the 21-cm line in the direction at which the telescope was pointed to the distance of the hydrogen causing it. Consequently, we can construct a map of how neutral hydrogen is distributed within our Galaxy.

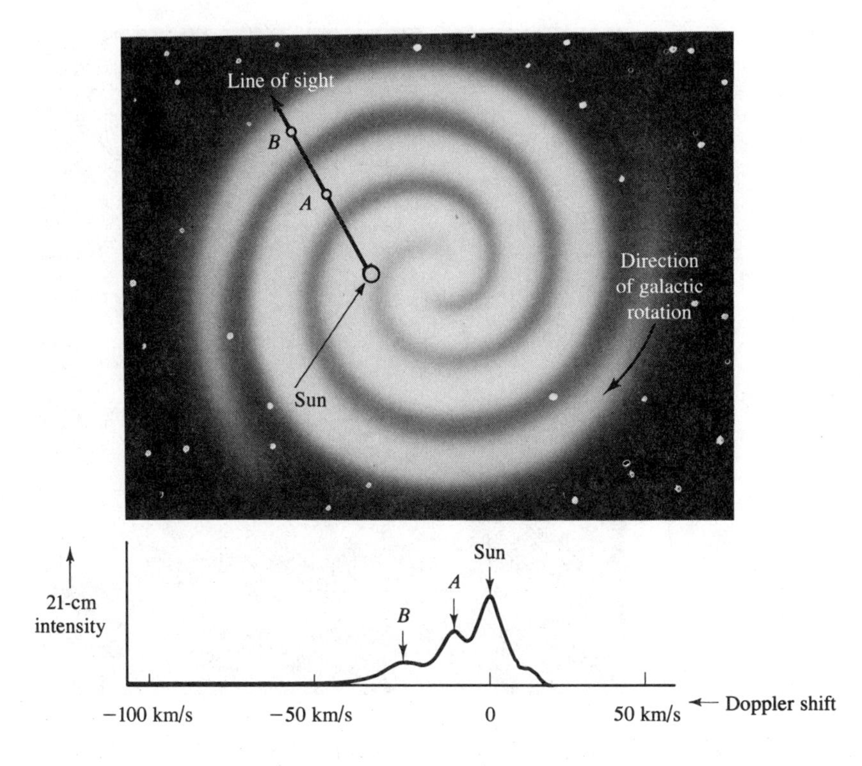

Profile of 21-cm line toward A and B

Figure 16–7 Diffferential galactic rotation measured by 21-cm observations.

A map of the Galaxy constructed this way is shown in Figure 16–8. The hydrogen is concentrated in long, narrow, spiral arms, not unlike those we find in some other galaxies. These arms, from radio measurements, agree quite well with the arm segments found from the O and B stars and nebulosities measured in visible light. The thickness of the arms is only about 300 parsecs; they therefore are imbedded within the disc of the Galaxy. The interstellar medium is concentrated within the arms. Estimates of the total mass of gas and dust in the form of interstellar matter indicate that it makes up about 5 percent of the total mass of the Galaxy. Most of the matter of the Galaxy has already formed into stars.

Stellar Populations

As our knowledge of the structure of the Galaxy developed, correlations between the characteristics of certain stars and objects like nebulae with their locations in the Galaxy become obvious. Walter Baade at Mount Wilson introduced the concept of stellar populations in 1944 while studying the Andromeda Nebula, a spiral galaxy that is frequently shown as an example of what our system probably looks like. Baade noted that the stars in the central bulge of the Andromeda galaxy were similar to the stars found in the globular clusters. However, these stars were not found in the spiral arms. Consequently, he

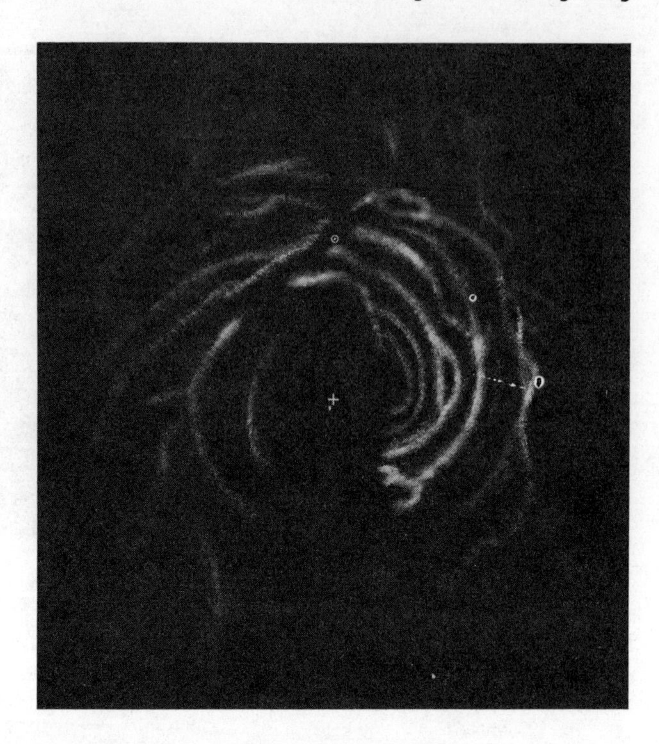

Figure 16–8 Radio map of the spiral arms.

proposed that there were two populations of stars. Population I contained the stars found in the spiral arms of Andromeda, while population II contained the stars of its central bulge and halo.

This same classification scheme works for our Galaxy. The spiral arms are marked by the **population I** objects, such as bright supergiants, O and B stars, young open clusters, and Type I cepheids. The **population II** objects consist of globular cluster stars, RR Lyrae stars, Type II cepheids, novae, and planetary nebulae. These objects are not restricted to the spiral arms but are found throughout the galactic disc. In the case of RR Lyrae stars, Type II cepheids, and globular clusters, they are members of the galactic halo. Today we recognize five populations. The *extreme population I* consists of the O and B stars, supergiants, T Tauri stars, young open clusters, Type I cepheids, and the interstellar dust and gas. This population is unevenly distributed in the spiral arms of the Galaxy. *Older population I* contains the sun, stars with high metal abundances, giants, and older open clusters. The members of this population show a patchy distribution somewhat restricted to the spiral arms. The *disc population II* stars contain the planetary nebulae, novae, stars in the galactic bulge, and RR Lyrae stars with periods less than four-tenths of a day. These stars are more or less smoothly distributed through the disc. Because the spiral arms are embedded in the disc, they are also found in the spiral arms. *Intermediate population II* stars, such as the long-period variables with periods less than 250 days, are strongly concentrated in an ellipsoidal distribution like a flattened sphere extending above and below the galactic disc. The last population is called the *halo* or *extreme population II*. It contains the globular clusters, Type II cepheids, RR Lyrae stars with periods longer than four-tenths of a day, and stars which are extremely metal-poor. These objects permeate the entire halo of the Galaxy.

The populations are related to the process of stellar evolution. The population I objects are young or middle-age stars or even the material from which stars may form. The population II stars, on the other hand, consist of stars which are old. Stars of all spectral types along the main sequence can be found among population I stars. Stars from the upper part of the main sequence—stars hotter than about spectral type F—are absent from population II objects. These stars will already have evolved away from the main sequence in the time since the population II stars were formed. A large proportion of the population II stars consists of horizontal branch stars, which we have seen are well evolved.

The chemical composition of the atmospheres of stars is well correlated with the populations. In the population I stars, the chemical composition is around 98 percent hydrogen and helium and about 2 percent of all other elements by mass. In the extreme population II stars, the elements other than hydrogen and helium account for less than 0.1 percent of the mass. As we have seen, this variation in compositions is a direct result of stellar evolution. Stars which have formed more recently contain more heavy elements, most likely because of the enrichment of the interstellar medium by the supernovae. The old population II stars are representative of the original chemical composition of the Galaxy, while the younger stars are representative of the present interstellar matter polluted by the supernovae which have already taken place.

The supernovae not only change the chemical composition of the interstellar medium, but they also are thought to provide the energy that drives the random turbulent motions in the interstellar matter. It is this turbulence that causes the multiple interstellar lines seen in some stellar spectra. Although supernovae are relatively infrequent, the tremendous energy released by their explosions is enough to send the interstellar matter into its chaotic motion.

Another feature of the interstellar medium which is not yet properly explained is the phenomenon of the high-energy **cosmic rays.** These are electrons and nuclei of atoms, mostly hydrogen and helium but including even heavier elements, which are moving at great speeds close to that of light. The observed characteristics of the cosmic rays indicate that they originate within our Galaxy, but their precise sources are not yet known.

In the last decade, radio astronomers have discovered the presence of complicated molecules in interstellar matter by observations made in the microwave part of the radio spectrum. This is exciting for two reasons. First, the complicated molecules strewn through some of the interstellar clouds are the basic building blocks from which even more complicated molecules found in living matter are made. But possibly more significant from the viewpoint of stellar evolution in our Galaxy, these molecules indicate that there are regions within the interstellar medium that are cold enough to permit gravitational condensations of material and hence lead to the formation of new stars.

KEY TERMS

Milky Way
the Galaxy
galactic disc
interstellar gas
emission nebula
stellar associations
spiral arms
interstellar lines
interstellar dust
interstellar reddening
reflection nebula
galactic rotation
21-cm radiation
population I
population II
cosmic rays

Review Questions

1. What observation led to the conclusion that stars in our Galaxy are generally in a flat system?
2. What two methods did Shapley use to find distances to globular clusters?
3. How are the globular clusters distributed in space?
4. Why did Herschel's star counts fail to reveal the galactic center?
5. On a clear, dark summer evening in the Northern Hemisphere, if you view the sky away from city lights, the Milky Way overhead shows up as two luminous bands separated by a dark band. What causes the dark band?
6. What is an O-association?
7. Emission nebulae are generally seen around O and B stars. Why aren't they seen around stars of other spectral types?
8. Why do O stars outline the spiral arms of our Galaxy so much better than G stars do?
9. How might you explain why a given spectral line shows up several times in a star's spectrum with slightly different Doppler shifts, with one line being much broader than the others?
10. In T Tauri stars, Mira variables, and infrared nebulae, we find objects which appear unusually bright at infrared wavelengths. What causes this excess infrared emission?
11. Why do globular clusters, on the average, have higher radial velocities than nearby stars?
12. Approximately how many trips has the sun made around the Galaxy?
    (1) 0.1
    (2) 1
    (3) 25
    (4) 5000
    (5) 1,000,000
13. How can the sun's motion be used to estimate the mass of our Galaxy?
14. Why has the 21-cm line of hydrogen been so useful in astronomy?
15. What have we used to trace nearby spiral arms?
16. Is most of the mass of our Galaxy in the form of stars or of interstellar clouds?
17. What is the primary cause of the differences between population I and population II stars?
18. List some of the ways population I stars are different from population II stars. To which group does the sun belong?
19. What apparently causes the random turbulent motion of interstellar gas?
20. At one time, astronomers expected that the gas atoms would not form into molecules in the extremely low-density, cold, interstellar gas clouds. How well has this expectation been confirmed in recent years?
21. What are cosmic rays?

Discussion Questions

1. What are two reasons why spectral lines from interstellar gas clouds are so narrow?
2. If a star appears rather red, how could you decide if it is a cool star, a hot star reddened by interstellar dust, or a hot star reddened by the Doppler shift of an incredibly rapid motion away from us?
3. List the ways that interstellar gas can be detected.
4. List the ways that interstellar dust can be detected.
5. In some reflection nebulae, the nebula close to the illuminating star is bluer than the star; but farther from the star it may actually be redder than the star. Why?
6. Interstellar clouds are thought to contain both gas and dust. Why are bright emission nebulae seen only around the very hottest stars, while reflection nebulae are discovered only around somewhat cooler stars?
7. If a star is completely surrounded by an opaque shell of dust, will the energy radiated by the dust shell be more than, the same as, or less than the energy radiated by the surface of the star?
8. How can we find the distance to an interstellar cloud from 21-cm radiation emitted by it?
9. Population I stars are confined to the galactic plane, while population II stars form a more spheroidal distribution. What does this imply about past changes in the shape of our Galaxy?
10. If a nearby star is observed to have an unusually large space velocity, is it more likely to be a population I or population II star?
11. Would spiral arms exist in a galaxy that did not have interstellar gas and dust clouds?

17

The Universe of Galaxies

Telescopic observations of the eighteenth century revealed large numbers of faint patches of light unlike stars in appearance. These objects were collectively called nebulae because of their veily, nebulous appearance. As it has turned out, some of them are nebulous concentrations of gas which are part of the Milky Way. Others, however, are actually star systems like our own Galaxy lying far beyond its confines. But before 1924, there was no conclusive proof that anything existed outside our own Galaxy. Even so, as far back as 1755, Immanuel Kant, a philosopher of great reknown, wrote that the nebulae are island universes—galaxies of stars like our own Milky Way.

In the nineteenth century, it was shown conclusively with the aid of photography that a number of nebulae are gaseous and are associated with stars in our Galaxy. As a consequence, there was a strong tendency to believe that all nebulae are part of the Milky Way system. Kant's idea fell out of favor.

The nebulae had been classified into two distinct groups according to their distribution in the sky. One class was called galactic nebulae because these objects, irregularly shaped and varying considerably in form, are distributed in the plane of the Milky Way. The other class was called extragalactic nebulae because they are distributed out of the plane of the Milky Way. The extragalactic nebulae appeared to have rather regular compact shapes, with spiral structure in some cases. In fact, until the early part of this century, all extragalactic nebulae were commonly referred to as spirals.

Around the turn of the century, bits of evidence began to crop up which indicated that the spirals are star systems like the Milky Way. Kant's idea was resurrected, at least in application to the spirals. These objects became the focus of great interest to astronomers, who separated into two camps regarding their true nature. One camp argued that they were merely parts of our Galaxy, while the other argued that these objects were themselves galaxies. The arguments were hardly

The Whirlpool Galaxy

quiet, but rather heated and emotional. Controversy over the nature of the spirals culminated in the "Great Debate" of April 26, 1920, at the National Academy of Sciences. The contestants were two California astronomers, H. D. Curtis of the Lick Observatory and Harlow Shapley of the Mount Wilson Observatory, who, because of his long tenure at Princeton and later at Harvard, came to represent the Eastern establishment. Shapley defended the more conservative view that the spirals are "truly nebulous objects" belonging to the Milky Way, while Curtis defended the more awesome, humbling idea that they are "island universes like our own Galaxy."

The debate was lively, but hardly conclusive. Certainly, neither Curtis nor Shapley persuaded the other to change his view, and the astronomy community remained divided in opinion. If you were to pick an underdog in the controversy, it would probably have been Curtis, who represented West Coast astronomy. Almost any idea of the time had to be endorsed by the Eastern establishment, centered on Harvard, to gain acceptance. Time changes things, however, and the West Coast has become at least as strong a center for the endorsement of ideas as the East.

Eddington, the greatest astrophysicist of the time, was neither East Coast nor West Coast, but English. His opinion concerning the nature of the extragalactic nebulae is expressed in a book, **Stellar Movements and the Structure of the Universe**. His clearness and intuitive brilliance are exhibited in his writing:

> The distribution of spiral nebulae presents one unique feature: they actually shun the galactic regions and preponderate in the neighborhood of the galactic poles. Indeed the mere fact that spiral nebulae shun the galaxy may indicate that they are influenced by it. The alternative view is that, lying altogether outside our system, those that happen to be in low galactic latitudes are blotted out by great tracts of absorbing matter similar to those which form the dark spaces of the Milky Way.
>
> If the spiral nebulae are within the stellar system, we have no notion what their nature may be. This leads to a full stop.
>
> If, however, it is assumed that these nebulae are external to the stellar system [Milky Way], that they are in fact systems coequal with our own, we have at least an hypothesis which can be followed up, and may throw some light on the problems that have been before us. For this reason the "island universe" theory is much to be preferred as a working hypothesis; and its consequences are so helpful as to suggest a distinct probability of its truth.[1]

Eddington's intuition was very sound, but intuition does not constitute proof. Regarding the true nature of the extragalactic nebulae, the issue was resolved by a man, a telescope, and a particular type of star. The significance of the issue is obvious. If the spirals are part of our Galaxy, then the Milky Way quite likely encompasses the entire universe. This is a large universe to be sure, being 100,000 light-years in size. But, if the spirals are themselves galaxies, then they are strewn through space at distances as great as tens of millions of light-years from their nearest neighbors. The size of the Milky Way is minute compared to the size of a universe which contains galaxies. Each person's ideas of the extent of the universe differed

---

[1] London: Macmillan and Company, Ltd., 1914, pp. 242–43.

drastically, depending upon the particular view held concerning the nature of the spirals.

The man involved in the resolution of the issue was Edwin Hubble, who, like Shapley, was an astronomer at the Mount Wilson Observatory in southern California. In 1917 the 100-inch telescope, which Hubble was to use, was completed. Southern California was even wilder and more uncivilized in those days, and construction equipment and materials had to be carried by mule train from Pasadena to the observatory in the San Gabriel Mountains. Drivers of the mule trains were too busy fighting off rattlesnakes to contemplate that they were carrying objects which, when assembled, would provide humanity with a true and mind-boggling picture of the universe.

With the aid of the 100-inch telescope, Hubble was able to show clearly that M31, the largest appearing spiral, contains stars and that some of these stars are cepheid variables. By employing the period-luminosity relation, Hubble showed conclusively that M31 is far beyond the confines of our Galaxy. The issue was resolved.

We have started several of our discussions by mentioning the work of Sir William Herschel. This displaced German musician, living in England and making his own telescopes (the best and largest of the time), was an amateur astronomer. Called the *father of double-star astronomy*, he made the first discovery of a planet, carried out the first determinations of the solar apex, gave the name to planetary nebulae, and began the study of the Galaxy. We must refer to him and his son John once again because it was their observations that began the discovery of large numbers of what were once called *nebulae*, which we now know are galaxies of stars—some smaller and some larger than the Milky Way Galaxy.

With his large telescope and during his systematic survey of the heavens, Herschel saw many nonstellar objects with greater clarity than anyone before him. Some objects, which earlier observers with inferior telescopes had taken to be nebulae, Herschel identified as star clusters. Some nebulae near the plane of the Milky Way were not resolved into stars; they became known as *diffuse nebulae*—in other words, the emission and reflection nebulae. A third group of objects was composed of compact nebulae which only appeared on the sky outside the confines of the Milky Way. They were called *extragalactic nebulae*, not because they were thought to lie beyond our Galaxy but because of their apparent positions away from the plane of the Milky Way. At one time Herschel convinced himself that these compact nebulae were external systems; he even referred to several of the larger ones as "telescopic Milky Ways." He later reversed himself and considered the extragalactic nebulae to belong to our Galaxy. He thought they probably were nebulae in various states of condensation, in the process of forming into a star or a cluster of stars.

By the early 1900s more than 10,000 extragalactic nebulae had been catalogued. But the argument over their nature still continued, with astronomers almost equally split in their support of the proposals that they were glowing gas or collections of stars. About this time Edwin Hubble, using the 100-inch telescope at Mount Wilson, was able to resolve stars in the outer part of the Andromeda Nebula, which is shown in Plate 15A. He identified several cepheid variables among these stars. Hubble was able to apply the newly discovered period-luminosity relation for cepheid variables to estimate the distance to the Andromeda system. He concluded that the system was 750,000 light-years away, which placed it well outside our Galaxy. With the recalibrated period-luminosity relation available today, we find it to be roughly 2 million light-years away. The Andromeda nebula is no ordinary nebulosity. It is in fact "a milky way" in its own right—a galaxy of stars which is even larger than our own. Hubble was also able to do the same analysis for two other extragalactic nebulae, showing them also to be far removed from our system. With Hubble's announcement of his measurements, at the winter meeting of the American Astronomical Society in 1924, the argument was finally settled. The **extragalactic nebulae** are galaxies, separate and distinct from our Galaxy.

Types of Galaxies

Hubble found that he could classify the **galaxies** into three major categories. One type is a spiral galaxy. The **spiral galaxies** show evidence of spiral arms, not unlike those which exist in our system. The spirals are subdivided into two groups. A **normal spiral** has arms which begin at the nucleus or central bulge of the galaxy, like those seen in the spirals shown in Figure 17–1. A **barred spiral**, however, has spiral arms beginning at the ends of what appears to be a bar extending through the galactic nucleus. Examples of some barred spirals are shown in Figure 17–2. They are less frequent than the normal spirals, accounting for less than one-third of the known spirals.

A galaxy in the next category was called an **elliptical galaxy** by Hubble. This type of galaxy appears to be an ellipsoidal collection of stars with no spiral arms. Figure 17–3 shows an example of an elliptical. By far the most numerous galaxies found, ellipticals contain both the least and most massive of all galaxies.

Hubble designated a galaxy of the third type as an **irregular galaxy**. These galaxies have no definite form or structure. Figure 17-4 shows both the Small and Large Magellanic Clouds, which many take as prime examples of irregular systems. A color photograph of the Large Magellanic Cloud is shown in Plate 15B. Recent work on the Large Magellanic Cloud has led some astronomers to the conclusion that it may be a barred spiral. The barred structure is not obvious, however; and it is still usually regarded as irregular. Galaxies in this type are quite rare; they make up less than 3 percent of all galaxies which have been classified.

Hubble refined his classification scheme for galaxies by further subdividing his types. He designated normal spirals by the capital letter S. This is followed by lower case a, b, or c, depending on the tightness of the spiral arms. The letter a designates a tightly wound arm system; b, moderately tight; and c, a loose arm system. The designation S0 was added later to indicate galaxies with a shape appropriate to the nucleus of a normal spiral, but with no arms present. The barred spirals are designated SB with a following a,b, or c, depending on the tightness of the arms. An SB0 type is also included to indicate a system showing a nucleus and bar but no spiral arms. The elliptical galaxies are designated by capital E followed by a whole number from 0 to 7. The number is the apparent ellipticity of the galaxy (the difference between its largest and smallest dimension on a photograph divided by its largest dimension) times ten and rounded to the nearest whole number. The observed ellipticity of an elliptical galaxy is generally not its true ellipticity because of the chance orientation of the galaxy to our line of sight. Statistical

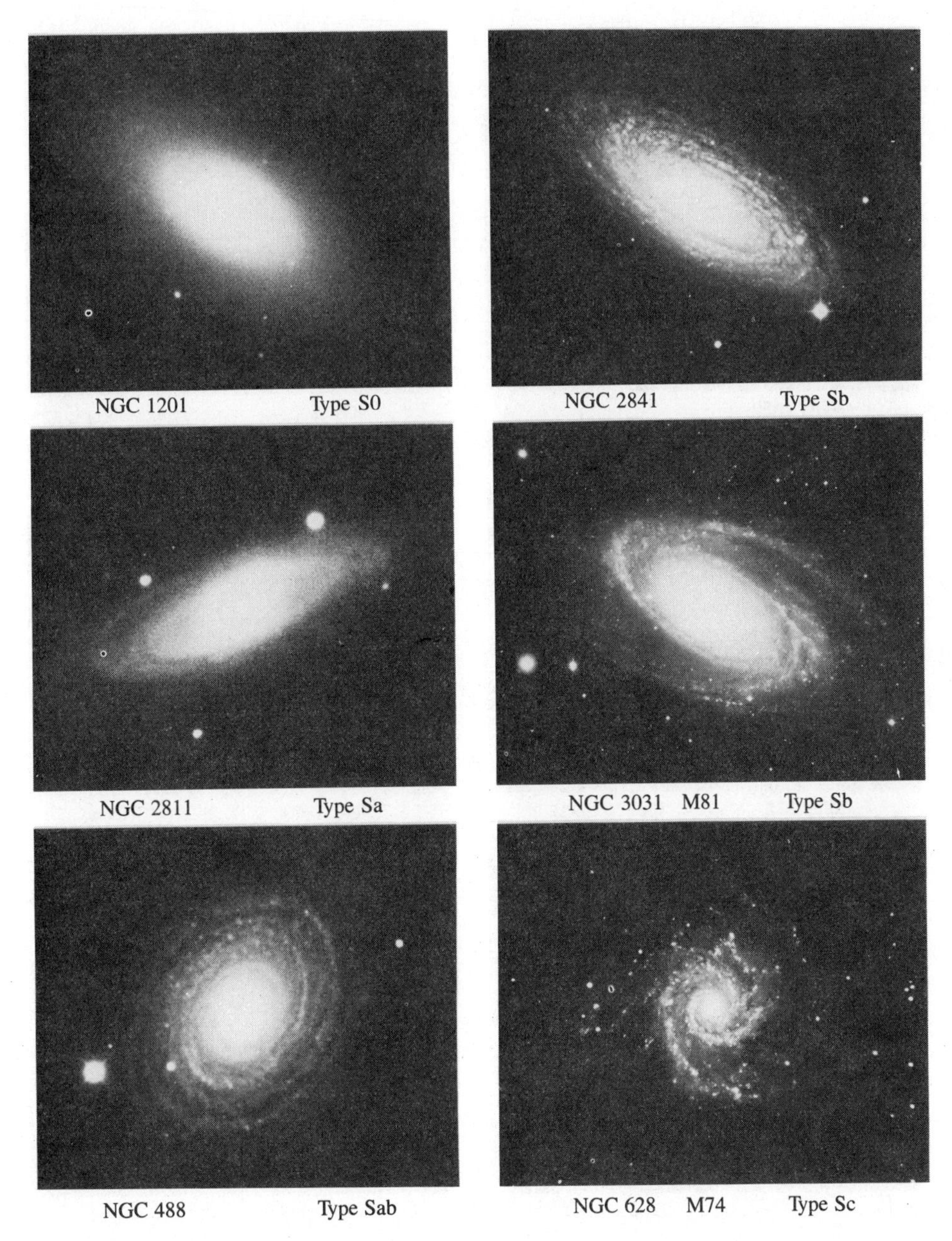

Figure 17–1 Sequence of normal spiral galaxies.

studies of the ellipticals, assuming they are oriented at random, indicate that all ellipticities up to 7 actually exist and that the distribution observed does not arise from a single shape viewed from different angles.

The Hubble classification of galaxies is nicely summarized by a diagram he proposed, which is shown in Figure 17–5. It is frequently called the *tuning fork diagram* because of its appearance. The ellipticals, in order of increasing ellipticity, are drawn along the

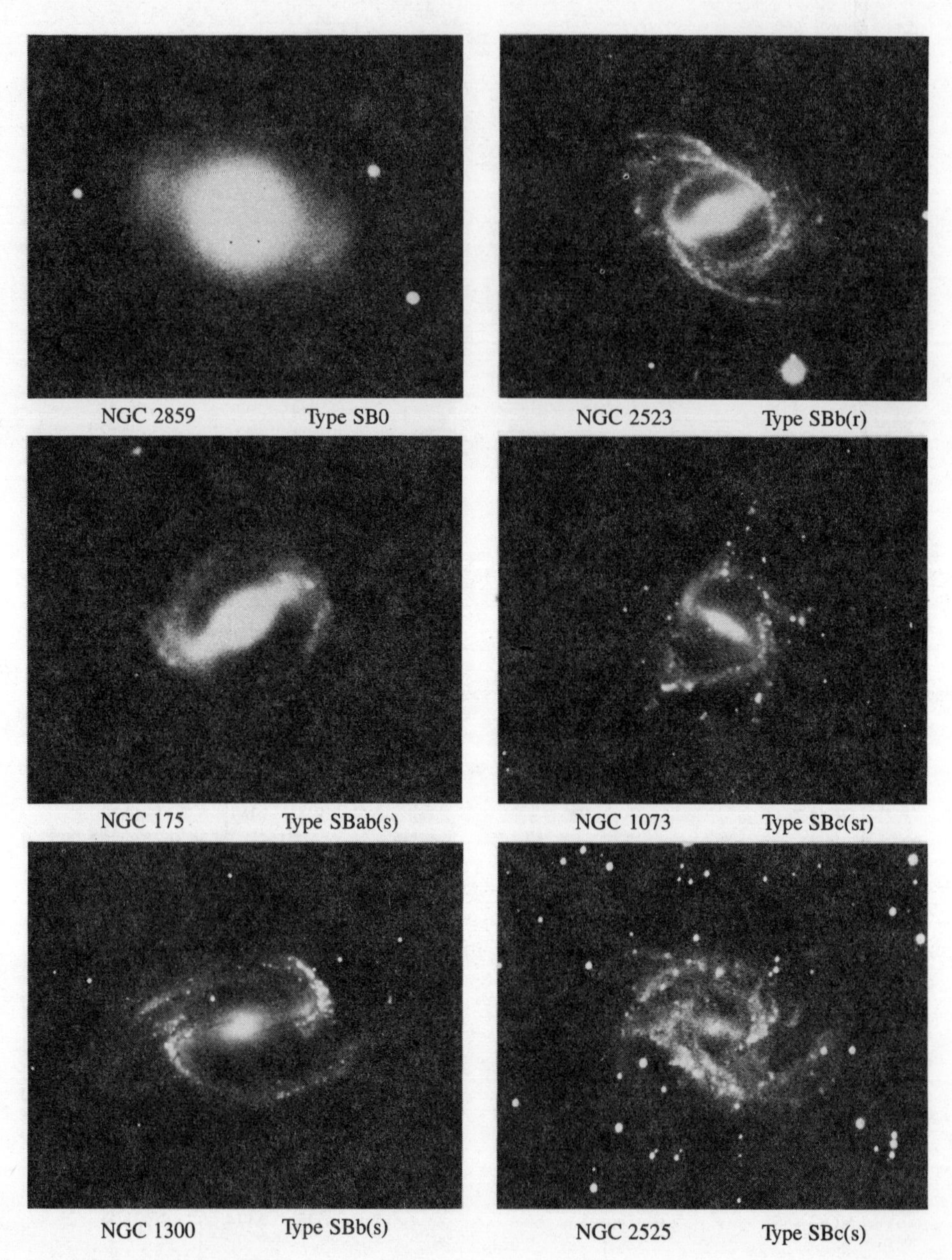

Figure 17–2 Sequence of barred spiral galaxies.

handle of the tuning fork. The S0 (and SB0) galaxies occupy the place where the handle and the branches of the tuning fork join. The normal spirals are drawn on one branch, from tight to loose, and the barred spirals, similarly, on the other branch. The irregulars do not appear on the diagram. When Hubble first presented his diagram, he thought it might represent an evolutionary sequence, with E0 ellipticals evolving to the right, eventually becoming Sc or SBc spirals. This idea is not accepted today, and Hubble's classification

Figure 17–3 NGC 4486, an elliptical galaxy.

Figure 17–4 The Magellanic Clouds.

scheme is taken to be simply morphological, that is, based only upon the observed shape or structure of the galaxies.

Some of the spiral galaxies, like the one shown in Plate 16A, happen to be seen edge-on. When this occurs, the galactic nucleus is seen to be spheroidal, with the spiral arms restricted to a rather thin disc surrounding the nucleus. The dark streak running through the disc is caused by the interstellar dust which is concentrated in the spiral arms.

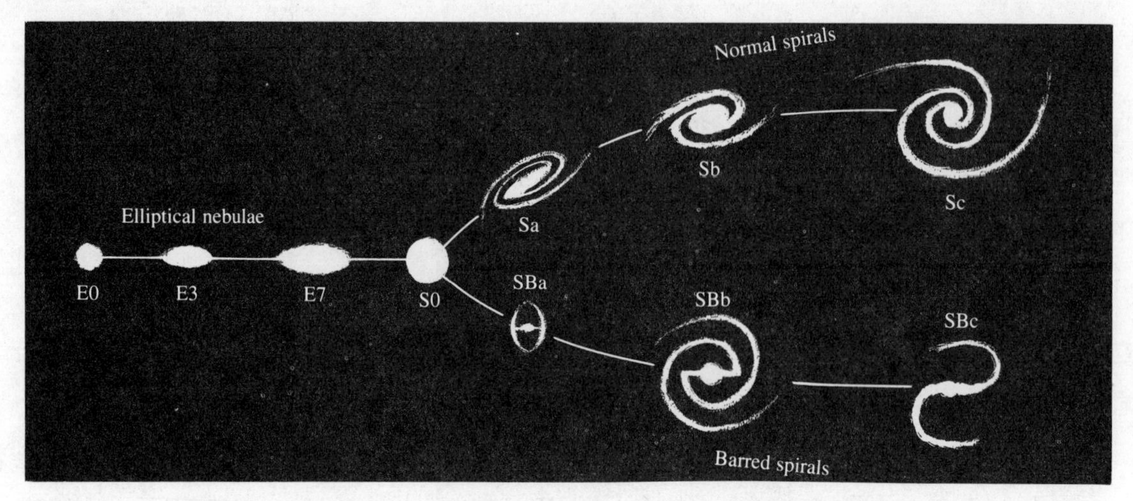

Figure 17–5 Hubble's tuning fork diagram.

Galaxies seen edge-on, like this one, and spirals, like the Andromeda Nebula, seen more face-on, supplied important clues to the interpretation of the structure of our own Galaxy.

Galactic Masses

The first determination of the mass of another galaxy came by applying essentially the same technique used to estimate the mass of our Galaxy. The highly flattened shapes of spiral galaxies suggest that they undergo rapid rotation. For well-resolved galaxies like Andromeda, we can measure the Doppler shifts of various parts of the galaxy. Figure 17–6 shows the **rotational velocity curve** for the Andromeda galaxy. The plot gives the radial velocity due to the galactic rotation as measured along the long axis of the galaxy, as we see it in the sky. Notice that for the inner regions, except near the very center, the speeds increase with distance outward in the galaxy. In other words, the inner regions of the galaxy rotate like a solid disc. In the outer regions, however, the speeds decrease with distance in the manner expected for Keplerian motion.

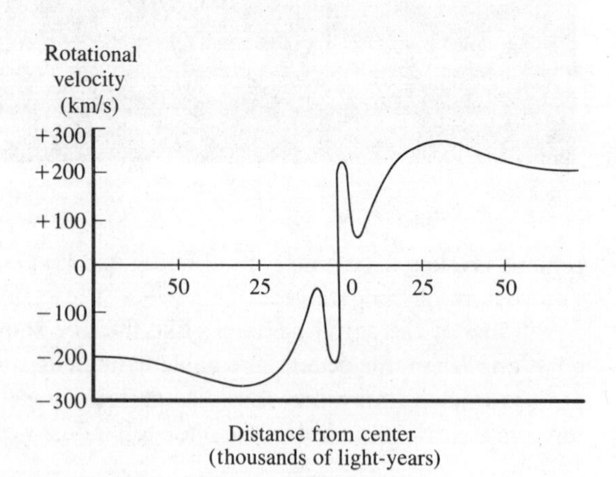

Figure 17–6 The rotational velocity curve for the Andromeda galaxy.

Unfortunately, the radial velocities measured are not the actual rotational velocities, because the plane of the galaxy is inclined to our line of sight. If the basic shape of the Andromeda galaxy is circular in the plane of its spiral arms, its observed shape implies that it is only 15° from being viewed edge-on. Correcting the velocities for this projection and using Kepler's harmonic law, we obtain an estimate of about 300 billion solar masses—50 percent more massive than our system.

An interesting feature of the Andromeda rotational velocity curve is exhibited by the sharp peaks seen on either side of the galactic center. The peaks are caused by a small core, 2400 light-years in diameter. It is in rapid rotation, taking only several million years per rotation, while the outer regions of the galaxy take several hundred million years.

There are only several dozen spiral galaxies which can be seen with sufficient detail to apply the rotational method for determining mass. The masses found by this technique range from ten billion to several hundred billion solar masses.

Elliptical galaxies cannot be studied by the rotational velocity curve method because they rotate too slowly. The absorption lines in the spectra of elliptical galaxies are much broader than those found in the spectra of single stars, for the broadened lines are composites of the lines of the individual stars in the galaxy. The fact that they are so wide indicates that the stars are moving within the galaxy with a large range of motions. Measuring the width of one of the galactic lines permits us to estimate the average velocity of the stars. The average velocity is related to how tightly the stars of the galaxy are bound together by their mutual gravitation. The total gravitation, in turn, depends upon the total galactic mass. Analyses of this nature have shown some ellipticals to have masses as small as a billion solar masses. On the other hand, there are large ellipticals which have masses 1000 billion times, or more, than that of the sun.

A number of galaxies are found in pairs. We think that most pairs are binary systems because they are found in much greater numbers than we would expect if they merely resulted from accidental alignment of randomly spaced galaxies. In principle, we can study these pairs as we did visual binary stars. Unfortunately, their periods of revolution are around several hundred million years, so there is no chance of finding an apparent orbit. Some pairs can be observed as spectroscopic systems, but we have no knowledge of the inclination of their orbits and obtain only one point on the velocity curve of each system. Such a measurement by itself is not very useful; but by measuring many pairs and assuming random orbital orientations, we can obtain a statistical estimate of the masses. Thornton Page has studied nearly 100 pairs this way. In forty-five pairs involving only S0 or elliptical systems, he found an average mass of 400 billion solar masses. In twenty pairs involving only spiral galaxies, the average mass was 30 billion solar masses.

Galaxies not only occur in pairs, but also in clusters. If a cluster is gravitationally bound together, the motions of the cluster members will be related to the total mass present in the cluster. We can, in essence, apply the same technique to find the mass of a cluster of galaxies as we did to find the masses of elliptical galaxies. By measuring the radial velocities of the cluster galaxies, we can estimate the total gravitational field, which, in turn, leads to an estimate of the cluster's total mass. Dividing the total mass by the number of galaxies in the cluster gives an average mass for a single galaxy. Masses found this way are usually larger—up to tens of times greater—than masses determined by other methods. There is no reason to believe that the average galaxy in a cluster of galaxies should be so much larger in mass than the other galaxies for which we have been able to estimate masses. There are three major hypotheses to explain this discrepancy, but none of them has been verified as yet. First, the observed velocities might not be gravitational equilibrium velocities, which would mean that our total mass estimate would be wrong from the start. Second, there could be a large contribution to the mass of the clusters of galaxies in the form of intergalactic matter, which by our procedure would unknowingly be assigned to the galaxies. Third, there might be black holes which would contribute to the gravitational field; but, being invisible, they would not be accounted for when calculating the mass per galaxy, thus making the galaxies appear more massive.

The lower limit to galaxy masses is found for the dwarf ellipticals. Their mass is estimated by counting stars and measuring the total light from the systems. The mass estimates usually turn out to be around several million solar masses. The dwarf ellipticals are so faint that they can only be seen when relatively nearby, say several million light-years away. Although few have been observed because of their low luminosities, they are thought to be the most common of galaxies. For example, of the twenty known galaxies (including our own) within a distance of two million light-years, nine are dwarf ellipticals. If this same ratio holds at distances where the dwarfs cannot be seen, it would mean then nearly half of all galaxies are dwarf ellipticals. The actual fraction may not be that high, but it is certainly large.

The different types of galaxies are compared by mass and size in Table 17–1. The greatest range occurs for the elliptical galaxies. The giant elliptical systems can contain nearly 100 times as much mass as a spiral, while the dwarf ellipticals are not much more massive than the largest globular clusters found within other galaxies. Our own system is a moderate to large spiral, by no means the largest either in mass or size. The irregular galaxies are on the small side of the range of spiral masses and sizes.

Table 17–1
Mass and size ranges of galaxies

Type	Mass (solar units)	Size (1000 light-years)
Spirals	10^9 to 4×10^{11}	20 to 150
Ellipticals	10^6 to 10^{13}	2 to 500
Irregulars	10^8 to 3×10^{10}	5 to 30

There are two other worthwhile comparisons of the types of galaxies: the kinds of stars they contain and the presence of gas and dust. The spiral galaxies invariably contain gas and dust in their arms as well as the hot, young stars associated with interstellar matter. The nuclei of the spirals contain the older, cooler, more evolved stars of population II. The ellipticals, however, except for one possible case, show no clear evidence of dust. However, from both radio and optical observations, some are known to include interstellar gas. The stars in these galaxies are characteristic of the old stars of population II.

The irregular galaxies divide into two groups. Both groups show interstellar gas, but one group has a lot of dust and the other has very little or no dust. Hot, blue stars are seen in the irregular galaxies.

In Plate 15A, compare the two companion elliptical galaxies with the Andromeda spiral. The color of the ellipticals is very similar to the central region of the spiral but quite different from the spiral arms, easily confirming the difference in the stellar population noted previously. Plates 15B and C show the Large Magellanic Cloud and an elliptical galaxy. The variation in color, while present, is not as valid as in Plate 15A because of difficulties in maintaining proper color balance in different reproductions.

The Distribution of Galaxies

After Hubble had conclusively shown that the extragalactic nebulae were in fact external galaxies, he turned to a survey with the 100-inch telescope to see how they are distributed on the sky. Ideally, he should have photographed the entire sky visible from Mount Wilson, but this would have taken many years. Instead, he photographed 1283 sample regions systematically spaced to give a good statistical sample. Hubble was able to identify more than 44,000 individual galaxies on his photographs.

He presented his results by preparing a map of the sky like that shown in Figure 17–7. The map shows the entire sky but is drawn so that equal areas on the sky are preserved as equal areas on the map. The map is centered on the Milky Way, the straight horizontal line representing the plane of the Galaxy. The cross-hatched areas are the portions of the sky which cannot be seen from Mount Wilson. The small filled circles are locations of photographs with more or less average galaxy counts. The large filled circles indicate high counts; and the large open circles, low counts. The squares indicate where very few galaxies appeared on the plates. The dashes show where no galaxies were counted.

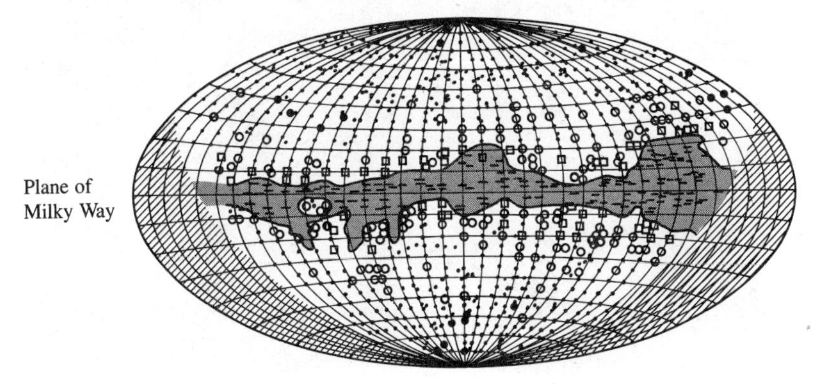

Figure 17–7 Hubble's plot of the observed distribution of galaxies.

The number of galaxies counted increases towards the poles of the Milky Way, that is, 90° from its plane. The irregular region where no galaxies were counted was called the **zone of avoidance** by Hubble. It agrees quite well with the boundaries of the visible Milky Way. There are a few holes in the zone, like the one to the left of center, where some galaxies are seen. Hubble interpreted the fall-off in the count of galaxies toward the Milky Way as the result of obscuring dust in the spiral arms of our Galaxy. The zone of avoidance is simply where absorption by dust is so effective that essentially no light from outside our Galaxy could reach his telescope. When he made allowances for this problem, he found that the distribution of galaxies is roughly the same over the entire sky. Based upon his sample counts, he estimated that there were nearly 100 million galaxies that should have been within the range of the 100-inch telescope. With modern photographic plates and the 200-inch telescope, there should be over one billion galaxies that could be photographed if we could remove the obscuring effect of dust in our system.

Even in the unobscured regions, near the poles of the Milky Way, Hubble's diagram does not show uniform counts per unit area. There is a clumping or clustering of galaxies. The clustering far exceeds the accidental statistical grouping that would occur if the galaxies were distributed randomly in space, so it must be an actual physical grouping. **Clusters of galaxies** can be one of two types. The irregular clusters, likc the one in Figure 17–8 from the constellation of Hercules, show no symmetry or central concentration in the distribution of the member galaxies. These clusters contain all types of galaxies—spiral, elliptical, and irregular. An irregular cluster can contain from a few tens to over a thousand galaxies. The regular clusters, on the other hand, show a spherical symmetry, usually with a strong central concentration of galaxies. They are made up almost entirely of elliptical and S0 galaxies, with very few or no spirals. The regular clusters are very rich, usually containing more than a thousand galaxies.

Figure 17–8 The Hercules cluster of galaxies.

George Abell, using the 48-inch Schmidt telescope at Palomar, surveyed the sky for clusters of galaxies much as Hubble had done for galaxies. He found over 2700 rich clusters, the only ones he could identify with certainty at large distances. His conclusions for clusters were quite similar to those of Hubble for single galaxies. He confirmed the zone of avoidance and found that the number of clusters appeared to increase as the poles of the Milky Way are approached. Allowing for the obscuration, he concluded that, on the average, the count of clusters of galaxies should be more or less the same no matter in what direction you look in the sky. Abell was also able to estimate the distances to the clusters. The nearest one he detected was about 300 million light-years away and the most distant ones between 2 and 3 billion light-years away. Combining his distance estimates with the observed directions to the clusters, he concluded that the clusters, on the average, are uniformly distributed. On a small scale, there is a statistically significant clumping. In other words, the clusters form clusters of clusters. These super-clusters have dimensions of the order of 100 to 500 million light-years and contain around one million billion solar masses.

Our Galaxy belongs to an irregular cluster, sometimes called the *local group of galaxies,* which contains twenty known members. The spatial arrangement of the local group is shown in Figure 17–9. Three of the members are spirals: our own system, Andromeda, and the open spiral in Triangulum. Thirteen of the members are ellipticals, of which nine are dwarf ellipticals. The remaining four galaxies are irregular systems and contain both the Large and Small Magellanic Clouds. The local group occupies an ellipsoidal volume of space, roughly 2.5 million light-years along its largest dimension. There also are two subclusters, one around our Galaxy and another around the Andromeda galaxy. Recently two galaxies, one a giant elliptical and the other a large spiral, have been discovered by P. Maffei on infrared plates taken in the plane of the Milky Way. The infrared light was able to penetrate the zone of avoidance. The distance to these galaxies is still uncertain, so we do not yet know whether they are in the local group.

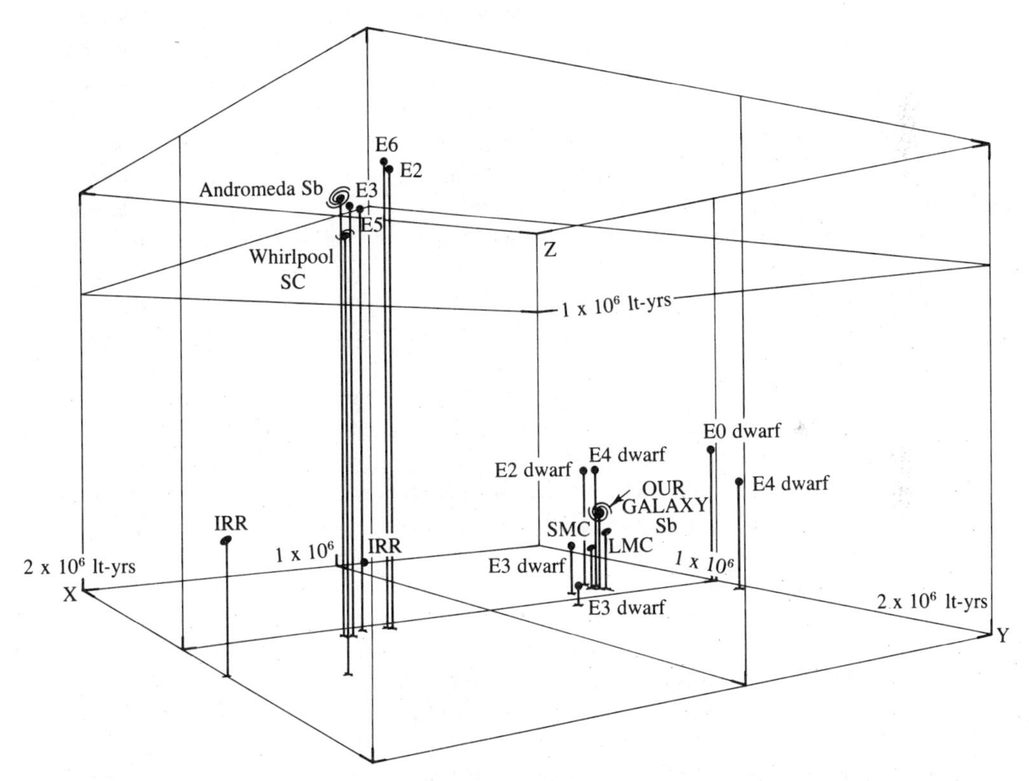

Figure 17–9 Spatial distribution of the local group of galaxies.

Distances of Galaxies

Measuring the distance to a galaxy is not easy. All galaxies, even the nearby Magellanic Clouds, are so far away that measuring trigonometric parallaxes is completely out of the question. The only method we can apply is to observe some object in the galaxy, maybe even the whole galaxy itself, for which we can obtain an estimate of absolute magnitude.

Comparing this to the apparent magnitude of the object gives the distance modulus of the galaxy, which can then be converted to distance in parsecs or light-years.

The most accurate distance indicators for galaxies are still the cepheid variables used by Hubble in his pioneering work. The most luminous classical, or Type I, cepheids are bright enough to be used for distance determinations up to 20 million light-years, using the 200-inch telescope. Beyond that distance the cepheids are too faint to be detected.

The RR Lyrae variables have been used in some of the nearby dwarf ellipticals, but their luminosity is not great enough to make them useful much beyond one million light-years. In all, there are thirty systems in which cepheids and RR Lyrae stars have been used to obtain distances.

Some of the bright blue stars in our Galaxy are sixteen times brighter than the brightest cepheids. In some of the spirals and irregulars for which cepheids have been used to obtain distances, we find that the brightest blue stars are roughly as luminous as the ones in our Galaxy. If we assume that the brightest blue star seen in a galaxy has an absolute magnitude equal to the average of those in our system and other nearby galaxies, we can compute distances to galaxies four times as remote as those we obtain using cepheids. Since there are no hot, blue stars in elliptical galaxies, this method of distance measurement will only work for irregulars and spirals.

There are other objects that are almost as bright as the bright blue stars. Novae, when at peak light, are nearly one-quarter as bright as the blue stars. Furthermore, because of their sudden changes in brightness, we can often identify them when individual stars are not well resolved. The novae are therefore considered to be good distance indicators. They suffer from one major fault, however; we must be fortunate enough to photograph a galaxy just when the nova event occurs. Bright globular clusters have nearly the same luminosity as a nova at maximum, so they can also be used as distance indicators. They are particularly helpful in estimating distances to the closer elliptical galaxies.

Emission nebulae are also useful as distance indicators. They convert enough of the ultraviolet light of their exciting stars to visible light to make them brighter than the exciting stars themselves. By comparing a yellow photograph of a galaxy with one taken in the light of H-alpha (the strongest line of hydrogen in the visible), we can identify the bright nebulosities in a galaxy. Such a comparison is shown in Figure 17–10. The nebulosities, being composed mostly of hydrogen, radiate strongly in H-alpha, while the illuminating stars do not. Nebulosities like these once caused confusion in distance measurements because in moderately distant galaxies they appear starlike. Thus, they are easily taken to be bright stars rather than nebulosities in an ordinary photograph. A nebula around a bright blue star is about five times brighter than the star. Using such nebulosities, therefore, extends our distance-measuring capability by slightly more than two times that possible with the bright-blue-star method.

Supernovae are excellent distance indicators because of their great luminosity at maximum light. They can outshine 10,000 ordinary novae. In fact, a supernova can outshine an entire galaxy. Figure 17–11 shows just such a supernova. In the photograph at the top, taken near maximum light, only the supernova appears. In the middle photograph, taken with double the exposure about a year later, the fading supernova is still easily seen, but the brighter stars of a galaxy are just apparent. The last picture, three and one-half years later, was taken with the exposure redoubled. The supernova has faded to obscurity, but the galaxy is now clearly seen. For a short while the supernova had a luminosity greater than the combined stars in its galaxy. As an aside, the supernova in the

Figure 17–10 Identification of nebulosities in a galaxy—Hα on the left, yellow on the right.

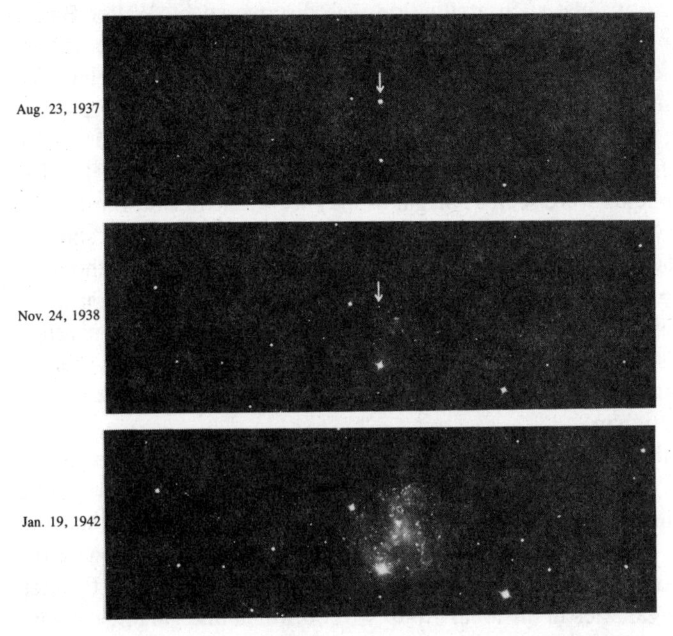

Figure 17–11 A supernova in a distant galaxy.

Andromeda galaxy seen in 1885 actually favored the argument that extragalactic nebulae were not external systems. The object became bright enough to be seen with the unaided eye, and the assumption that it was an ordinary nova led to the conclusion that the Andromeda system would have to lie well within the boundary of our Galaxy. This was an easy mistake to make, since supernovae had not yet been recognized as being different from ordinary novae. However, it did supply a compelling false clue to the distances of the extragalactic nebulae until the supernova's true nature was realized. Supernovae can be used to measure distances over a billion light-years. Unfortunately, they are seen at a frequency of only one every three or four hundred years in most galaxies. Of course, they suffer from the same drawback as ordinary novae: The galaxy must be observed at just the right time or the event is missed.

The accidental occurrence and observation of a supernova are not reliable enough to make supernovae useful for finding the distance to a particular galaxy. To get around this, the distances to the more remote galaxies are estimated in yet another way. By pooling all the information for galaxies for which distances are known, we have found that the most luminous galaxies are about ten times brighter than supernovae. In clusters of galaxies, the brightest galaxy turns out to be about this luminosity. Therefore, if we assign to the brightest galaxy in a cluster the average luminosity of the leading galaxies in clusters for which we have already estimated a distance, we can estimate the distance to that cluster. The brightest-cluster-member technique is good to distances in excess of three billion light-years.

Needless to say, finding the distances to galaxies, particularly the remote ones, is precarious at best. Each of the techniques we have mentioned is based on a prior method, which in turn is based on a prior one. Any error in an earlier calibration, such as the period-luminosity relation for the cepheid variables, is transmitted through the other distance methods which use it. Just such a situation was discovered by Walter Baade during his work on the Andromeda galaxy using the 200-inch telescope in the 1950s. Baade had calculated that RR Lyrae stars in the Andromeda Nebula should be detectable on the 200-inch photographs, using the distance for Andromeda obtained from the period-luminosity relation for cepheid variables. But they were not recorded on the photographs. The galaxy had to be farther away for Baade not to photograph the RR Lyraes. Thus, since the cepheids were seen as bright as they were, the period-luminosity relation must have been incorrectly measured. With some further work, Baade showed that the classical cepheids were about four times brighter than previously thought. Therefore, every distance based ultimately on the cepheid relation was twice as far away as had been thought. Overnight, so to speak, Baade had doubled the size of the universe. We do not expect any more surprises like this. The various distance measurements seem to be reasonably consistent, but a prudent person still treats distance estimates to galaxies with great care and a bit of skepticism.

When reasonably reliable distances to galaxies became known, a startling discovery concerning them was made. V.M. Slipher at Lowell Observatory had obtained the spectra of a number of galaxies in the early 1900s. He noticed that they invariably showed absorption lines displaced to the red and that the fainter galaxies usually showed the greater shifts. Hubble followed this work with an exhaustive study using the 100-inch telescope. Except for some galaxies in the local group, he confirmed that galaxies always show spectral lines that are shifted to the red.

By this time, enough distances had been measured that Hubble was able to compare the observed red shifts with the distances to the galaxies. To his surprise, the red shifts in the spectra of galaxies increased in proportion to the distances to the galaxies. This meant that if the red shifts were Doppler shifts, the galaxies are all moving away from us. Furthermore, the more distant a galaxy, the faster it recedes. Although there was considerable argument at the time of Hubble's announcement, today we accept the red shifts as Doppler shifts. Figure 17–12 shows some representative spectra for galaxies at known

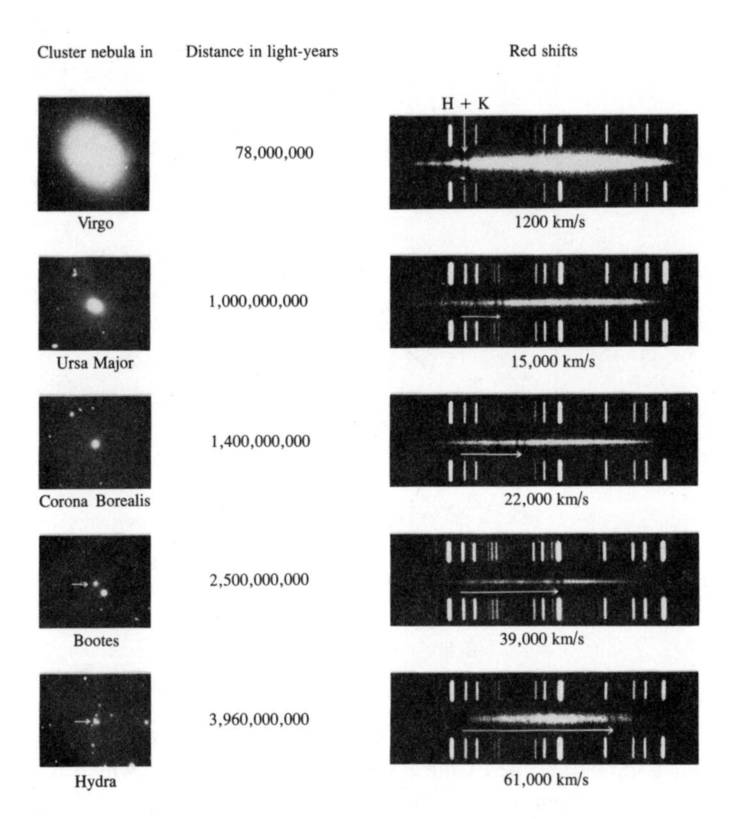

Figure 17–12 Examples of the change in red shift with distance.

distances. Notice how the prominent lines of ionized calcium are increasingly shifted to the right—that is, to longer wavelengths—as the distances to the galaxies increase.

Hubble summarized his results in a formula, now called Hubble's **law of red shifts**. The formula states that

$$V = Hd$$

where V = the velocity of recession of a galaxy
H = Hubble's constant
d = the distance to the galaxy

When V is given in kilometers per second and d in **megaparsecs** (1 million parsecs = 3.26 million light-years), **Hubble's constant** H is found to be between 30 and 100 kilometers per second per megaparsec. The presently accepted value of H is 55 kilometers per second per megaparsec. Hubble's law predicts that the plot of the galactic recessional velocities against the distances to the galaxies gives a straight line, as shown in Figure 17–13. The uncertainties in the determination of H arise mainly from the difficulties in measuring the distances of galaxies used to calibrate Hubble's law. It is not uncommon, now that the law is rather well established, to use it to find the distance of a galaxy once its velocity of recession has been measured. More important than this convenient property of Hubble's law is the impact that it has had on our concept of the universe. We will address this in the next chapter, where we discuss cosmology.

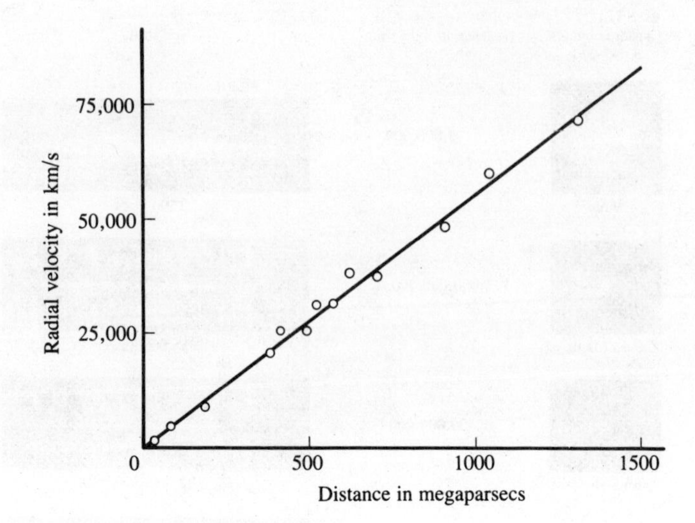

Figure 17–13 Hubble's law of red shifts.

Peculiar Galaxies

The galaxies we have discussed so far have been, for the most part, normal. There are many abnormal or peculiar galaxies. Some of these are pairs of galaxies in which the shapes are severely distorted. The Toomre brothers, using high-speed electronic computers, have constructed mathematical models of galaxies gravitationally interacting. One of their computer-generated diagrams is shown in Figure 17–14. Their models have reproduced many of the features seen in the peculiar pairs, and we now think that the distorted shapes arise from tides mutually induced by the members of the pairs.

Some galaxies are observed to be strong radio sources. Although most normal galaxies are weak to moderate sources, peculiar galaxies are frequently strong sources. One of the first radio sources discovered was Centaurus A, so named because it is the brightest radio source in the constellation of Centaurus. This source was subsequently identified with the optical object shown in Plate 16B. This is indeed a peculiar galaxy, since it is the only elliptical galaxy definitely showing the presence of dust. Two photographs of another peculiar galaxy, the giant elliptical M87, are seen in Figure 17–15. They were obtained by taking a long and a short exposure of the galaxy. This galaxy is also a very strong radio source. The long exposure shows what appears to be a rather normal elliptical system. But in the short exposure we see a bright starlike nucleus from which a jet of material seems to stream. Not all ellipticals that are strong radio emitters show obvious optical peculiarities like M87. We think that their radiation is associated with large amounts of gas present in these galaxies but absent in the usual elliptical galaxy.

In the 1940s, Carl Seyfert noticed a few spiral galaxies that had very bright, blue nuclei. The spectra of the nuclei show the presence of very high temperature gas displaying high turbulent velocities measured in thousands of kilometers per second, indicative of a violent disturbance in the center of these galaxies. The **Seyfert galaxies**, as they came to be known, were not given much attention. They were treated as interesting curiosities. But later, radio observations proved many to be strong radio emitters, and this

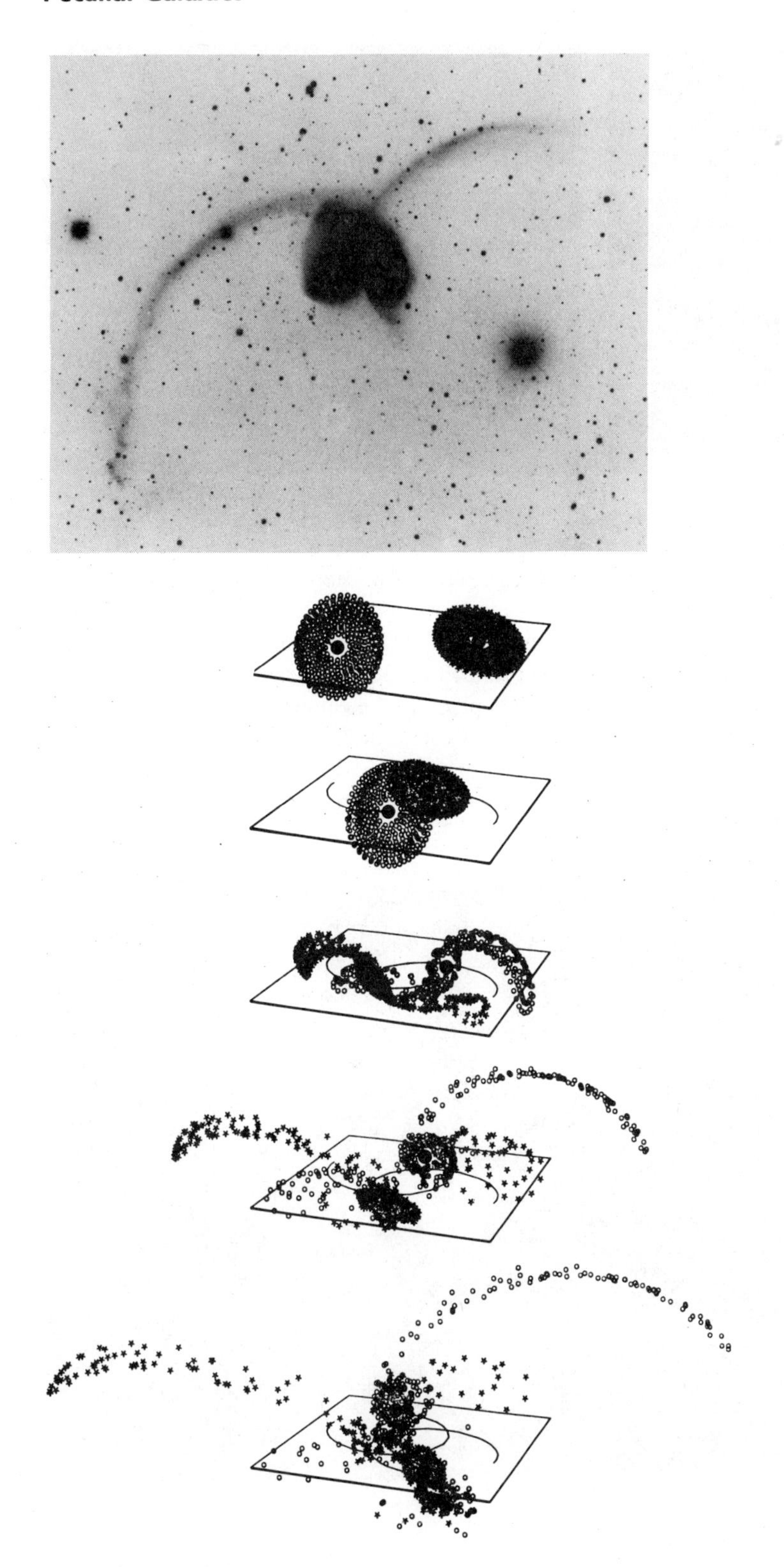

Figure 17–14 Interacting galaxies.

Figure 17–15 A peculiar galaxy, M87.

led to a search for Seyfert galaxies. Enough have now been discovered to lead some astronomers to think that maybe as many as one percent of all spiral galaxies are Seyferts. Figure 17–16 shows the Seyfert galaxy labeled NGC 4151. Not only has it been seen to be a strong radio source, it has also recently been identified as an X-ray source. Some people have proposed that the violence which takes place in the center of a Seyfert also takes place to a lesser extent in our own Galaxy, to explain the point source of radio signals observed in the direction of our galactic center.

Possibly the most extreme example of galactic violence is the irregular galaxy M82, shown in an H-alpha photograph in Figure 17–17. Spectra of this nebula show Doppler

Figure 17–16 NGC 4151, a Seyfert galaxy.

Figure 17–17 M82, an exploding galaxy.

shifts indicative of material streaming outward from its center at velocities up to 1000 kilometers per second. Some of the filaments of material seen in the photograph extend as much as 12,000 light-years above the galactic plane of M82. The material must have been violently discharged millions of years ago. This galaxy is a strong radio source, showing a radio spectrum characteristic of electrons moving near the speed of light in a magnetic field. The color photograph of this object shown in Plate 16C, while suggesting the violence indicated in the H-alpha photograph, is not nearly as convincing as Figure 17–17. This is because the displaced hydrogen in the outer regions has fallen to such a low density that it essentially radiates only in the emission lines of hydrogen and hence is difficult to see except in H-alpha photographs. We do not yet have a satisfactory explanation of cosmic explosions like the one that must have taken place in M82 or of the violence in the Seyfert galaxies. It is safe to say that we cannot claim to understand galaxies until we can account for these violent outbursts.

KEY TERMS

extragalactic nebulae
galaxies
spiral galaxy
normal spiral
barred spiral
elliptical galaxy
irregular galaxy
rotational velocity curve
zone of avoidance
cluster of galaxies
law of red shifts
megaparsec
Hubble's constant
Seyfert galaxy

Review Questions

1. On what basis did Herschel differentiate between galactic and extragalactic nebulae?
2. How did Hubble first establish that an extragalactic nebula was outside our own Galaxy?
3. What are the major types of galaxies?
4. Which type of galaxy is the most common? The most massive? The least massive? The least common?
5. On what basis are both normal and barred spiral galaxies classified?
6. How are elliptical galaxies classified?
7. How can we find the mass of a well-resolved spiral galaxy?
8. Why can't we derive the masses of two galaxies orbiting about each other in the same way as we found the masses of the components of a binary star?
9. How do masses of galaxies found for single and binary galaxies compare with galactic masses found in clusters of galaxies?
10. What is the difference between a cluster of galaxies and a galactic cluster?
11. How do the masses of irregular galaxies compare with masses of spirals?
12. What are some differences between regular and irregular clusters of galaxies? To which type of cluster does our Galaxy belong?
13. What causes the zone of avoidance for galaxies?
14. Do clusters of galaxies themselves appear to cluster together?
15. What are some of the objects observed when finding distances to galaxies?
16. Why has the use of cepheid variables for distance measurements been limited to relatively nearby galaxies?
17. Why can emission nebulae usually be seen at a distance greater than the stars that illuminate them?
18. What difference is found when we compare the spectra of nearby and distant galaxies?
19. Are there any galaxies for which blue shifts have been measured?
20. If one cluster of galaxies is twice as far away as another, how will their red shifts compare?
21. What are Seyfert galaxies? What does our Galaxy have in common with the Seyferts?
22. Is there any evidence that huge explosions can occur in the nuclei of galaxies?

Discussion Questions

1. What might determine whether a galaxy too massive to be irregular would be a spiral or an elliptical galaxy?
2. What type of galaxy has stars that are arranged and that move in a manner similar to the arrangement and the movements of the globular clusters in our own Galaxy?

Plate 15A The great Andromeda spiral galaxy.

Plate 15B Large Magellanic Cloud.

Plate 15C Elliptical galaxy.

Plate 16A Spiral galaxy seen edge-on.

Plate 16B Centaurus A, a peculiar galaxy.

Plate 16C M82, a peculiar galaxy.

3. Under what circumstances is it impossible for us to obtain the mass of a nearby spiral galaxy, using the rotational velocity method?
4. In systems where there is little or no apparent rotation (such as globular clusters, elliptical galaxies, and clusters of galaxies), how can the mass be found?
5. What reason can you give for the fact that bright, hot, main-sequence stars are usually not found in elliptical galaxies?
6. Given the size of the earth's orbit, list all of the intermediate distance determinations which you would have to make before you could find the distance to a cluster of galaxies. (Hint: Cepheid variables have been found in galactic clusters.)
7. If the Hubble constant is 55 kilometers per second per megaparsec, how fast would a cluster of galaxies 100 megaparsecs away be receding from us? If it has maintained this velocity of recession, how long would it have taken to travel from us out to its present distance? (One hundred megaparsecs is roughly 3×10^{21} kilometers, and there are about 3×10^{7} seconds in a year.)
8. Would a cluster of galaxies 200 million parsecs away take more, the same, or less time to travel to its present distance than a cluster 100 million parsecs away?

18

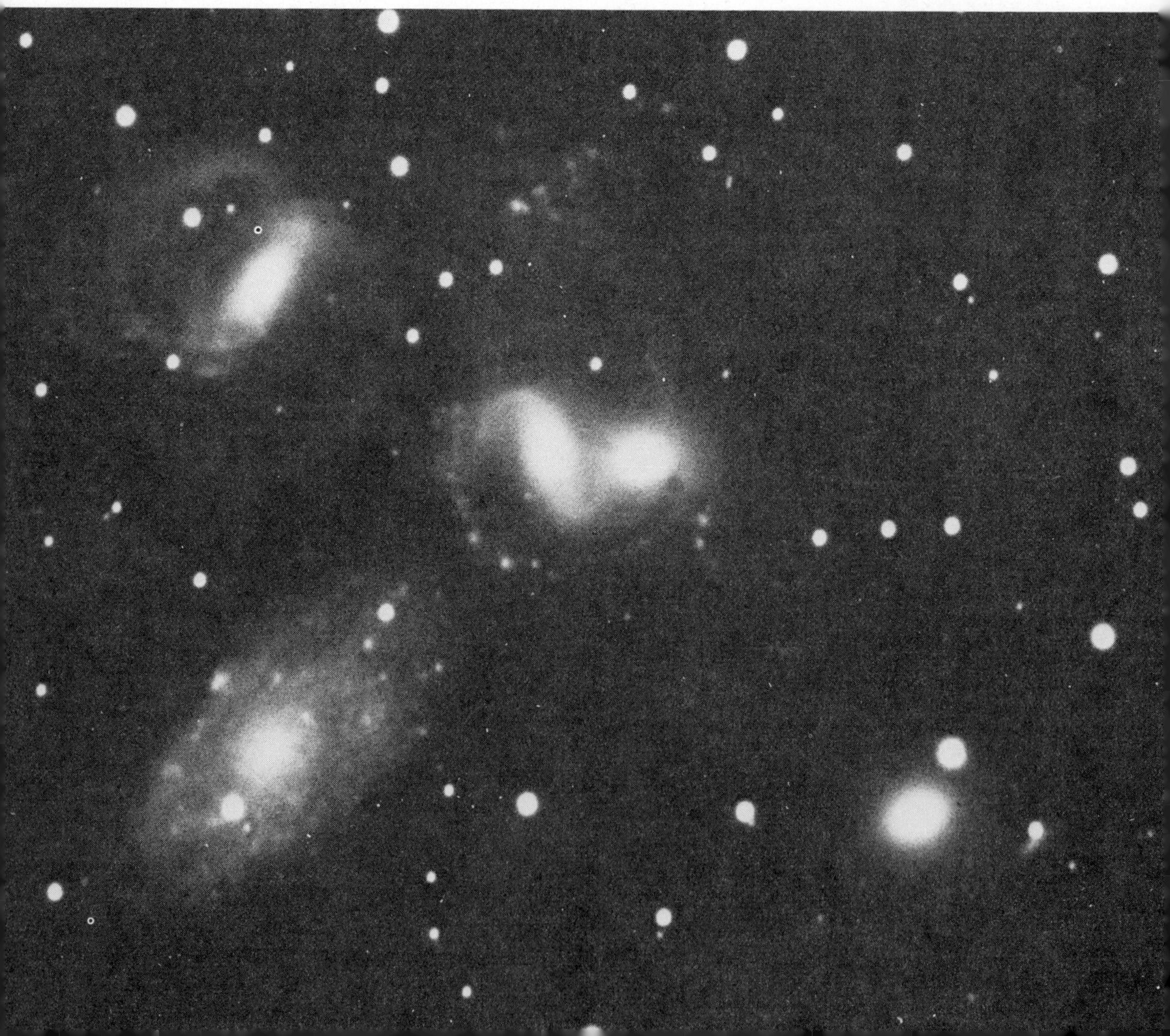

Cosmology—The Origin, Structure, and Evolution of the Universe

At one time or another, many parents are confronted by a child's question, "How did everything begin?" Patiently, according to the belief of the parent, a reasonably satisfactory explanation can be presented. However, that is not always the end of it. More often than not, the child proceeds to the next question: "What happened before that?"

It is a quite universal human quality to wonder how things did come about. Ideas concerning the origin of the universe and life on the earth provide the most basic premises of religions and philosophy. It has been said at times that scientists are people, too; as you might expect, one of the aims of science is to learn if and how the universe originated. Even though it is true that understanding the purpose of the universe and the life in it is outside the capabilities of what is normally called science, it is impossible for scientists to build a picture of the universe that does not form as a result of the human qualities of those who practice science. One of these qualities is to have some emotional opinion—or as we might say, "gut feeling"—concerning the nature of the universe.

The two major, basic ideas proposed for the universe are that it originated at some particular time in the distant past and that it in fact never did begin but instead always existed. A logician could hardly be surprised at such a division in concepts. Either the universe had a beginning or it didn't. There seems hardly any room for another alternative. The idea of a universe with a definite beginning, often called the creation hypothesis, together with the concept of a creator, is the foundation of Judaeo-Christian thought. Among the faiths of Eastern peoples are religions which teach that the universe has no beginning or end, a sort of steady-state hypothesis.

In view of the rather limited number of choices, it should not seem strange that astronomers seriously considered both hypotheses as candidates for the true explanation of the universe and its origin or lack of origin. Until the 1960s, one candidate

Stefan's quintet of galaxies

seemed as likely a choice as the other. Before that time, all we knew regarding the evolution of the universe was that it seems to be expanding. Extrapolating back in time, it is logical to conclude that everything in the universe was packed together at some point in space and time—a beginning of sorts when the universe was "created." However, a "steady-state" explanation can also be made for the expansion. Extrapolating back in time, we reach not one, but an infinity of creation points. Following the history of one atom leads us to its creation at a particular point in space and time. Following another atom leads us to another creation point located elsewhere in space and time. If the rate of creation of atoms is just right, going back, we would find the universe to be no more crowded than it is now; going forward, the steady-state universe would be no less crowded than it is now, because newly created matter would replenish the matter which is already present and thinning out.

You might say that the steady-state universe is not actually steady state because creation is occurring continuously. But for all practical purposes, a steady-state universe as described here is a universe that, overall, does not appear to change during an infinite period of time.

During the 1960s, however, evidence began to accumulate that supports the creation hypothesis and not the steady-state hypothesis. In this, our last presentation, we describe the origin and evolution of our universe based on the creation hypothesis. We hope that you won't be disappointed to find that we don't say anything about the why's—that is, the purpose of this huge and wondrous universe. Such a discussion would be beyond the realm of science. However, being human, we can understand how you might want to contemplate it.

Photons, atoms, the earth, the solar system, stars, nebulae, clusters of stars, the Milky Way system, galaxies, and clusters of galaxies—these are some of the things we have talked about so far. Now we address the **universe**—the totality of all the observed, and not yet observed, objects. How is the universe arranged? Does it change? What is its past and its future? These are the questions that we try to answer through **cosmology.** The dictionary defines the word *cosmos* as an orderly, harmonious, systematic universe. So *cosmology* can also be taken as the process of trying to understand the order and system of the universe.

The usual procedure in cosmological studies is to devise a theory or model, called a **cosmological model**, which gives a description of the universe that can be checked against observations. Frequently, the observational tests of a model are at the very limits of our observational ability, and it is difficult to verify the model. Most models are based upon the hypothesis that the same laws of physics which are valid in our vicinity also hold throughout the universe. Some models have been proposed in which physical constants, like Newton's constant of gravitation or the mass of the hydrogen nucleus, gradually change with time. These models have not been well accepted because most astronomers are reluctant to invoke a hypothesis which is not subject to direct experimental verification, particularly as long as adequate progress is apparently being made with more conventional approaches.

There have been basically two types of cosmological models. One type is a *steady-state model* in which the universe is pictured as being, on the average, unchanged with time. It would have neither beginning nor end. This does not imply that small changes at some localized position, like the evolution of a star or a galaxy, could not occur, but it does require that new stars and galaxies would have to appear at the same rate that others are dying out. This process would keep the mix of old and young systems the same in the future as we observe it now. To generate the matter needed to form the new, young systems, a never-ending supply of fresh hydrogen would have to be created continuously. For a number of years, the continuous creation model was considered to be a possible cosmological model; but recently it has failed several observational tests. Possibly its greatest failing is its inability to account for the weak radio energy we detect coming from all directions in the universe. This same radiation, as we will see, supplies one of the most compelling arguments in favor of the major competitor of the steady-state theory. The continuous creation hypothesis has now been set aside by many as observationally unverified.

The other type of cosmological model is an *evolutionary model*. In this kind of model, creation occurs at one specific time, and the universe thereafter changes with time; that is, it evolves. We will restrict our discussion to the evolutionary model that appears at this time to account best for the observed properties of the universe.

Any cosmological model must satisfy the observed fact of Hubble's law of red shifts. If we accept the red shifts as Doppler shifts, this means that the model must predict that distant galaxies will recede from our Galaxy with speeds proportional to their distances. In addition, most astronomers accept what is considered to be a fundamental hypothesis, the **cosmological principle**. This principle states that the general properties of the universe as observed from our location are the same as those observed from any other location in the universe. In other words, we do not ascribe any special features to our region of space.

The Expanding Universe

The recession of distant galaxies seems to imply that our Galaxy is at the *center* of an **expanding universe**, as depicted in Figure 18–1. If that were true, it would be a violation of the cosmological principle because it would give our Galaxy a special property. Instead, we say that the expansion is a general property of the space in which a galaxy or cluster of galaxies is imbedded. In this way, observers inhabiting any other galaxy would see all other galaxies outside their own cluster to be receding from that galaxy as if it were located at the center of the universe.

At first glance, this seems self-contradictory; but fortunately, there is a simple analogy which can clarify the basic idea of an expanding space. Consider a loaf of raisin bread. We can think of each raisin as a galaxy (or a cluster of galaxies). If the original dough of the bread fills a space ten centimeters in each dimension, then as the yeast in the dough causes it to rise, the dough will expand uniformly. The expansion causes its length, width, and height to become larger. Assume the dimensions of the loaf double in one hour. Then, as illustrated in the cross section of the loaf shown in Figure 18–2, the raisins are carried along by the expanding dough and the distances between raisins increase, although the raisins themselves remain the same size. If raisin A represents our Galaxy, then the neighboring raisins (galaxies) would be twice as far away one hour later. Thus raisin B, originally two centimeters distant, would be found at four centimeters. Raisin E

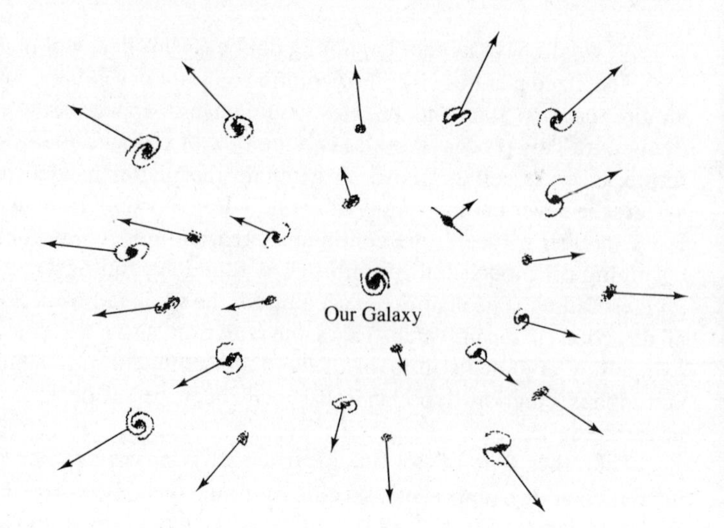

Figure 18–1 The expanding universe.

would move from seven to fourteen centimeters. But each raisin has accomplished its motion in one hour. The speed of recession of B is four minus two, or two centimeters per hour. For E it is fourteen minus seven, or seven centimeters per hour. Raisins C and D at intermediate distances give similar results. Thus, the greater the velocity of recession of a raisin, the greater is its distance.

Of course, the choice of raisin A to represent our galaxy was arbitrary. We could just as easily have chosen any other raisin. We would, however, have come to exactly the same conclusion. Every raisin moves away. The more distant it is, the faster it seems to move. The motion of the raisins is not a property of the raisins; instead, it is a property of the dough in which they are located. Similarly, the observed recession of actual galaxies from our own is taken as evidence that the universe is expanding. Just as in the raisin bread, this expansion is a property of space. Neither galaxies nor clusters of galaxies have to change in size as the system expands.

If the universe is expanding, the galaxies must have been more crowded together in the past than now. The Belgian priest Abbé Lemaître carried this reasoning to its ultimate

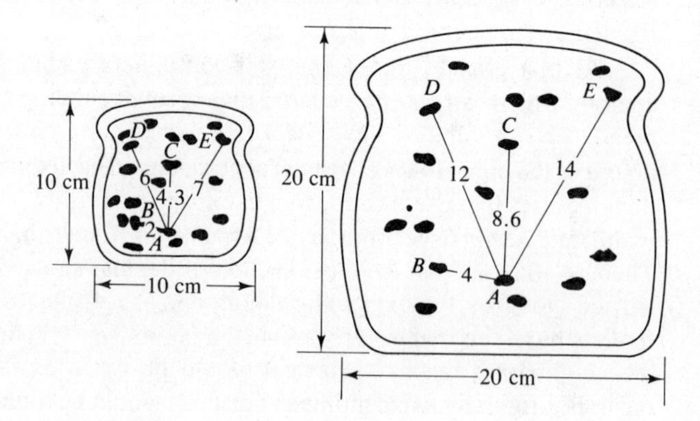

Figure 18–2 The raisin bread analogy.

conclusion in 1931 and proposed that at some time in the past the entire universe was crowded into a very small volume. Lemaître referred to this state of the universe as the **primeval atom** and assumed that it was instantaneously created. Subsequent workers, particularly George Gamow, showed that the primeval atom would have been extremely hot—hot enough to explode in a "big bang." This big bang would have expelled the material of the primeval atom outward. The expansion we see today is the residual motion of this violent event which took place at the beginning of time. The model of the universe stemming from the proposal of Abbé Lemaître, as modified later by others, is commonly called the **big-bang universe**. It automatically incorporates the red shift law and, as we shall see, also satisfies the cosmological principle. We will first see how such a universe would be expected to evolve, then demonstrate that it satisfies the cosmological principle, and finally see whether there are any observational tests which tend to verify it.

Evolution of the Universe

Because of the extremely high temperature, the primeval atom would have consisted of elementary particles normally bound together in nuclei of atoms, immersed in an intense sea of radiation. Just after the big bang, the universe would be reminiscent of the fireball that occurs after a nuclear explosion. Consequently, it is often referred to as the **primeval fireball**.

While the universe was extremely young, the fireball would have quickly expanded and cooled to the point that helium nuclei produced through hydrogen fusion could survive. But the cooling would have been extremely rapid. Thus, by the time about 25 percent of the hydrogen had been transmuted, the temperature would have fallen below that required for fusion to continue. Essentially no nuclei of elements heavier than helium would have formed. The temperature would still have been high enough that electrons would not have been able to associate with the protons and helium nuclei to form neutral atoms. Just after the big bang, then, the matter in the universe consisted of an ionized gas which was about 75 percent hydrogen and 25 percent helium by mass.

With continued expansion, the universe would have cooled relatively quickly. In about one million years following the big bang, the temperature would have become low enough for protons and helium nuclei to capture electrons and for the matter of the universe to be converted to a normal gaseous state. The temperature at this time would have been only about 3000K; so, in accord with Planck's black body radiation law, only an extremely tiny fraction of photons in the universe would be energetic enough to interact with hydrogen and helium atoms. We say that the radiation *decoupled* from the matter in the universe when the neutral atoms formed.

Before the radiation decoupled, both the pressure of the hot ionized gas and radiation pressure due to strong coupling of photons with the freely moving electrons would have prevented any condensations of matter in the universe. With the decoupling, however, the pressure would have dropped to the point where condensations in the gas could take place. Furthermore, these condensations would then have grown by gravitation to form huge clumps of gas that eventually would form into galaxies. Less than 100 million years after the big bang, the universe would have expanded sufficiently for the temperature to fall to only a few tens of degrees above absolute zero—certainly cool enough to permit gas clumps to condense gravitationally if the universe were static. But the expansion of the universe would have tended to disperse the gas and prevent gravitational condensation, unless large, well-defined regions with densities at least one percent greater than the average gas density existed prior to the decoupling of radiation from the gas. If the big-bang model is correct, density variations must have arisen by some as yet unknown process.

Some recent calculations show that if there were a tendency for gas to condense in the early stages of the universe, then clusters of condensations would have formed. Thus, not only would galaxies have ultimately come into being, but clusters of galaxies would also have developed. The precise mechanisms by which galaxies and the stars within them formed has not been completely solved, at least not by people; but nature obviously found the solution.

The Apparent Edge of the Universe

As we have already seen, galaxies and clusters of galaxies are, on the average, uniformly distributed in the sky after we correct for the obscuration by dust in the Milky Way Galaxy. By the cosmological principle, the sky should have this same property viewed from any other galaxy. We say that the universe should appear to be *isotropic*, which means it should look the same in all directions from any place in the universe.

It is even more difficult to conceive of this isotropic property than the expansion of the universe. For example, our raisin bread analogy completely breaks down. As shown in Figure 18–3, if we were located at the center of the loaf, an all-sky map would show the more or less uniform distribution required. But if we were located near the edge of the loaf, only half the sky would show galaxies. The only way to correct the raisin bread model would be to assume the loaf to be infinite in all directions.

But the big-bang model predicts a finite universe, because the universe has only been expanding for a finite time. It is possible for a big-bang model to have the property of an expanding, isotropic, finite universe, no matter where an observer is located, but we cannot draw a three-dimensional diagram of such a system. We shall again try to

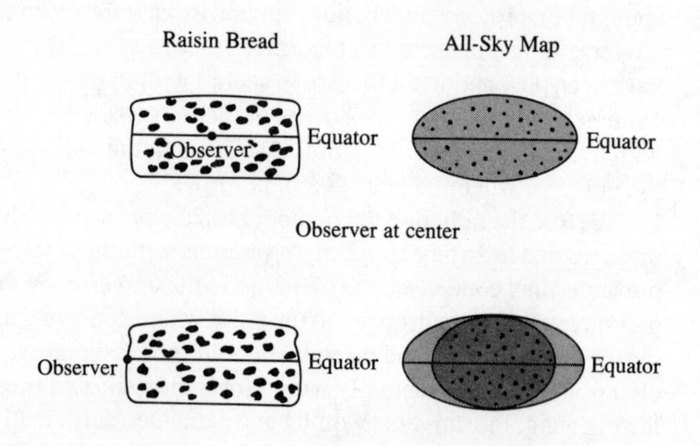

Figure 18–3 The view of the sky in the raisin bread analogy.

understand our dilemma by resorting to an analogy, in this case a two-dimensional analogy. In Figure 18–4, we see two two-dimensional surfaces. The one on the left is the curved surface of a sphere and the one on the right is the flat surface of a thin disc. In the curved two-dimensional universe represented by the spherical surface, no point on the surface has the unique property of a central location. Also, there is no apparent edge. In the finite flat-disc universe, there is a center and there is an edge. And just as in our raisin bread analogy, the particular view will depend upon the location of the observer.

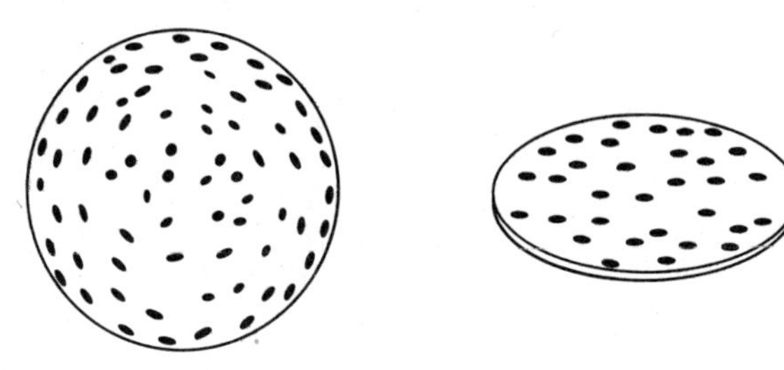

CURVED: Sphere

FLAT: Disc

Figure 18–4 Curved and flat two-dimensional models.

We can understand the properties of the spherical two-dimensional universe by referring to Figure 18–5. A spherical surface has the property that the shortest distance between two points on the surface is always an arc of a great circle. As long as the two points do not lie on a diameter of the sphere, there is only one possible great circle through any given points. A great circle is any circle on the surface of the sphere having a radius equal to the radius of the sphere, as shown on the left. Since light always travels by the shortest distance, light from "galaxy" A would travel along the great circle arc to "galaxy" B. If you were an observer positioned as shown on the right, your locale would appear flat just as our neighborhood on the earth's surface appears flat. Thus, light from other galaxies would reach you from different directions and distances in the apparent disc centered on your location. You would interpret your view of the curved two-dimensional universe in terms of an apparent flat-plane universe centered upon you.

If you saw light from the most distant point in your curved universe, you would be able to see the point in any direction, as demonstrated on the right in Figure 18–6. The

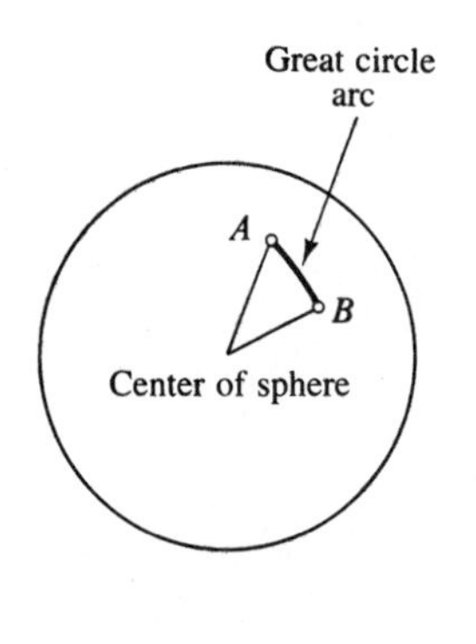

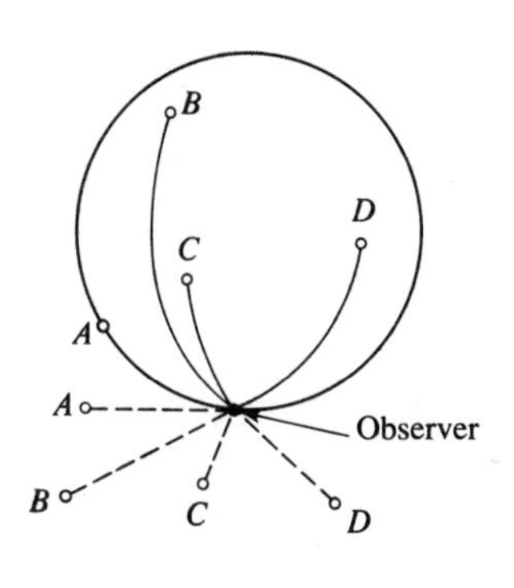

Figure 18–5 Light paths in a spherical-surface universe.

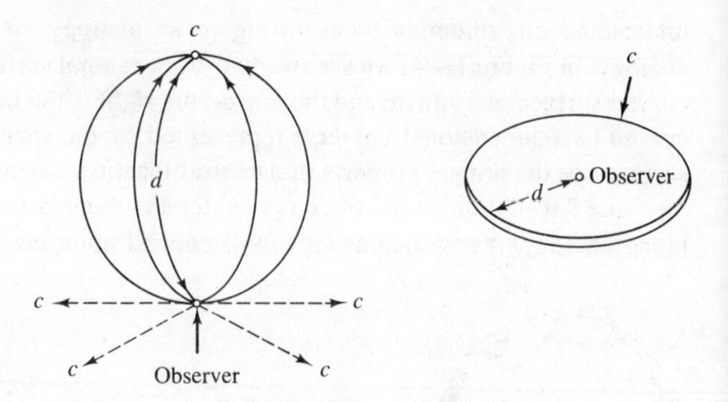

Figure 18–6 The apparent disc universe seen by an observer in a spherical-surface universe.

point would appear to be a circle centered upon your location. If we assume that light could not travel more than halfway around the sphere, the circle—in reality a point—would be interpreted by you as the edge of your apparent universe.

If we now suppose that the sphere expands uniformly, you would see the apparent disclike universe expanding away from you, as illustrated in Figure 18–7. Your expanding universe has a handy property: The more distant a galaxy, the faster it has to travel away from you in order to keep up with the expanding space of the spherical-surface universe. In other words, Hubble's law would hold.

Light, however, travels with a finite speed—300,000 kilometers per second. So you would not see a distant galaxy as it is at the moment you receive its light, but as it was some time in the past. The more distant a galaxy, the longer its light takes to reach you. The more distant galaxies appear as they were in earlier times, and the closer galaxies more nearly as they are at the present. You can never look into space any farther than light would have traveled since the origin of the universe. Thus, the edge of your apparent universe is actually its origin. It completely surrounds you, regardless of your position in the curved two-dimensional space.

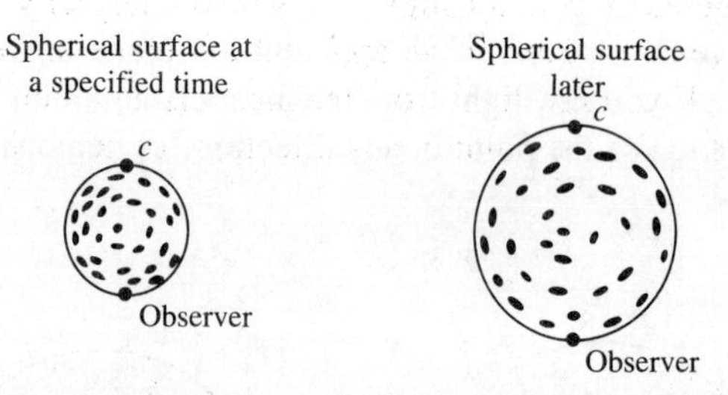

Figure 18–7 The observed expansion for a spherical universe.

Our apparent universe in the big-bang model is the *three-dimensional analogue* of the two-dimensional disc. We interpret the apparent sphere around us as if we are at its center. Thus, as illustrated in Figure 18–8, if we could see the surface of the apparent three-dimensional sphere, we would be looking at the origin of our universe. In the same manner that the third dimension was needed in the spherical-surface universe to explain the apparent flat-disc universe, we need a fourth dimension to explain the apparent three-dimensional volume of our spherelike universe. The fourth dimension is time. Einstein's general theory of relativity incorporates time and the three usual dimensions of space into a four-dimensional space to describe our universe. We can manipulate the mathematics of this four-dimensional space, but we cannot draw a diagram of it any more than the hypothetical observer in the two-dimensional space could have drawn a diagram of the spherical-surface universe.

In the big-bang universe, the matter was packed so tightly at the origin of the universe that it was opaque to radiation. Shortly after creation, the primeval fireball expanded enough that radiation decoupled from matter and it became transparent. The boundary of our apparent spherical universe is consequently the surface of the primeval fireball at the time it became transparent. We can never hope to see completely back to the actual moment of creation.

Some objects have been observed to have red shifts so great that the Hubble law places them nearly 10 billion light-years away. Consequently, we see them now as they were nearly 10 billion years ago. They must be closer than the primeval fireball, so the fireball started out at an even earlier time. The big-bang model predicts that if the universe did in fact begin with a primeval atom, the origin took place more than 10 billion years ago.

At the time the primeval fireball became transparent, it was still hot, around 3000° above absolute zero. The radiation now reaching us from the fireball, almost at the apparent edge of the universe, would be tremendously red-shifted because of its expansion. If the big-bang theory is correct, the fireball radiation because of the red shift would appear to us now like radiation coming from a black body only a few degrees above absolute zero.

Arno A. Penzias and Robert W. Wilson, of the Bell Telephone Laboratories, were testing a very sensitive radio receiver to be used with a satellite communication system in

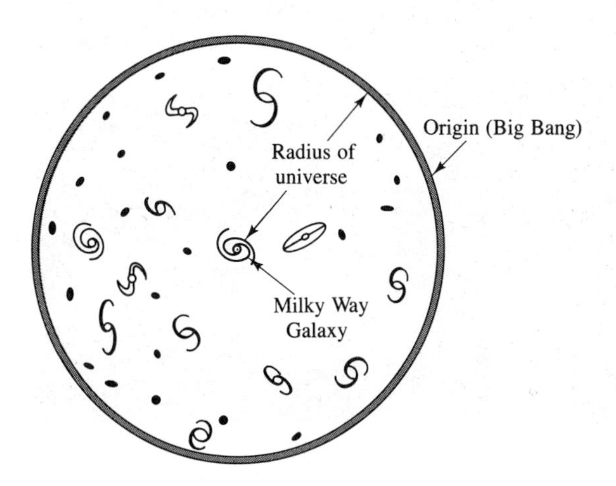

Figure 18–8 The observed spherical universe in the big-bang model.

1965. To their surprise, much as had happened to Jansky thirty years earlier, they detected an unaccounted-for radio signal. Their signal was a very weak one that was present no matter where they pointed their antenna, and it corresponded to radiation which would be received from a black body at 3K. Their observations have been confirmed by measurements at different wavelengths. It is now generally accepted that we are receiving energy from the primeval fireball just as the big-bang model had predicted. Jansky had detected the center of our galaxy; but Penzias and Wilson detected the center of the universe, the primeval fireball which surrounds the apparent universe. The **three-degree background radiation** is considered the most positive evidence so far uncovered in favor of the big-bang model.

From the time a galaxy first formed, there is almost certainly a sequence of evolutionary stages it must have undergone to reach its present stage. Consequently, since we are essentially looking backward in time, we might expect to see differences between the more distant galaxies and the close ones. But it seems that normal galaxies reach the later stages of their evolution in a relatively short time. As yet, we do not have instruments quite sensitive enough to study in detail the extremely faint light of galaxies so remote that they should show the evidence of evolutionary differences.

In the early 1960s another type of object was discovered. These were called **quasi-stellar sources** (popularly, *quasars*), because they are strong radio sources which appear starlike in an optical telescope. In 1963, Maarten Schmidt found that the strange appearance of the spectra of these objects came about, in great part, from extremely large red shifts. Figure 18–9 shows the spectrum of 3C273, the quasi-stellar source with the greatest apparent brightness. A red shift like the one in 3C273 could in principle be caused by strong gravitational fields. However, the general consensus is that the large red shifts

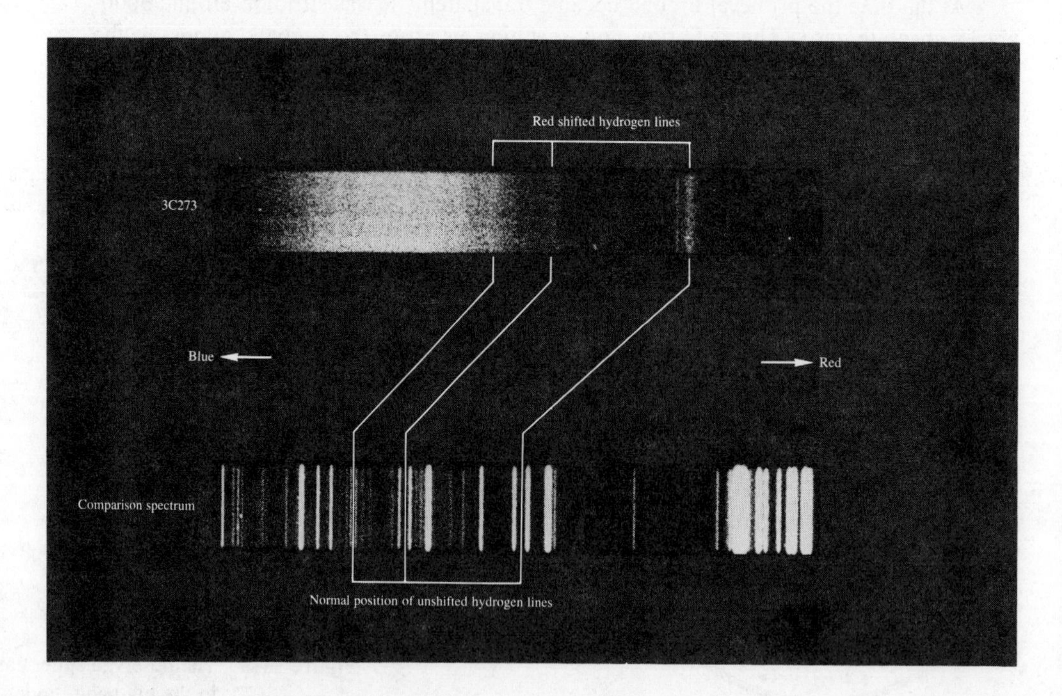

Figure 18–9 Spectrum of the quasi-stellar source 3C273.

are due to high speeds of recession. Subsequently, another class of objects was identified which has all of the characteristics of quasi-stellar sources except that these objects are not strong sources of radio emission. They are called *quasi-stellar* objects.

The largest known red shifts have been found in quasi-stellar source spectra. If these objects are moving with the general expansion of the universe, they are among the most distant objects known. This would imply that the quasi-stellar sources were formed by a process that took place in the early periods of the universe. The process must also be short-lived, because no nearby sources of this type have been observed.

Taking the distances for the quasi-stellar sources predicted from the Hubble law by their red shifts, we find that, based on their apparent brightness, they must be the most luminous sources known. They would emit, in some cases, thousands of times as much visible energy as our Galaxy.

Besides the largest red shifts, the spectra of quasi-stellar sources and quasi-stellar objects display bright emission lines not found in the spectra of normal galaxies. Similar bright lines are seen in the Seyfert galaxies, however. These emission features, as seen in Figure 18–10, are extremely broad, implying that mass motions may be taking place with speeds which are a few percent the speed of light. The emission lines of the Seyferts arise in the nuclei of these galaxies. The nucleus of a Seyfert is quite small, usually less than ten light-years in diameter. The entire mass of a Seyfert is about the same as that of a normal spiral. Why and how the great energy rate of the nucleus can exist, being much greater than that of the rest of the galaxy, is not yet known.

The quasi-stellar objects and quasi-stellar sources share much in common with the Seyfert galaxy nuclei. Some Seyfert nuclei are radio quiet, like the quasi-stellar objects, while some are strong radio emitters, like the quasi-stellar sources. All present a starlike image and are very blue in color. Another common characteristic is their rapid variation in brightness with time. The time of the variations is so short that the dimensions of the source

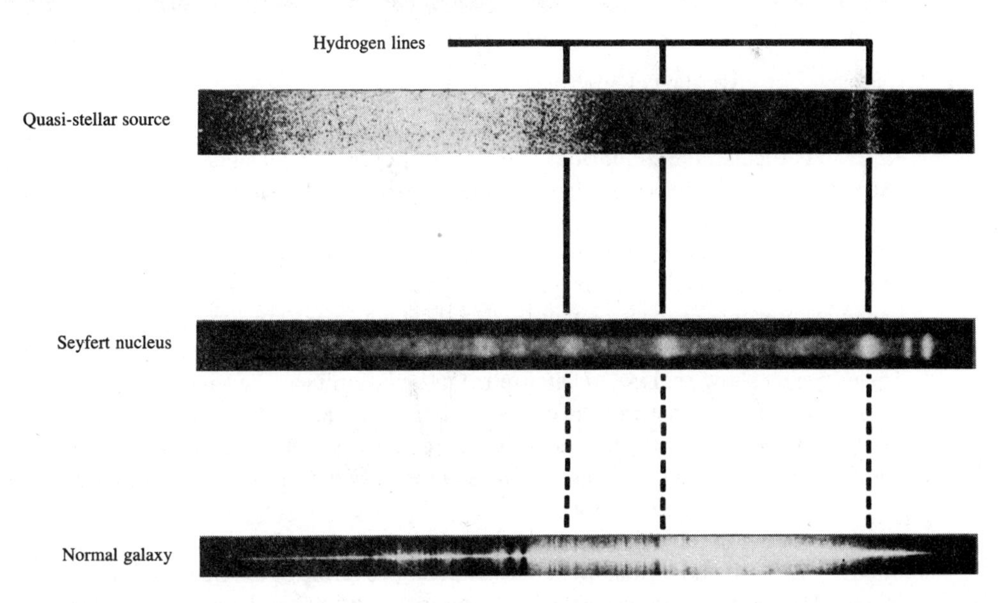

Figure 18–10 Spectra of a quasi-stellar source, a Seyfert galaxy, and a normal galaxy.

of their radiation cannot be much larger than about one light-year for the quasi-stellar sources or objects, up to about ten light-years for the Seyfert galaxy nuclei. Much bigger sizes would cause the light travel-time across the region to be large enough that the observed brightness variations would broaden and average out with time. In other words, if the region were large, by the time energy was received from a more distant part of the region, its nearest part would have already diminished in brightness, thereby reducing the apparent abruptness of any variation. Thus, for all of these objects, we are driven to the conclusion that a tremendous supply of energy is being generated in a very restricted region of space. Figure 18–11 illustrates dramatically the rapid change in the luminosity of one of the quasi-stellar sources.

July 17, 1966 August 8, 1966 September 23, 1966

Figure 18–11 An example of the variation in brightness of a quasi-stellar object.

In particular, the similarities between the Seyfert nuclei and the quasi-stellar sources suggest that the sources may also be nuclei of galaxies. Recent observations, in fact, seem to reveal faint wisps of emission to be surrounding some quasi-stellar sources. There is not yet enough confirmation of these observations to let us be sure that the sources are really nuclei of galaxies. In any case, if the quasi-stellar sources are anything like Seyfert galaxies, it would be extremely difficult to detect a galaxy surrounding them at their large distances. Possibly the best evidence of matter surrounding the sources comes from radio maps which show regions of intense radio emission near the visible object. The radio radiation is characteristic of the emission from a hot, ionized gas, so it is still too early to say the material also contains stars. The radio maps are not unlike those of radio galaxies, which favors accepting the quasi-stellar sources as being part of a larger but fainter galaxy.

Figure 18–12 shows a plot of the range in total luminosity of normal galaxies brighter than our own, Seyfert galaxies, and quasi-stellar sources, plotted in units of the brightness of our Galaxy. It shows that there is considerable overlap of the Seyfert luminosities with those of the normal galaxies at the low-luminosity end and with the quasi-stellar sources at the high-luminosity end of the Seyfert distribution. It is almost as though the Seyferts are a bridge between the quasi-stellar sources and normal galaxies. Because of their many

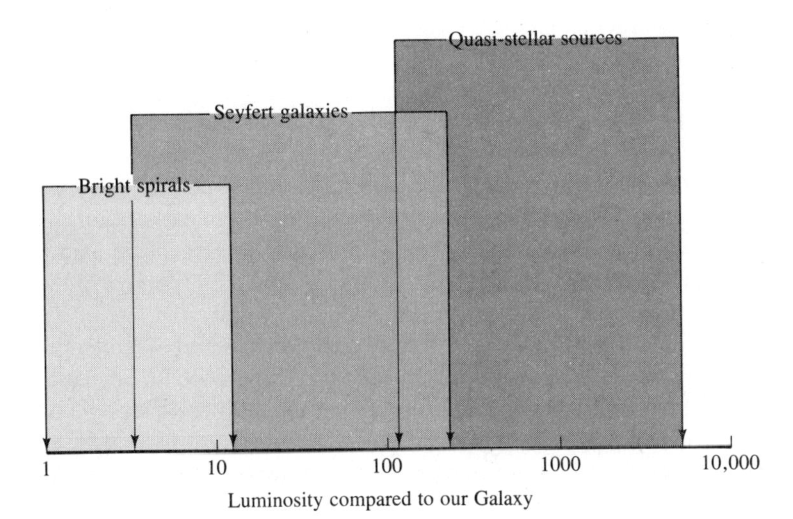

Figure 18–12 Luminosity ranges of normal galaxies, Seyfert galaxies, and quasi-stellar sources.

similarities to Seyfert nuclei, we think that quasi-stellar sources, and quasi-stellar objects as well, may in fact be very distant, luminous objects which are associated with the early stages of a galaxy. They may represent a long-past feature of galaxies during the early evolution of a universe that commenced with a big bang. The quasi-stellar sources and objects may well be objects from the distant past.

The Universe of the Future

What kind of future appears to be in store for the universe in the big-bang model? It will depend on the total mass which the universe contains. The gravitational forces of the various parts of the universe acting upon each other will slow down the expansion. If there is not sufficient total mass, the total gravity will be too small; the universe, although slowing down somewhat, will continue to expand. The distances between the galaxies will increase forever, and the stars within the galaxies will use up their energy sources. At some time in the future the amount of interstellar matter in the galaxies will have been diminished by star formation, so there will not be enough matter left for new stars to form. The universe will become dark, cold, and empty.

There is one alternative to this rather foreboding possibility, assuming the universe contains enough mass that the gravitational forces will be great enough to halt the expansion. Then the matter of the universe under its own gravitational field will begin a collapse. Eventually, all the matter in the universe will again be compressed into a small volume, heating up immensely to form another fireball in which all of the matter will again be reduced to its elementary forms. The primeval atom of Lemaître will be formed again. Two possibilities then arise. The universe may remain a small, dense object forever, or perhaps the reconstituted object will bounce back into an expanding universe with another fireball and formation of new galaxies and stars. Some time later the expansion will again be halted, and contraction will set in once again. The universe will continue its expansion and contraction indefinitely—it will be an "oscillating universe." In the big-bang model, time starts with the creation of the universe when the fireball takes place. In the oscillating universe, time is endless and beginningless. But as far as observers are concerned, the observations available will be limited by the last fireball; they would not be able to detect events outside the oscillation in which they exist.

Calculations put the critical mass—that is, the smallest possible value if the universe is to cease expanding—at about 10^{55} grams. Current estimates of the actual mass of the universe, obtained from Galaxy counts and estimates of how much intergalactic matter there may be below our current ability to detect it, turn out to be about one percent of the critical mass, that is, only 10^{53} grams. At the moment then, we believe the fate of the universe is one of a slow, cold death. But we have overlooked things before—remember how long it took to recognize interstellar matter in our Galaxy. There may still be matter unaccounted for because of the limitation in our observational capabilities. It may yet be that we will eventually find enough mass to make it possible for the universe to be an oscillating one. Then, instead of a cold finish, the universe may have a succession of hot endings and beginnings.

Whether the universe goes through many bangs or not, the model of the universe expanding outward from an initial fireball seems to agree reasonably well with the time scales of a number of objects we have studied. The age of the oldest star appears to be around 10 billion years, the sun is about 5 billion years old, and the ages of the moon and earth are quite similar to that of the sun. How old could the universe be? We can estimate the possible age from Hubble's constant. Taking an object at some distance, Hubble's law predicts its velocity of recession. Dividing its distance by its velocity should give how long it took the object to reach its present distance from us. But, because Hubble's law is a proportionality, we get the same answer at any distance we choose our sample object.

With the currently accepted value of the Hubble constant, the age would be about 18 billion years. This age, of course, represents the maximum which the universe could have at this moment since the fireball occurred. If the universe is slowing down because of gravity, its age would be less. Some people argue that because we find the maximum age of objects in the universe to be around 10 billion years, the universe itself cannot be much older, and it must be slowing down. Our observational capability now permits us to reach almost to the distances where differences in the density of matter in space would be evident in a universe that is slowing down. Hopefully, in the very near future we will have the information to decide just what the universe is doing. Of course, as has happened before, it could well be that some new evidence will show up that makes the big-bang model unacceptable. Until that occurs, it appears to be the best model we have available to help us interpret our observations.

KEY TERMS

universe
cosmology
cosmological model
cosmological principle
expanding universe
primeval atom
big-bang universe
primeval fireball
three-degree background radiation
quasi-stellar sources

Review Questions

1. What is cosmology?
2. What is the big-bang theory? When it was first proposed, what observation was it intended to explain?
3. Why were hydrogen and helium the only elements made in any significant amount by the big bang?
4. Early in the big bang, the matter was opaque. When and why did it first become transparent?
5. Why does the big-bang theory require a curved four-dimensional space? Why doesn't it consider the big bang to be a normal explosion expanding about its center?
6. Suppose you wish to discover a galaxy that could be seen as it was one million years ago. Where in space would you expect such a galaxy to be?
7. If the big bang exploded somewhat over 10 billion years ago, what would be the greatest distance away that an object could conceivably be seen?
8. At what temperature did the matter in the universe cease being opaque? What color would a nearby star be if it had this temperature? Why do we not see the early universe having this same color?
9. Why is the discovery of the 3K background radiation coming from the entire sky so important?
10. What is a quasi-stellar source? What characteristics does one have?
11. What is the main evidence that quasi-stellar sources are very distant? Why do we conclude that they are the most luminous objects in the universe?
12. What evidence is there that quasi-stellar sources are small—less than one light-year in diameter?
13. How can we explain why many quasi-stellar sources appear at great distances and none appears nearby?
14. What similarities are there between the quasi-stellar sources and the nuclei of Seyfert galaxies? What differences are there between Seyfert galaxies and quasi-stellar sources?
15. If quasi-stellar sources really are as far away as their red shifts indicate, what is the great unexplained problem they pose?
16. What is the difference between quasi-stellar sources and quasi-stellar objects?
17. Much effort has been spent to determine the total mass of the universe. Why is there so much interest in this quantity?
18. If the expansion of the universe eventually stops and is replaced by a contraction, what are two possible results of the collapse?
19. What is the present best guess as to the ultimate fate of the universe?

Discussion Questions

1. Suppose two galaxies appear close to each other in the sky. One galaxy is 100 million parsecs away, receding from us at 5500 kilometers per second, and the other is 300 million parsecs away, receding three times as fast at 16,500 kilometers per second. If an observer on the nearer galaxy observed both our galaxy and the distant galaxy, what distances and recession speeds should be measured? Would he or she derive the same value for the Hubble constant as we do?
2. Suppose an explosion occurs and various bits of debris are ejected at different speeds. If we looked at the debris a short time later, what relation should we find between the distance each bit traveled compared with its speed?
3. What observations can be explained by the big-bang theory?
4. Suppose an explorer living at the North Pole of the earth were not aware that the earth is round. He sets off in one direction, traveling in what he thinks to be a straight line over the earth's surface, until he reaches a stake set in the ice at the South Pole. He then retraces his path to the North Pole and sets out in another direction, again arriving at the South Pole. What would he conclude about the stake at the South Pole? Does it make sense to say we are completely surrounded by the point where the big bang occurred?
5. Many centuries ago, those who did not believc the conclusions of Aristarchus argued that the earth was not infinitely big and must have an edge over which one risked falling if he ventured too far from home. How do we avoid worrying about this difficulty now? Today some people argue that if the universe does not extend forever, it must have an edge similar to that envisioned for a flat earth—a barrier we might approach but which could not be crossed. How do we counter the argument that a barrier of this sort exists at the outer regions of the universe?
6. Since the universe became transparent about one million years after the big-bang explosion, where would the point of the explosion be located as seen from earth? Why can't we see this point?
7. What evidence is there that quasi-stellar sources may be nuclei of galaxies?
8. If the expansion of the universe has not been slowing down, how can we estimate how long ago the big bang occurred?
9. If the expansion of the universe is slowing down, would we underestimate or overestimate the age of the universe if we neglected the slowing down? Why?
10. The Hubble constant derived from relatively nearby clusters of galaxies is presently taken to be about 55 kilometers per second per megaparsec. If the expansion of the universe is slowing down, would distant galaxies show a smaller, the same, or a larger Hubble constant? (Remember that looking at distant objects is equivalent to looking back in time.)

Answers to Review Questions

Chapter 1

1. As was noticed in ancient times, the celestial sphere appears to rotate from east to west. This motion causes the sun, for example, to rise in the east and cross the sky to the west.
2. The ecliptic is the path through the stars that the sun appears to take in the course of a year.
3. The sun is farthest north about June 21.
4. The planets slowly moved through the "fixed stars" in the sky. The word *planet* is taken from the Greek word for wanderer.
5. Retrograde motion is the occasional east to west, or backwards, motion of a planet through the stars.
6. No.
7. The Mesopotamian explanation of planets and their motions was that the planets are actual gods in the heavens. By contrast, the Jewish explanation was that the planets were created by a god, but are not gods themselves.
8. Constellations are various groupings of stars which are often associated with various animals or objects.
9. The ecliptic passes through the twelve constellations of the Zodiac.
10. Since the moon is a sphere, the sun shines on only one-half of the moon at any one time. At different times, people on earth see different fractions of the sunlit side.
11. The moon goes through its cycle of phases in a little less than thirty days.
12. The classical Greeks.
13. In a solar eclipse, the moon's shadow falls on the earth. This requires the moon to be closer to earth than the sun is. Hence, for at least the sun and moon, the sky has depth, with one object being farther away than the other.
14. The shadow of the earth on the moon during a lunar eclipse is always circular. The only object which always casts a circular shadow is a sphere.
15. Careful observations with the unaided eye failed to reveal any parallaxes of stars, showing either that the earth was stationary or that the stars were very far away indeed. Many people in ancient times felt it easier to believe that the earth was stationary than that the universe was so large.

16. Aristarchus deduced, from observations of the sun, moon, and lunar eclipses, that the sun is many times larger than the earth. It seems easier to picture a smaller sphere (the earth) going around a larger sphere (the sun) than the other way around.
17. Because *geo* is derived from the Greek word for earth, a geocentric universe is one in which the earth is at the center of the universe.

Chapter 2

1. Ptolemy's explanation of the motion of the sun and moon was that they move in circular orbits, with the earth near the center of those circular orbits.
2. Each planet occasionally undergoes retrograde motion, while the sun and moon never stop their easterly motion through the stars.
3. In Ptolemy's system, a planet does not move along a deferent as the sun and moon do. Rather, the planet moves on an epicycle, a circular path, about a point which itself moves along the deferent. The planet's speed along the epicycle is faster than the motion of the whole epicycle along the deferent, with both motions being eastward, or counterclockwise. When the planet is closer to the earth, it thus has an apparent clockwise motion *as seen from the earth*.
4. The centers of the epicycles of Mercury and Venus are always lined up with the sun, as seen from earth, because Mercury and Venus never get a great distance away from the sun in the sky.
5. An inferior planet has an orbit about the sun smaller than the earth's orbit. A superior planet has a larger orbit than the earth's orbit.
6. Mercury and Venus.
7. The smaller the planet's orbit, the faster it moves in its orbit.
8. Although Mars is going west to east, near the time of opposition the earth passes by it, going even faster in the same direction. As we see Mars from our faster earth, it is being "left behind" near the time of opposition and therefore appears to be going backwards.
9. When an inferior planet passes between earth and sun, it is in inferior conjunction. (Usually at this time it does not lie exactly on a direct line from the sun to the earth, so it is not seen silhouetted against the sun.) When an inferior planet is in the same direction but on the far side of the sun, it is in superior conjunction.
10. It would appear to be half-illuminated.
11. The angle between the line from the planet to the earth and from the planet to the sun is 90°, and the earth is west of the sun as seen from the planet. Therefore, the earth is at western quadrature as seen from the planet.
12. You would look in the direction opposite to the sun. Jupiter would rise in the east at sunset, cross the sky, and set in the west at sunrise.
13. The sidereal period is the time it takes a planet to complete an orbit as seen from the sun, while the synodic period is the time it takes a planet to complete one full cycle of configurations.
14. Tycho made the observations of planetary positions, and Kepler used these observations to deduce his three laws of planetary motion.
15. An ellipse, with the two tacks at the two foci. The sum of distances from the foci to any point on an ellipse must stay the same. In this case, the sum of the distances is the length of the string.
16. The eccentricity is smaller, and hence the ellipse is more like a circle.
17. Kepler's harmonic law states that $P^2 = a^3$. Since a is the semi-major axis of the orbit, the value of a for Saturn is a little less than 10 AU, and the value of a^3 therefore is somewhat less than

1000. The period is therefore close to thirty years, since thirty squared is also somewhat less than 1000.

18. Galileo.

19. Galileo discovered four moons around Jupiter, the phases of Venus, the mountains and craters on the moon, and spots on the sun. (Incidentally, large sunspots can occasionally be seen with the naked eye; but Galileo used the telescope to show the spots really were on the sun and do not result from some object between us and the sun, silhouetted against the surface.)

Chapter 3

1. Velocity is the speed and direction of motion of a body— for example, ten meters per second in the northerly direction.

2. Acceleration is the rate at which velocity changes. An acceleration may result in either a change in speed, a change in direction of motion, or a change in both speed and direction of motion.

3. 200 kilometers per hour.

4. No. In both cases the acceleration was the same—100 kilometers per hour for each second of firing. The longer the acceleration lasts, the greater is the rocket's increase in speed.

5. The *gas pedal,* often called the *accelerator pedal,* can be used to increase the car's speed. The *brake pedal* can also be used to change the car's speed. Often called "deceleration," this decrease in speed is another example of an acceleration. The *steering wheel* can be used to change the car's velocity by changing its direction of motion.

6. (1) A body has a constant velocity unless acted upon by a force.
   (2) The force on a body equals its mass multiplied by its acceleration, with the acceleration being in the same direction as the force.
   (3) For every force acting on a body, there is an equal and opposite force exerted by that body.

7. The full-sized car, with twice the weight of a subcompact car, will have twice the mass. Hence it will require twice the force to accelerate it, as compared with the subcompact model.

8. There are two basic ways to find the mass of a body by using the law of gravity. We can measure the force resulting from one object's gravitational pull on another, or we can measure the acceleration resulting from that force. For instance, we can measure the gravitational *force* of the earth pulling on objects at rest at the earth's surface, which is usually called "weighing" the objects. *Acceleration* of a freely falling body also measures the strength of the gravitational force. The masses of stars, planets, moons, and the sun are found by the accelerations they produce in other objects.

9. We can find the mass of an object by measuring how much acceleration it undergoes for a given force. This technique has been used by astronauts in space, where "weightless" conditions prevent the use of normal doctor's scales to find mass. The astronauts measure how rapidly they are accelerated by a spring. The faster their acceleration, the less is their mass.

10. (3) The same (by the law of reaction)

11. (1) Eighty-one times greater (by Newton's law of force)

12. The force of gravity would double.

13. The force of gravity between the earth and moon would become four times stronger than it is now if both bodies doubled in mass.

14. Since the square of the distance would become four times greater, the force would become one-fourth as great.

15. The side of earth away from the moon has less of an acceleration toward the moon and is somewhat "left behind" as the rest of the earth accelerates more toward the moon. Therefore, the tides there are "higher."
16. Tides on earth depend on the *difference* between the acceleration on one side of the earth and that on the opposite side. The nearest point on earth is 3.3 percent closer to the moon than the farthest point. But the point on earth nearest to the sun is only 0.01 percent closer than the farthest point. So even though the moon's gravitational pull is less than the sun's, the difference in pull across the earth is greater.
17. The earth rotates on its axis.
18. Observing aberration of starlight and the parallax of the closer stars.
19. Rotation of the earth on its axis.
20. The gravitational pull of the moon and sun on the equatorial bulge of the earth.
21. (1) Revolution about the sun
    (2) Rotation on its axis
    (3) Precession of its axis of rotation
    A fourth motion of the earth, of which the effects are not as obvious, is the motion of the earth about the center of mass of the earth-moon system.

Chapter 4

1. A star on the celestial equator would rise exactly due east, reach its highest point when on the celestial meridian, and set due west.
2. A star at the celestial pole would not appear to move at all. All the other stars would appear to go around it, while it remained stationary.
3. At the equator.
4. Since your latitude equals the altitude of the celestial pole, you would be at about latitude 40°N.
5. (1) Celestial equator
   (2) North Celestial Pole
   (3) Declination
   (4) Right ascension
6. Longitude 0° is defined by the meridian through Greenwich, England. Zero hours right ascension is taken as the vernal equinox, or the intersection of the celestial equator and ecliptic where the sun goes from the Southern to Northern Hemisphere.
7. The celestial sphere appears to rotate 360° in 24 hours, or 15° per hour. This rate corresponds to 1° in 4 minutes.
8. Moving about 1° in 4 minutes, the sun will take about 2 minutes to move its own diameter.
9. Sidereal time is time measured directly from the stars. (It is defined as the local hour angle of the vernal equinox.)
10. Time is defined by the rotation of the earth, which in turn is measured by observing stars as they transit an observer's meridian.
11. Local apparent solar time.
12. While completing one rotation relative to the stars, the earth moves along its orbit about one degree. The apparent position of the sun thus changes, and it takes about four minutes for the earth to rotate through this extra degree.

13. Whenever you moved east or west, you would have to reset your watch by four minutes for each degree of longitude, which would be downright inconvenient.
14. Your standard time is an approximate measure of how far ahead of or behind Greenwich you are as the earth rotates. And since the earth rotates west to east, time zone boundaries lie along meridians of longitude, west or east of Greenwich.
15. Nine o'clock in the morning. Since the earth rotates toward the east, California *follows* New York by three hours. Three hours after New York has noon, California will have noon.
16. Stars near the sun in the sky are not easily seen, so the constellations generally associated with each season are the ones roughly in a direction opposite to the sun in the sky. As the earth orbits the sun, the direction *toward* the sun, and hence the direction *away* from the sun, changes. Different constellations therefore appear as the seasons progress.
17. The tilt of the earth's axis by 23½ degrees means that when the earth is in one part of the orbit, the North Pole of the earth is tilted toward the sun. The Northern Hemisphere of the earth then receives more sunlight than the Southern Hemisphere. When the earth is on the other side of the sun, the situation is reversed; the Southern Hemisphere receives more sunlight and warmth.
18. Due north, because the sunlight comes to you over the North Pole at midnight.
19. If we measure how long it takes the sun to make two successive passages by a given star as the sun appears to travel along the ecliptic, we measure a sidereal year. If we use the time it takes the sun to make successive passages through the vernal equinox, we measure a tropical year.
20. It takes about 365¼ solar days for a tropical year. Every four years, the ¼ day adds up to an extra day.

Chapter 5

1. True.
2. Yes. When the moon is new as seen from the earth, the earth would be full for someone on the moon. At time of first quarter moon, the earth would appear in last quarter phase for a lunar observer. The earth would, in every case, have just the opposite phase that the moon had at that moment.
3. Waxing crescent phase, almost to first quarter. Since the first quarter moon follows the sun in the sky by about 6 hours, the first quarter moon would transit six hours after the sun, or about 6:00 P.M., assuming standard time is used. For the situation posed in this question, the moon transited five hours after the sun, and is therefore five hours or 75° to the east of the sun. This places it 15° short of first quarter.
4. Since a full moon is opposite to the sun in the sky, it will be highest above the horizon when the sun is farthest below the horizon, which is at midnight.
5. No, a narrow crescent moon cannot be far from the sun in the sky, so it would be too close to the sun to be above the horizon at midnight in temperate latitudes. (At the North or South Pole, just after the sun has set for the first time in six months, a narrow crescent moon could be seen continuously as it moved around the sky just above the horizon.)
6. Since the moon is low in the east soon after the sun sets in the west, the moon is opposite to the sun in the sky. This position corresponds to full phase, not crescent phase, so the moon must be undergoing an eclipse.
7. A node of the moon's orbit is a point where the moon's path in the sky crosses the ecliptic.

8. The nodes of the moon's orbit slowly move westward along the ecliptic, so it takes the sun less than a full year to go from one node back to the same node.
9. An *umbra* is that part of a shadow where all direct light from the light source is cut off, while in the *penumbra* only part of the light is blocked.
10. True. Although the nodes slowly regress along the ecliptic, at any one time they are exactly opposite to each other. Incidentally, this explains why if the sun is at one node and the moon at the other, the sun, earth, and moon line up, and we have an eclipse of the moon.
11. It takes the moon about two weeks to go from new moon to full moon, the two phases of the moon when solar and lunar eclipses can occur.
12. It takes the sun about six months to go from one node of the moon's orbit to the other.
13. The moon's location in its eccentric orbit. When the moon is near its closest approach to the earth, its umbra reaches the earth, giving a total eclipse. When the moon is near its greatest distance from the earth, it is too far away for its umbra to reach the earth, and an annular eclipse takes place.
14. The color in the earth's umbra results from red light being bent by our atmosphere into the umbra. Because the moon does not have an atmosphere, the same effect cannot occur in the moon's umbra.
15. A lunar eclipse can be seen from the entire night side of the earth, while a solar eclipse is visible from only a small part of the day side of the earth.
16. After the entire surface was heavily cratered, molten rock flowed over some of the surface and covered the craters in these regions.
17. Lunar rills are valleys cutting across the surface of the maria. (Some may have been formed by channels of flowing lava.)
18. Apollo astronauts found several layers of rock on the side of a lunar rill, indicating successive flows of lava formed that mare.
19. The dense core of the earth causes the earth to have an average density of 5.5 grams per cubic centimeter. Since the moon's average density is only about 3.3 grams per centimeter^3, it can't have a significant dense core.
20. Moon rocks are depleted in those elements which vaporize at temperatures somewhat hotter than those normally occurring on the earth's surface. Hence we conclude that the moon material went through a higher temperature phase, at which time those elements were vaporized and escaped into space.
21. The earth, being a larger target for objects to run into, must have suffered many more collisions than the moon. There are only a few craters which can still be seen on earth, which shows the power of wind and water erosion to erase features on the earth's surface.
22. The three theories of the origin of the moon are (1) the *fission* theory, which describes the moon as having come from the earth; (2) the *accretion* theory, in which the moon formed from material left orbiting the earth after the earth had formed; and (3) the *capture* theory, according to which the moon was formed in another part of the solar system and ultimately captured in the earth's gravity.

Chapter 6

1. Most objects in the sky are so far away that the only way of understanding them is by studying light from them. Other scientists can often do experiments, manipulating whatever is being

studied and seeing how it responds. We have no means of manipulating a star to see what effect our efforts have on it.

2. Diffraction is the spreading of waves around an obstacle.
3. If light is directed at a group of slits, the light waves from each slit interfere with light from the other slits. The regions where light from all slits is *in phase* (where the wave crests add together) depend on the wavelength of the light. Hence, interference phenomena can be used to measure the wavelength of light.
4. Red light has a longer wavelength than yellow light.
5. In the wave model of light, color depends on the wavelength of the light.
6. The longest-wavelength visible light is red, and the shortest is violet.
7. The highest-frequency visible light is violet.
8. Since X-rays have very short wavelengths compared with radio waves, X-rays have a much higher frequency.
9. A photon is a particle of light.
10. The photoelectric effect is the process in which light is absorbed by a metallic surface, and electrons are given off as a result. The effect occurs only for light of less than some critical wavelength, depending on the type of metal. Its significance is that it led to Einstein's confirmation of the particle behavior of light.
11. Einstein described light as consisting of photons, with the energy of a photon proportional to its frequency. A long-wavelength, low-frequency photon does not have sufficient energy to overcome the forces holding the electrons in the metal.
12. Compared to blue light, a photon of red light has longer wavelength, lower frequency, and less energy.
13. The photon of twice the wavelength would have half the frequency and, hence, half the energy of the shorter-wavelength photon.
14. A continuous spectrum shows light of all wavelengths, with no spectral lines visible. An emission spectrum has bright spectral lines, with lines appearing only at certain discrete wavelengths. An absorption spectrum is like a continuous spectrum except that at certain wavelengths there is less energy, giving dark spectral lines.
15. When an atom absorbs a photon, an electron absorbs the energy of the photon, causing the electron to go into a larger, higher-energy orbit. Unless the photon has exactly the right energy (or wavelength) to place the electron in one of the allowed electron orbits, the electron cannot make the jump. Hence the atom cannot absorb the photon.
16. The electrons in helium are subjected to a stronger pull from the nucleus, which has two positive charges, thus making the orbits smaller than in hydrogen. In addition, the two electrons in helium repel each other, and this extra force further changes the electron orbits. Because the orbits are different, jumps between them require different amounts of energy. These result in different wavelength spectral lines.
17. The Doppler effect is the apparent change in frequency or wavelength of a signal which occurs because the distance between the observer and source of the signal is changing.
18. Yes, if the observers are in different places. An object might be between the two observers, moving toward the one and away from the other. Also, the object could be stationary, with one observer traveling toward it and the other moving away from it.
19. The star and the observer are getting closer together.
20. The Doppler shifts will be the same because the speeds of approach are the same.
21. Radial velocity is the rate at which the distance from an observer to an object is changing.

22. The sun is hot, so its emitted radiation is largely in the visible region of the spectrum. The planets, being comparatively cool, emit much less radiation. What they do emit is mostly at longer wavelengths, in the infrared and radio regions of the spectrum.
23. A blue star must be a relatively hot star to radiate so much blue light.
24. It gets brighter at every wavelength, and the point where it reaches peak brightness moves to a shorter wavelength.
25. (5) 81 times more, since $3^4 = 81$.

Chapter 7

1. The objective of a refracting telescope is a lens, while a mirror is used to form the image in a reflecting telescope.
2. Focal length is the distance behind the lens at which the image of a distant object is brought to a focus.
3. The lens with the longer focal length will produce the larger image.
4. False. Some telescopes are used only for taking photographs and are never used for visual observations.
5. The magnifying power is greater for a longer focal length objective and a shorter focal length eyepiece. The amount of magnification equals the focal length of the objective divided by the focal length of the eyepiece.
6. On most telescopes, by exchanging the eyepiece for one of a different focal length.
7. A very large telescope has great light-gathering power, which lets us study very faint and distant objects.
8. The eyepiece for the Newtonian focus is on the side of the telescope, toward the end where the light enters. For the Cassegrain focus, the eyepiece is located behind the objective mirror.
9. The resolving power of a high-quality telescope is determined by the diffraction of light. The larger the telescope, the less diffraction occurs, so a large telescope would have better resolving power.
10. Use an interferometer to collect radiation at two or more places and then combine the signals. The resolving power depends on the separation of the collectors. It can be made much larger than the diameter of a single telescope, giving improved resolving power.
11. Since diffraction is greater at longer wavelengths, and radio wavelengths are *much* longer than wavelengths of visible light, the radio telescope would have much poorer resolving power.
12. Since the number of electrons ejected from a metal surface depends on the number of photons striking the surface, the electrons released due to a star's light can be counted to measure the star's brightness. Filters can be used to let just one color fall on the metal surface at a time, so the brightness of each color can be found. Comparing the brightness of a star in different colors will usually indicate the star's temperature.
13. A reflecting telescope.
14. Since the planets shine by reflected light and are too cool to emit much visible light of their own, we must use longer wavelengths to measure radiation emitted by the planets. The hotter the planet, the more intense are the infrared and radio waves. The strength of the emitted radio waves can thus be used to find a planet's temperature.
15. The time delays and Doppler shifts in the echo.
16. False. Objects outside our solar system are too distant to be detected with radar.

17. The most accurate way to measure the mass of a nearby planet is by measuring the gravitational pull it exerts on a spacecraft. The acceleration of the probe is detected by changes in the Doppler shift of radio signals it transmits to the earth.
18. If the spacecraft goes behind the planet (as seen from the earth), radio signals transmitted to the earth as the spacecraft disappears must travel through successively deeper layers of the atmosphere. The radio signals slow down as they pass through the atmosphere. The amount the waves are retarded measures the amount of gas at each layer of the atmosphere. The process is repeated as the probe reappears from behind the planet.

Chapter 8

1. 25,000. Distances of the planets are very much greater than the sizes of the planets.
2. Radio and radar telescopes and space probes.
3. Mercury never appears very far from the sun. Even at greatest elongation, Mercury is difficult to see without a telescope, appearing low in the twilight sky.
4. Venus is covered by opaque clouds.
5. By analyzing the range of Doppler shifts in radar signals reflected from each planet's surface.
6. The temperature of the gas and the velocity of escape from the planet.
7. The lighter particles in a gas move faster than the heavier particles, so hydrogen and helium atoms can more easily reach the velocity of escape.
8. The strength of the radio emission from Venus' surface showed how hot it is.
9. A covering over the surface must allow some visible light to come in, but it will absorb more infrared radiation given off by the surface. This prevents the emitted radiation from escaping easily and effectively traps the heat.
10. The terrestrial planets are Mercury, Venus, Earth, and Mars. The Jovian planets are Jupiter, Saturn, Uranus, and Neptune. Pluto is sometimes included in the list of terrestrial planets.
11. Of the terrestrial planets, Venus has the most atmosphere, while Mercury has essentially none.
12. Yes. Faulting in the crust of Mars was discovered by *Mariner 9* on the same region of the planet where many volcanoes occur.
13. *Mariner 9* photographed meandering valleys apparently formed by running water.
14. Martian craters are generally much more eroded. Also, no craters on Mercury or the moon have been found showing the evidence of fluid flow in the ejected matter that sometimes is found on Mars.
15. The ridges may be debris that collected at the edge of thick ice caps during previous ice ages on Mars.
16. The Jovian planets are larger, more massive, and lower in density than the terrestrial planets. The Jovian planets also contain a smaller percentage of rock and metals but more light gases and ices. Their rotation rates are greater, and they have more satellites.
17. First, Jupiter has a sizable equatorial bulge. Second, there is a strong belted structure in the planet's atmosphere. Finally, the rapid rotation presumably is important in generating Jupiter's strong magnetic field.

18. Short radio waves which are generated by electrons trapped in the magnetic field. The bursts of radiation at longer wavelengths are related to the moon Io as it interacts with the magnetic field. Fast electrons apparently accelerated by Jupiter's magnetic field have been detected near the earth.
19. The Great Red Spot is a cloud formation, apparently resulting from a fairly stable circulation pattern in the clouds, somewhat like a hurricane on Earth.
20. Venus rotates backwards, or retrograde. Uranus is the only planet with its axis of rotation in the plane of its orbit about the sun.
21. Uranus was discovered by accident, when an astronomer noticed that an apparent star had a disc appearance when viewed through a telescope. The location of Neptune was discovered by its gravitational perturbation on Uranus' orbit. Pluto was discovered through a long systematic search for a ninth planet. The search was based upon erroneous calculations involving Uranus' motions.
22. When Uranus passed between the earth and a star, observers on earth saw the star dim briefly each time one of the rings blocked some of the starlight.
23. Pluto's orbit is highly inclined to the orbit of Neptune; so even when Pluto gets closer to the sun than Neptune, it will not get very close to Neptune's orbit, let alone to Neptune itself.

Chapter 9

1. True.
2. Between the orbits of Mars and Jupiter.
3. The easiest way to pick out an asteroid from among the stars is to take a long time-exposure photograph. The motion of the asteroid among the stars causes it to move across the photographic plate. You can easily pick out the trail of the asteroid from the pointlike images of the stars.
4. The visible head of a comet, made up of gases flowing from the central ices, is much larger than the largest asteroids. However, the "iceberg" itself is much smaller than a large asteroid.
5. Due to their irregular shapes, asteroids reflect light unevenly as they rotate. The rotation period can be found by timing how long one complete cycle of light variation takes.
6. Comets are predominantly icy, while asteroids are not.
7. A comet as close to the sun as are the asteroids would eventually be destroyed as its ices evaporated.
8. The solar wind carries ions from the comet away from the sun, forming the gas tails. But atoms which are not ionized are not transported by the solar wind and, in general, do not flow into the gas tail.
9. The pressure of sunlight shining on very tiny solid particles from the comet forces them away from the sun, forming the dust tail.
10. The increased solar heat causes the ices to evaporate faster, so more gas and dust is given off; and the intensity of sunlight, which excites the gas and reflects off the dust, also increases as a comet approaches the sun.
11. Many newly discovered comets have incoming orbits, before planetary perturbations affect them, which extend out to the order of 50,000 AU.
12. Comets arriving from the Oort comet cloud come from any point in the sky, with orbits which are not at all confined to the ecliptic.
13. A meteor is a fast moving trail of light resulting from a solid body entering our atmosphere. If the body reaches the ground, it is called a *meteorite*.

14. False. The particles causing meteor showers are normally too small to reach the earth's surface as meteorites.
15. Meteor showers occur when the earth passes near the orbit of a prior comet, and the earth sweeps up particles given off by the comet.
16. (2) A few postage stamps. Even fairly bright meteors have masses of less than one gram.
17. The zodiacal light is a faint band of light along the ecliptic, caused by sunlight reflecting off dust particles orbiting the sun, mostly inside the earth's orbit.
18. If two objects collide very gently, because they are on very similar orbits, they will probably stick together. If the objects are on considerably different orbits, they will collide much more violently and will probably break into smaller pieces.
19. The gases in the sun are so hot that if some of the gas could be removed, the gas would rapidly expand and soon be lost to interstellar space. The material from which the planets formed had to be much cooler than the temperature in the sun.

Chapter 10

1. The sun is much closer than the stars seen in the night sky, which makes it appear larger and brighter.
2. False. The photosphere is a very thin layer surrounding the interior of the sun.
3. Photospheric granules are regions where hotter gases from deeper layers have risen to the photosphere and are radiating away their excess heat.
4. Every point on a solid ball would rotate with the same period, but the equatorial regions of the photosphere are observed to have shorter rotation periods than the polar regions. The photospheric granulation results from small-scale motion, demonstrating a nonrigid nature. Also, only a gas could produce the many absorption lines seen in light from the photosphere.
5. An H− ion is a normal hydrogen atom with an extra electron attached.
6. Wiggly lines result from rising and descending volumes of gas in the photosphere.
7. By photographing the sun using only the light of a strong absorption line.
8. Spicules are thin jets of chromospheric gas extending upward at the top of the chromosphere.
9. At the center of the disc, the chromospheric gases are silhouetted against the brighter light from the lower photosphere. At the limb, the same gases look brighter compared with the blackness of outer space behind them.
10. The chromospheric network is the pattern of large cells outlined with bright lines usually seen in photographs of the sun taken in the light from ionized calcium atoms.
11. Throughout the photosphere, the temperature decreases at higher altitudes. In the chromosphere, however, it rises at higher altitudes. At the top of the chromosphere, the temperature sharply increases to the very high temperatures of the corona.
12. The identification of coronal spectral lines from the often-ionized iron and other elements confirms the high temperature.
13. A black body absorbs all light incident upon it, so it must be opaque. The corona is almost perfectly transparent, so it is far from being a black body.
14. X-rays from the sun are emitted from the corona because of its extremely high temperature.
15. False. Individual sunspots occasionally last as long as a few months, but never for many years.
16. The umbra is darker, and hence cooler, than the penumbra.

17. The strength of the magnetic fields in the photosphere is found by the amount that spectral lines are split into components; the larger the splitting, the stronger is the field.
18. The faster rotation of the equatorial regions of the sun causes the field to wrap around the sun many times in a few years. This wrapping of the field makes it more concentrated.
19. The magnetic field emerging like a kink in a rope from beneath the photosphere has the direction of the field coming up in one spot and returning in the other spot. This gives the spots opposite magnetic polarities.
20. Sunspots, the corona, and apparently the chromosphere were all much more difficult (or impossible) to observe during that time, and aurorae were very rare on the earth. Possibly this decrease in solar activity helped cause the unusually cold weather on the earth during the Maunder Minimum.
21. Prominences and filaments are both clouds of relatively cool gas suspended in the corona, but they are simply viewed from different angles. A prominence is seen above the limb of the sun, and a filament appears as a dark cloud silhouetted against the disc of the sun.
22. A plage is not as bright as a flare; and a plage changes brightness or shape slowly (over a period of days), while a flare brightens rapidly (within a few minutes).
23. The northern lights result when ionized particles originally from the sun strike the earth's upper atmosphere, causing it to glow faintly.

Chapter 11

1. Parallax is the apparent change in the position of an object due to the change in position of the observer.
2. A parsec is a unit of distance, equal to about 3×10^{13} kilometers or 3.26 light-years.
3. By definition, a parsec is the distance away a star would be to have a parallax of one second of arc.
4. The distance to a star in parsecs is found by dividing 1 by the parallax of the star (in seconds of arc). One divided by 0.01 is 100, so the star is 100 parsecs away.
5. For distant stars, the parallax angle is too small to be measured accurately.
6. Proper motion is the apparent change in the position of an object in the sky which does not result from the earth's orbital motion around the sun. Proper motion therefore results from the motion of the object relative to the sun; it is measured in seconds of arc per year.
7. Doppler shifts are just a measure of velocity along the line of sight and are not affected by distance, so the stars would have the same Doppler shift if they have the same velocity with respect to the sun.
8. Both proper motion and tangential velocity are measures of the motion of the star relative to the sun across our line of sight; but tangential velocity is the speed in kilometers per second, and proper motion is the apparent change in the star's position, measured in seconds of arc per year.
9. Since the stars have the same parallax, they are the same distance away. Hence, the only way to explain why one has twice as much proper motion as the other is to conclude that it has twice as much tangential velocity.
10. The peculiar velocity of a star is its velocity through space relative to a local standard of rest, while its space velocity is its motion through space relative to the sun.
11. Obviously, the other train is overtaking you and heading west faster than you are. To find its motion, simply add its motion relative to you (10 kilometers per hour) to your own motion (40

kilometers per hour). Thus, the other train is moving west at 50 kilometers per hour, relative to a local standard of rest (the surface of the earth, in this case).

12. (1) Since the solar apex is the direction of the sun's peculiar motion, on the average we are getting closer to stars near the apex; so, on the average, they should show a Doppler shift to shorter wavelengths.
13. (4) Stars 90° away from the solar apex are, on the average, stars that the sun is "passing by" on its motion toward the apex. Their average proper motion away from the apex is just a reflection of the sun's peculiar velocity.
14. The local standard of rest is any point moving with the average velocity of the stars in the neighborhood of the sun. Usually, this neighborhood is taken as about 100 parsecs around the sun. Since the sun really is no different from other stars in the neighborhood, its velocity is included in the average.
15. Apparent magnitude is a measure of the brightness of a star as it actually appears in the sky, while absolute magnitude is a measure of how bright it would appear if we could place the star at the standard distance of 10 parsecs away and make sure the space between us and the star is perfectly transparent.
16. Since the smaller the magnitude, the brighter is the star, the third-magnitude star is brighter than the fourth-magnitude star.
17. Since the stars have the same absolute magnitude, they are emitting the same amount of light each second. The one that looks much brighter looks that way because it is closer to us.

Chapter 12

1. The Draper classification was based on classifications made with photographic plates, which allowed the spectra to be recorded and thus examined in much greater detail.
2. Photographic spectra are a permanent record for detailed study. Time exposures can record much fainter light than the eye can detect, so photographic spectra can be used for fainter stars and for higher dispersion (where the light is more spread out) to show finer detail. Finally, with an objective prism, large numbers of spectra are obtained at one time.
3. O, B, A, F, G, K, and M.
4. In cool stars, almost every hydrogen atom has its electron in the first orbit, and collisions between atoms are usually too weak to excite the electron to a higher orbit. Since the visible spectral lines of hydrogen, the Balmer lines, occur when a photon is absorbed by an electron in the second orbit, very few atoms in the atmosphere of a cool star are able to absorb Balmer lines.
5. In very hot stars, collisions between atoms, ions, and electrons are relatively violent, often causing an electron to be ejected from any hydrogen atom involved in the collision. Without an orbiting electron, a hydrogen ion cannot produce spectral lines.
6. Hydrogen and helium are the two abundant elements in stars. About three-quarters of the mass is hydrogen and about one-quarter is helium.
7. The hottest stars are hot enough that virtually all the calcium in the atmosphere is ionized twice; that is, two of the normal quota of electrons are removed. The strong spectral lines of doubly ionized calcium are outside the visible spectral region (occurring at short ultraviolet wavelengths).
8. The differences in the lines in stellar spectra result from temperature differences. But R stars have temperatures similar to K stars; N and S stars are about as hot as M stars. The unusual

appearance of the spectra of R, N, and S stars must therefore result from a difference in the amount of some of the elements present in these stars.

9. The star with four times the luminosity must have four times the surface area (or twice the radius) of the other star.
10. The A star is hotter, so in one second it emits more light per unit area. But since it emits the same total amount of light each second as the G star, the A star must be considerably smaller.
11. Absolute magnitude is plotted vertically, while spectral type is shown on the horizontal axis. Absolute magnitude corresponds to luminosity; spectral type, to surface temperature.
12. White dwarfs are found in the lower left part of the H-R diagram. Supergiants are at the upper right. Dwarfs, or main-sequence stars, lie on a diagonal band from upper left to lower right. Giants lie below the supergiants but above the main sequence.
13. The most luminous stars are visible to the naked eye at great distances, so the naked-eye stars include very luminous stars in a large volume of space. However, the low-luminosity stars can be seen only if they are in a small volume of space close to the sun.
14. Th G refers to the spectral type of the sun; the 2 means the spectrum is two-tenths of the way from G0 to K0; and the V is the luminosity class of the sun, a dwarf.
15. The color of a star is usually given as the difference between its magnitude in blue light and its magnitude in yellow light. These magnitudes can be found from photographs (where a combination of the wavelength sensitivity of the photographic plate and a color filter determines whether blue or yellow light is being recorded) or by photoelectric cells (where filters select the wavelength being recorded).
16. Open, or galactic, clusters have fewer stars, show a smaller concentration of stars toward the center, and lack the round appearance of globular clusters. Open clusters are generally found closer to the Milky Way in the sky, and their H-R diagrams usually have main sequences extending to hotter, more luminous stars than the main sequences of globular clusters. More red giants appear in globular clusters.
17. (2) The cooler the stars at the upper end of the main sequence, the more giants there are in the cluster.
18. Distance modulus is defined to be apparent minus absolute magnitude. A distance modulus of zero means that absolute and apparent magnitudes of the star are equal. Since absolute magnitude is the apparent magnitude a star would have if it were ten parsecs away, the star's distance would be ten parsecs.
19. The apparent magnitude and color index (or spectral type) of stars in the cluster.
20. The rotational velocity of the star and the angle between the star's rotational axis and the line of sight.
21. The splitting of spectral lines of atoms in a magnetic field, called Zeeman splitting, is large enough to be measured in a small number of stars. The larger the splitting, the stronger the field.

Chapter 13

1. (4) soon after the American Revolutionary War.
2. Visual binaries are two stars close together which over a period of time show orbital motion about each other. Spectroscopic binaries are identified by the periodic changes in their Doppler shifts. An eclipsing binary appears as a single star which periodically changes in brightness.

3. Separation of the two components of a binary and the direction from the brighter star to the fainter.
4. The brighter star is the primary.
5. If the primary appears exactly at the center of the circular orbit, the true orbit is circular; otherwise, it is an ellipse viewed at an angle.
6. From observations of the apparent orbit of a binary star, we can derive the orbital eccentricity, inclination of the orbit, semi-major axis in seconds of arc, and period. If we know the star's distance, we can find the semi-major axis in astronomical units and the mass of the system.
7. The amount of motion of each component relative to the center of mass of the system gives the ratio of the masses; the greater the motion, the less is the mass.
8. Since neither component is orbiting either toward or away from the earth, neither would show any Doppler shift due to orbital motion and the lines would appear single.
9. The velocity curve is a measure of the orbital speed of the components. By adding up the velocities at various times, we can calculate the total distance traveled toward and away from the earth. If the inclination of the orbit is also known, we can then find the orbit size.
10. The masses may be found if we measure the velocity curves of each component, the inclination of the orbit, and the period.
11. If a spectroscopic binary is also an eclipsing binary, the orbital inclination can be found from an analysis of the light curve.
12. During a total eclipse the light from the binary does not vary, so the light curve will have a flat minimum. For a partial eclipse, the light will steadily fade until mid-eclipse, and then it will start brightening. Hence, the minimum will be more pointed than flat-bottomed.
13. Tidal distortions and reflection effects.
14. A visual double can be noticed on a single photograph of the sky. A spectroscopic binary requires good quality spectra to be noticeable (which limits us to brighter stars). An eclipsing star will not be discovered unless someone measures the star's brightness at several times, including one time when an eclipse takes place.
15. False. More stars exist in binary systems than as isolated stars.
16. The mass-luminosity relationship shows that, in general, the more massive the star, the more luminous it is. The range of luminosities among stars is much bigger than the range of masses.
17. Densities of stars are found to fall in a much greater range than stellar masses.
18. Main-sequence stars follow the mass-luminosity relationship, while white dwarfs do not. Giants generally do not follow it very well.
19. The components must be relatively close together, and one star must expand enough to exceed the critical surface.
20. Most, and presumably all, novae occur in close binary systems. The current theory describes the outburst as resulting when matter from the larger star impacts on the smaller star, which is probably a white dwarf.
21. Some stars have been observed to suffer nova outbursts more than once.

Chapter 14

1. The pressure forces are balanced by the weight of the overlying matter.
2. The nuclear force.

3. Protons must be moving very fast (and hence have a high temperature) if they are to overcome the repulsive force caused by their like charges.
4. In the equation $E = mc^2$, the mass converted to energy is multiplied by the speed of light squared, which is a huge number—approximately 9×10^{20} $(cm/s)^2$.
5. Massive stars consume hydrogen so much faster that they exhaust their larger hydrogen supplies in a shorter time.
6. Its mass.
7. Only by calculating a model which fits the sun's observed luminosity, radius, and spectral type.
8. (2) All main-sequence stars are undergoing hydrogen fusion in their cores.
9. By calculating how long the present hydrogen supply in the sun's core will last with its current luminosity.
10. A zero-age main sequence is the one expected for a group of stars having the same chemical composition and deriving their luminosities by hydrogen fusion. It is called zero-age because the composition would not yet have been changed by nuclear fusion.
11. Because eventually it must consume all of the hydrogen in its core.
12. Hydrogen fusion in a shell around the core. As the temperature in the shell is raised by the contracting core, the fusion rate in the shell will increase.
13. The central temperature of the sun is too low for helium to react.
14. A red giant. The density will increase in the core but decrease at the surface.
15. The massive O-type stars evolve from the main sequence in about a million years, so those that we see on the main sequence now must have formed recently.
16. About ten billion years, of which half has already expired.
17. About ten billion years. The surface chemical composition is representative of the initial composition of the cluster stars. If they had formed with a different composition, the lifetime of a star with the sun's spectral type would be different, and the cluster would have a different age.
18. The cluster turn-off points for globular clusters almost always occur at stars with lower surface temperatures.
19. Gravitational contraction.
20. The energy released by the onset of hydrogen fusion.
21. A massive protostar reaches the main sequence in a very short time compared to a low-mass star.
22. The upper mass limit is set by radiation pressure and the lower mass limit by the inability to reach a high enough central temperature for hydrogen fusion.
23. A neutrino is a particle that moves with the speed of light and has no charge and essentially no mass. It interacts with matter so rarely that a neutrino emitted in the sun's core almost always escapes.
24. Fewer neutrinos than theory predicts have been detected.

Chapter 15

1. T Tauri stars appear to be newly formed stars, just approaching the main sequence. They appear on the H-R diagram where stars approaching the main sequence are expected to be and are found in young clusters or in locations where there is considerable gas and dust, out of which stars are born.

2. Dust particles in a shell surrounding a T Tauri star are heated by the star to the point where they emit infrared waves.
3. Cepheid variables periodically expand and contract with the light variations.
4. The period-luminosity relation shows that the more luminous a cepheid, the slower it pulsates.
5. The period-luminosity relation is important in establishing distances to cepheid variables. The apparent magnitude and period of variation must be measured before we can use the relation to find the distance to a cepheid.
6. RR Lyrae stars are pulsating stars; but they are fainter, hotter, and pulsate faster than cepheids.
7. A Mira variable is a red giant whose period of variability is comparatively long and a bit irregular, typically around one year.
8. The excess infrared radiation comes from dust around the Miras—dust which condensed from the gas flowing from them.
9. Planetary nebulae are shells of gas, each expanding around a hot central star. Their name derives from their similarity in appearance to the planets Uranus and Neptune when viewed through a small telescope.
10. From the ultraviolet photons emitted by the central star.
11. Larger planetary nebulae tend to have smaller central stars. Since the nebulae get larger as time passes, the central stars must be shrinking.
12. Planetary nebulae appear to be the transition stage between red giants (especially Mira variables) and white dwarfs.
13. On the main sequence, more massive stars are larger. For white dwarfs, more massive stars are smaller.
14. The residual heat left inside the white dwarf from when it was a red giant.
15. The largest mass a white dwarf can have is about 1¼ solar masses. The larger the mass, the smaller is the star; and a white dwarf of about 1¼ solar masses would have zero radius.
16. T Tauri stage, Main sequence, Red giant, Planetary nebula, White dwarf, Black dwarf
17. The helium flash is the very rapid onset of the nuclear reaction that converts helium to carbon in red giants.
18. Horizontal-branch stars are in the process of converting helium to carbon in their cores, becoming larger and cooler (but keeping about the same luminosity).
19. A Mira is thought to have a carbon core surrounded by a helium shell, which in turn is surrounded by original, hydrogen-rich matter.
20. Planetary nebulae and Mira variables apparently supply most of the hydrogen and helium, while supernovae supply the heavy elements.
21. Younger stars have considerably more heavy elements (carbon and heavier). These heavy elements must have been created in supernova explosions which occurred before the star formed, and the star formed in part out of the debris of the explosions.
22. A supernova apparently is the explosion of an entire star, while a nova is a surface phenomenon that probably leaves the star just about as it was before the explosion.
23. A pulsar is a very small star that emits radio waves which are received on earth as periodic bursts.
24. The supernova observed by the Chinese in A.D. 1054 apparently caused the Crab Nebula. A pulsar, thought to be a rapidly rotating neutron star, is located in the middle of the Crab Nebula.
25. A neutron star would typically have roughly the volume of a large mountain. (The radius of a neutron star would be around ten kilometers.)

26. A black hole is an object that has such a strong gravitational field that nothing, not even light, can escape. Hence, we could not see it.

Chapter 16

1. Counts of the number of stars in different parts of the sky visible with a telescope showed that stars are concentrated in the band visible as the Milky Way.
2. For clusters in which Shapley found RR Lyrae stars, he found the distance from the apparent magnitudes of these stars. For other globular clusters, he estimated the distance by the apparent size of the cluster.
3. In a roughly spheroidal shape, centered on the center of the Galaxy.
4. Because of obscuring clouds of dust, we see to about the same distance in any direction we look in the Milky Way. Hence, the number of stars visible in different parts of the Milky Way is roughly the same.
5. Interstellar dust clouds blocking out our view of more distant stars.
6. A group of O and B Stars in a localized region of space, usually immersed in a concentration of interstellar matter.
7. Only O and B stars are hot enough to produce large amounts of the ultraviolet light needed to make the nebulae glow brightly.
8. O stars have such a short lifetime that they must still be near the clouds of gas and dust in the spiral arms from which they formed.
9. The broad line is from the star itself, and the remaining lines are from clouds of interstellar gas moving with different velocities relative to the star.
10. Dust heated by embedded stars emits the infrared radiation.
11. Nearby stars are orbiting around the galactic center at about the same speed as the sun, so they have small space velocities. Since globular clusters do not share in the rotation of the galactic disc, the sun's own orbital motion is reflected as the large radial velocities seen in many globular clusters.
12. (3) 25
13. From the sun's orbital period around the Galaxy and its distance from the galactic center, we can estimate the mass of the Galaxy from Newton's form of Kepler's harmonic law.
14. The 21-cm line arises from the abundant cool hydrogen atoms of the gas clouds in interstellar space, so it is very useful for mapping out these clouds. Also, the 21-cm line is at a wavelength long enough to penetrate the dust clouds, allowing us to probe the Galaxy to greater distances than we can in visible light.
15. O-associations and interstellar gas clouds emitting 21-cm radio waves.
16. Stars. Only around 5 percent is in the form of interstellar material.
17. Age. Population II stars are older than population I.
18. Population I stars are younger, are never found very far from the plane of the Galaxy, include hot main-sequence stars but fewer giants, and have a higher metal abundance. The sun belongs to population I.
19. Disturbances caused by supernova outbursts.
20. Not very well. Molecules far more complex than expected have been discovered in abundance in interstellar gas clouds.

21. Cosmic rays are particles like electrons and nuclei of atoms moving through space with speeds close to the speed of light.

Chapter 17

1. Galactic nebulae were located in the Milky Way, while extragalactic nebulae were located outside this band.
2. Hubble observed cepheid variables in extragalactic nebulae and calculated how far away cepheids of their periods would have to be to have the observed apparent magnitudes.
3. Spiral (normal and barred), elliptical, and irregular.
4. For the first three questions, the answer is elliptical galaxies. The rarest are the irregular galaxies.
5. According to how tightly their arms are wound.
6. By the letter E followed by a number designating the shape of the galaxy. The smaller the number, the more circular the galaxy appears.
7. From the rotational velocities and Kepler's harmonic law.
8. The orbital periods are so long that we have been unable to detect the changing position angle of the components. Also, orbital velocities of a binary galaxy found from Doppler shifts do not give masses unless the inclination is known, and it is usually not possible to find it.
9. Masses of galaxies derived from clusters of galaxies generally turn out to be much larger than masses found for single or binary galaxies.
10. A cluster of galaxies is a collection of galaxies thought to be held together by mutual gravity. A galactic cluster is a cluster of stars contained in a galaxy.
11. Although both spirals and irregulars have a wide range in possible masses, the spirals are on the average about ten times more massive than the irregulars.
12. Regular clusters are more nearly round in appearance, more concentrated in the center, and usually have more member galaxies than irregular clusters. The galaxies in a regular cluster are almost all elliptical or S0 galaxies, while an irregular cluster contains all types of galaxies. Our Galaxy is a member of an irregular cluster.
13. Dust in the spiral arms of our Galaxy.
14. Yes, there seem to be clusters of clusters. (Any larger-scale clustering has not been found, but it would be very difficult to find it, if it should exist.)
15. Distances to galaxies are found by observing such objects as RR Lyrae and cepheid variables, bright blue stars, novae and supernovae, and entire galaxies.
16. Beyond about 20 million light-years, cepheid variables are too faint to be easily detected.
17. The stars illuminating emission nebulae are hot, emitting most of their energy in the ultraviolet. Since much of this ultraviolet light is converted to visible light by the nebulae, the nebulae emit more visible light than the stars; so the nebulae can be seen from a greater distance.
18. Spectral lines in distant galaxies are red-shifted compared with closer galaxies.
19. Yes. Some galaxies in our local group show blue shifts in their spectra.
20. The more distant clusters will have twice as large a red shift.
21. Seyfert galaxies are spiral galaxies with very small bright nuclei, some of which emit extremely intense radio waves. Our own Galaxy is a spiral and also emits radio waves from a

pointlike source in its nucleus, although the source is very much weaker than those usually found in Seyferts.

22. Yes. Several galaxies emit very strong radio waves and are ejecting large quantities of gas at high speeds.

Chapter 18

1. The study of the structure, origin, and evolution of the universe.
2. The big-bang theory proposes that the universe started as an explosion at one point. It was designed to explain the Hubble law of red shifts.
3. The expanding matter cooled before it had time to form heavier elements.
4. After about one million years, the matter became cool enough for electrons to be captured by nuclei of atoms. The transparent gases hydrogen and helium were found.
5. The universe apparently does not have a "center" the way a normal explosion does. Our universe does not seem to be made so that someone can stand near one edge and see most of the matter lying to one side.
6. One million light-years away.
7. A little over 10 billion light-years.
8. The universe first became transparent at around 3000K. A star of this temperature would appear red, but the 3000K surface from the early universe is so far away that its light is extremely red-shifted, primarily into the radio region of the spectrum.
9. The 3K radiation agrees well with the predictions of the big-bang theory and strengthens our belief that this cosmological model describes the universe well.
10. A quasi-stellar source is a starlike point of light in the sky, which varies in brightness, has a red shift in its spectrum, and emits radio waves.
11. The main evidence that quasi-stellar sources are very distant is the large red shifts in their spectra. If they really are that distant, they would have to be more luminous than any galaxy in order to have their observed apparent brightness.
12. The rapid light variations of quasi-stellar sources indicate that they are small. Objects over one light-year in diameter probably would not be observed to vary as rapidly as those sources do.
13. The greater the distance to an object, the further into the past we look when we see it. If quasi-stellar sources have a lifetime less than the age of the universe and they appeared early in the history of the universe, then none would have lasted to the present. Hence we could not see them nearby, but they would still be visible at great distances.
14. Quasi-stellar sources and Seyfert nuclei are very small and luminous, with very broad emission lines in their spectra. Both are blue in color and vary rapidly in time; frequently, Seyfert nuclei, like quasi-stellar sources, are strong radio sources. However, the nucleus of a Seyfert galaxy is usually less luminous than a quasi-stellar source. A Seyfert shows a spiral arm structure and is generally not as far away as the quasi-stellar sources.
15. Explaining how such enormous amounts of power can be generated in such a small volume.
16. Quasi-stellar sources emit strong radio waves, but quasi-stellar objects do not.
17. The ultimate fate of the universe, cold and empty or hot and dense, depends on whether there is enough mass that gravity can stop the expansion we presently observe and cause the universe to collapse.

18. If the universe eventually collapses, it may shrink down to a tiny point and stay that way. Or possibly there may be a new bang, followed by another expansion.
19. Based on our present estimates of the total mass of the universe, it appears as if the universe will expand forever. (If there is enough unseen mass, such as the apparent "missing mass" in large clusters of galaxies, it is possible the universe will eventually contract.)

APPENDIX I
Relativity

Newton's first law of motion, originally demonstrated by Galileo, is called the law of inertia. Simply stated, a body of given mass which is at rest or moving uniformly in a straight line will remain at rest or continue to move uniformly in a straight line unless acted on by an unbalanced force.

Just what is meant by "rest" has led to many discussions. If you were to stand on the platform of a train station while a train passed, you would have a different opinion of what constitutes rest than would a passenger on the train. Suppose you notice a passenger seated at a table in the dining car with a bottle of wine on the table. As you anxiously stare at the bottle, you observe it to move. On the other hand, the passenger, staring eagerly and with less futility at the bottle, happily observes it to be motionless. And if this were a very smooth riding train, moving uniformly on a straight, level track, the passenger would observe the wine bottle to remain at rest unless disturbed by the application of a force. On the other hand, you would observe the wine bottle to remain in uniform motion in a straight line unless it were to be disturbed by an applied force.

So the motion of the wine bottle is relative and depends on the viewpoint of the observer, but the law of inertia is absolute. That is, the bottle remains in the particular state perceived by either of the two observers unless disturbed by an applied force.

Now suppose that the passenger is totally irresponsible and after drinking the wine passes the bottle through the open window and allows it to drop. In the hypothetical case we are considering, there is no effect on the bottle due to air resistance. Consequently, the passenger would observe the bottle to fall to the ground along a vertical, straight-line path as a result of the earth's gravity. But you would see the bottle fall to the ground along a curved path, a particular type of curve called a *parabola*.

Again, there is an effect of relative motion. The path followed by the bottle will depend on the velocity of the bottle relative to the observer at the instant it is released. More specifically, the passenger and you would agree that the horizontal motion of the bottle is unchanged. To the passenger it is zero when the bottle is released, and it will remain zero until the bottle strikes the ground. To you, the bottle is moving horizontally with the same speed as the train when released, and it continues to move horizontally with that same speed. The vertical motion of the bottle, however, would be observed to change by you and the passenger. At the moment the bottle is released, it is not moving vertically.

But due to the earth's gravity, the bottle would begin to fall. After one second, both you and the passenger would observe it to have a vertical velocity toward the ground of 9.8 meters per second. So even though the path of the bottle depends on the viewpoint of the observer, the acceleration does not.

The acceleration of an object of a given mass is defined by Newton's second law of motion. The acceleration is equal to the force on the object divided by its mass. Notice that the acceleration of an object is not affected by its velocity. Acceleration is independent of the motion of the observer, a direct result of the absolute nature of Newton's second law of motion. That is, if Newton's second law holds for objects observed by any particular observer, it holds for any other observer moving uniformly in a straight line relative to the first observer. In this sense, even though motion is relative and not absolute, Newton's laws of mechanics are absolute. The relative nature of motion together with the absolute nature of the laws of mechanics constitute a principle of relativity.

Strictly speaking, Newton's laws of motion will not appear to hold for an observer who is stationary on the surface of the earth. The observer is, in fact, continually accelerating because of the earth's rotation. For example, the behavior of the Foucault pendulum is not what an observer would expect. Its apparently strange behavior is actually used to demonstrate the earth's rotation.

As another example, if you were to see an object that is truly moving in a straight line with uniform speed, the apparent motion of that body would change as a result of your own acceleration. For instance, suppose you were to hold a baseball in your hand. Furthermore, assume by some magic that you are able to make the ball unresponsive to gravity, air pressure, or any other force. Now suppose you release the ball. According to the law of inertia, under these magical conditions, you would expect the ball to remain suspended at the exact spot of release from your hand. But, in fact, the ball would move eastward on a straight line with the speed you were being carried by the earth's rotation when you released it. You would continue eastward on your circular path due to the earth's rotation. The ball would, at first, appear to rise and move westward in your sky. Eventually, as it moved ever farther away, it would dip toward the horizon. Finally, it would disappear below the western horizon. The reason for this strange behavior is the earth's rotation. Even though the ball's motion would appear to you to violate the law of inertia, its true motion would not. The apparent violation arises because you are continually accelerating in the circular path you follow on the spinning earth. Since you would see the ball as if it did not obey the law of inertia, you are said to be observing from a *noninertial frame of reference*. In contrast, a frame of reference free of acceleration, and in which the law of inertia holds, is called an *inertial frame of reference*.

Maybe somewhere in the universe there is a truly inertial frame of reference, but that is not the major issue. Given the existence of an inertial frame of reference, there are an infinite number of inertial frames of reference. Any frame of reference moving with a uniform velocity in a straight line relative to the given inertial frame of reference is itself an inertial frame of reference. In any such frame, the laws of mechanics hold true.

To establish the concept of an inertial frame of reference, we have stressed that the surface of the earth is not an inertial frame of reference. However, for motion occurring over relatively short distances and short intervals of time, such as the motion of the dropped wine bottle, the effects of the earth's rotation are small. Newton's laws of motion hold very well, if not strictly. You may have suspected that this is so, because results of experiments made on the surface of the earth led to the formation of Newton's laws.

By using the concept of an inertial frame of reference, we state in accordance with the principle of relativity, as outlined above, that (1) motion is defined relative to the observer's frame of reference and is in this sense not absolute; and (2) Newton's laws of mechanics hold in all inertial frames of reference and in this sense are absolute.

Suppose we apply this principle of relativity to an imaginary gun which shoots bullets with a speed of 300,000 km/sec. If you fired the gun, you would see the bullets traveling at 300,000 km/sec no matter where you aimed the gun.

Assume you have a friend with a similar gun moving away from you at 200,000 km/sec. According to our principle of relativity, a bullet fired toward you by this gun would approach at only 100,000 km/sec. If your friend were moving away at 300,000 km/sec, a bullet fired by the gun would never reach you.

An ordinary light bulb has a property similar to our imaginary gun. The bulb "fires" photons, particles of light, with a speed of 300,000 km/sec. Two scientists, Albert Michelson and Edward Morely, studied the motion of photons in a series of famous experiments at Cleveland, Ohio, in 1896. They showed that photons behaved like the bullets from our imaginary gun, traveling away from the light source at 300,000 km/sec in all directions. They found the same result at different times during the year. So the motion of the earth about the sun did not change the observed velocity of the photons. But this is what we might have expected. Their gun was at rest relative to the earth, so from their position of rest they should measure the same velocity in all directions.

However, in 1913 a Dutch astronomer, Willem de Sitter, found that the speed of light from two stars in orbit about each other was the same no matter where the stars were in their orbits. But the speed of an orbiting star toward the earth changes as it changes its orbital position. In other words, de Sitter found that the speed of light does not depend on the relative speed of the source toward or away from the observer. If the bullets from your imaginary gun behaved in this way, you would indeed have a strange gun. Bullets fired from the gun when at rest would travel at 300,000 km/sec. But bullets fired from the gun when moving away at 200,000 km/sec would still arrive at your location with a speed of 300,000 km/sec.

Surely, if the interpretations of the results of Michelson and Morely and of de Sitter are correct, our principle of relativity falls apart when light is considered. The speed of light is not relative at all; it is absolute. The speed of light is independent of the motion of both the observer and the light source. Albert Einstein, in 1905, had already shown that the apparent fault is not in the concept of relativity. Instead, it is in the basic assumptions which we made when we applied relativity. In his special theory of relativity, Einstein included the absolute nature of the speed of light. He also included the absolute nature of the laws governing two properties associated with light—electricity and magnetism.

According to the special theory of relativity:

1. Motion is defined relative to an observer's frame of reference and is, in this respect, not absolute except that the speed of light is absolute.
2. The laws of mechanics and the laws of electricity and magnetism hold in all inertial frames of reference and are in this sense absolute.

In his special theory of relativity, Einstein preserved the absolute nature of the laws of mechanics and established the absolute nature of the laws of electricity and magnetism. But a price had to be paid. Human intuition was severely shaken.

We have already remarked on the seemingly unreasonable idea that the speed of light is independent of both the motion of the source of light and the motion of the observer. Just as seemingly unreasonable are the implications concerning the nature of space and time made by this strange idea. For example, nothing seems so reasonable as to say that the time interval between two separate events or occurrences is absolute. That is, two observers in two different frames of reference should observe the same period of time elapse between the occurrence of two different events.

But consider this supposition in terms of Einstein's postulate that the speed of light is absolute. A train 600,000 km long is moving by us at 100,000 km/sec on a straight, level track. (This is a very fast, very long train.) The train is equipped with doors at either end that open when struck by beams of light emitted from a source located at the midpoint of the train.

Suppose that a passenger turns on the light source at the instant the midpoint of the train passes us. Clearly, after one second, since light travels at 300,000 km/sec, the light from the source would reach both ends of the train. The doors would open, and the passenger would say that they opened simultaneously.

But light also travels at 300,000 km/sec in our frame of reference. So after one second, in our frame, the light from the source would be 300,000 km up the track and down the track. In that same second, however, the rear of the train would move 100,000 km toward where the light source had been. In other words, the light would already have opened the rear door from our vantage point. The front of the train would be an additional 100,000 km farther away from the point where the light was first emitted. The front door would not yet have opened. Thus, we would not say that the doors open simultaneously as the passenger would. The interval between the two events, the activation of the two doors, cannot be considered equal in our frame of reference and the passenger's frame of reference.

In general, the interval of time between any pair of events is relative and not absolute. Speed is the measure of the distance traveled in a certain interval of time. Since the time interval between the arrival of light at two separate points is relative, then the distance between those points must also be relative in order that the speed of light be absolute. Because of the relative nature of distances and time intervals, the way in which the velocity of an object depends on the frame of reference of the observer is not what we expect. Let us return to the first train and its wine-drinking passenger. Suppose that the irresponsible passenger, instead of simply dropping the emptied bottle, had thrown it forward at 50 km/hr. Assume that the train was also traveling at 50 km/hr. Galileo, and your own intuition, would say that you should see the bottle travel horizontally at 100 km/hr. The velocity of the bottle relative to the train and the train relative to you are simply added. We can write this as

$$V = v + v'$$

where V = the combined velocity

v and v' = the velocities being combined

But Einstein, taking into account the relative nature of distance and time, would say something different. His rule for combining velocities is

$$V = \frac{v + v'}{1 + \dfrac{vv'}{c^2}}$$

where c = the velocity of light

In the example we have chosen, he would give the speed which you would see for the bottle as 99.9999999999998 km/hr. There is only an extremely small difference between the velocity of the bottle determined by the simple formula which adds the velocities and that determined by the more complicated formula from Einstein's special theory of relativity. Only if the speed of the train or the speed of the bottle relative to the train is close to the speed of light would the two formulae lead to very different results. For example, if the bottle were moving at the speed of light relative to the train, the simple formula would give

$$V = v + c$$

This result would have the bottle exceed the speed of light. However, Einstein's formula would give

$$V = \frac{v + c}{1 + \frac{vc}{c^2}} = c$$

This result is in accordance with the absolute nature of the speed of light.

If, somehow, the passenger could pitch the bottle forward at a speed greater than the speed of light, Einstein's formula predicts that you would observe the bottle to move slower than the passenger observed the bottle to move. This is absurd! How could the bottle move faster relative to the passenger than to you, when the passenger is already moving away from you when he pitches the bottle forward? The answer lies in the fact that the bottle could not move faster than the speed of light in the first place—not relative to you nor relative to the passenger.

The speed of light takes on tremendous significance. It is not only the absolute speed with which electromagnetic radiation travels, but it is the limiting speed of material objects. Nothing can travel faster than light. Anything traveling at the speed of light relative to one inertial frame of reference travels with the speed of light relative to all inertial frames of reference.

Now consider the acceleration of a body as it would be expressed using Newton's second law: $a = F/m$. Thus, in Newtonian mechanics, if a constant force were applied to a body, it would be continuously accelerated by the same amount. The longer the force were applied, the faster the body would move. If the force were applied long enough, Newtonian mechanics predicts that the body would eventually gain a speed greater than that of light. But we have just seen that the speed of light is the fastest that any object can travel.

Therefore, as the object's speed approaches the speed of light, the energy supplied by the force could not be utilized to increase the speed of the object very much. Where does the supplied energy go? Einstein showed that it turns up as an increase in the object's mass. In other words, mass and energy are equivalent. A careful analysis of the relation between mass and energy shows that even a body at rest in your reference frame has an energy equivalent for its mass. Einstein's famous equation states

$$E = mc^2$$

where E = the energy in ergs

m = the mass in grams

c = the velocity of light in cm/sec

An excellent confirmation of this mass-energy relation is observed in nuclear power generation. The mass of the original fuel in a nuclear power plant is greater than the final mass. The mass lost, expressed as energy by Einstein's formula, is just the energy released by the nuclear fuel. Stars like the sun generate the energy by which they radiate by converting mass to energy.

Certainly, the special theory of relativity ranks as a great achievement. But it is important to note that Einstein's theory by no means invalidates Newton's laws. Newton's laws work perfectly well for everyday phenomena and reasonably well for extreme conditions. Only when velocities near the speed of light are attained does Einstein's theory give much different predictions. The main significance of Einstein's theory is that it incorporates the laws of motion and the laws of electricity and magnetism into a principle of relativity.

Certainly, the special theory of relativity would have been accomplishment enough for a single individual, but Einstein also developed a theory of gravitation—a more general theory than Newton's. To gain some understanding of Einstein's theory of gravitation, first consider Galileo's law of falling bodies. If you were to hold two objects of different mass in your hands and release them from the same height, they would hit the ground at the same time. This happens because the force of gravity on an object and the object's resistance to motion depend on its mass in the same way.

Now suppose that there were no such thing as gravity and that the section of the earth upon which you are standing were being accelerated away from the earth's center. If you were to release the two objects once more from the same height, they would hit the ground at the same time.

There is really no way to distinguish between the behavior of falling bodies which are being accelerated by gravity in one direction or which are completely free of forces and are merely reflecting your own acceleration in the opposite direction. More specifically, when all objects are observed to accelerate similarly relative to a particular frame of reference, it is impossible to decide whether the frame of reference is inertial and gravity is present, or whether there is no gravity and the frame of reference is not inertial but is itself accelerating. This is essentially the principle of equivalence—the principle which postulates the equivalence of gravitation and acceleration.

What happens to a beam of light according to the principle of equivalence? Imagine you are in a spaceship far removed from all other bodies, so gravity does not matter for the moment. If the spaceship has a constant speed, when you shine a light across the spaceship, it will strike a certain point on the wall. The beam will travel in a straight line. Now, if the spaceship is accelerating at a right angle to the beam, the point where the light used to hit will have moved out of position before the light strikes it. The light beam will have a curved path relative to your accelerating frame of reference. Einstein reasoned, by the principle of equivalence, that you would also see the curved light path in an inertial reference frame in the presence of gravitation. Light itself is accelerated in the presence of gravitation.

The path of an object in the presence of gravitation depends not only on its acceleration but on its velocity. Suppose two objects are moving at some angle to the direction of the gravitational force. The path of the faster object will be more nearly straight. Since nothing travels faster than light, no object can have a more nearly straight path than light. Therefore, for motion from one point to any other point, the shortest path will be taken by light. Since light generally follows curved paths because of gravitation, what we usually think of as a straight line is not the shortest distance between two points.

Space itself is curved! The effect of gravitation is to cause a curvature of space; the stronger the gravitation, the greater is the curvature.

On the basis of the principle of equivalence, Einstein developed a theory of gravitation that involves the curvature of space. Because of the relative nature of space and time, this theory is formulated in four dimensions—three space dimensions and one time dimension. Einstein's theory of gravitation, called the general theory of relativity, translates to Newton's law of gravity when the gravity is not extremely strong or when the velocities of objects influenced by gravity are not close to the speed of light.

It is easy to see how Newton's law of gravity is inadequate under extreme conditions. According to this law, the force of gravity between two objects depends on their masses and the distance between them. This law becomes ambiguous when the objects are moving rapidly relative to one another or relative to the observer studying them. This is because mass depends on speed, and the distance between two objects is not absolute but instead is relative (it depends on the frame of reference in which the distance is defined). But is not sufficient merely to modify Newton's law of gravity to take into account the relative nature of mass and distance; indeed, the general theory of relativity does more than that.

Gravitation is the weakest of all known forces, and only massive bodies such as the earth cause easily observed accelerations. Therefore, if you are to find discrepancies between the actual behavior of objects in the presence of gravitation and that predicted by Newton's law, you must look for them in the behavior of objects or of light in the vicinity of extremely massive bodies. Mercury is fairly close to the sun and certainly, among the planets, is in the region of the strongest gravitation. The behavior of Mercury in its orbit is not exactly that predicted by Newton's law when only the sun's gravity is taken into account. The semi-major axis of Mercury's orbit changes direction in space, rotating slowly about the sun in the direction of Mercury's revolution. This is the behavior predicted for Mercury's motion when the gravitational perturbations of the other planets are taken into account. However, the semi-major axis of Mercury's orbit is found to rotate faster than predicted, by 43 seconds of arc per century. This is not a swift rotation by any stretch of the imagination. Nevertheless, it is observable, and it is not in compliance with Newton's law of gravity. On the other hand, Einstein's general theory of relativity predicts the extra 43 seconds of arc exactly.

In 1919, two groups of Britain's most distinguished astronomers set out to photograph the solar eclipse of May 29. One group traveled to Brazil; the other, to an island off the coast of West Africa. Their purpose was to test Einstein's general theory of relativity. Light from stars that passed close to the sun would supposedly be deflected from a straight-line path because of the gravitational influence of the sun. The general theory predicted that the deflection should be 1.75 seconds of arc. By comparing the relative positions of stars seen near the edge of the disc of the sun during the eclipse with their positions when the sun is in another part of the sky at a different time of year, the British astronomers were able to confirm Einstein's prediction. Space is curved in the presence of gravitation, just as postulated by the general theory of relativity.

Finally, Einstein's theory predicts that light emitted from the surface of a strongly gravitating body will lose energy in escaping from the gravitation of the body. The energy of a photon is related to its wavelength. Therefore, a decrease in energy would lead to an increase in wavelength. For visible light, the color red is at the long-wavelength end of the spectrum; so an increase in wavelength due to gravity is called a *gravitational red shift*. In accordance with Einstein's theory, this effect has been detected in the spectral lines of white dwarf stars, very compact high-density stars with very high surface gravities. By using very sensitive equipment, even the tiny gravitational red shifts which are produced in photons as they rise several floors in an elevator shaft on the earth have been detected.

Einstein realized that there is another way to visualize a gravitational red shift. If we receive photons that left a massive object at some known wavelength (for example, H-alpha), we would receive the photons shifted to a longer wavelength, and hence a lower frequency. It would appear to us as if the photons came from a region of space where time passed more slowly. We could claim that since one second on the massive body took longer to pass than one second on our own watch, successive wave crests in the light beam were generated more slowly on the massive body. This would account for the fact that we receive the waves at a lower frequency. We can define one second as the time it takes for a certain number of waves to be generated in a given spectral line, so a gravitational red shift quite validly can be thought of as a measure of the effect that mass has on the rate at which time passes.

It is tempting to think that changes in the length of one second, due to high speeds or to strong gravitational fields, are not "real," and that they apply to the real world only in some abstract and detached way. However, in 1971, scientists at the Naval Observatory in Washington, D.C., conducted an experiment to see if accurate clocks really show the effects Einstein predicted. A set of atomic clocks was kept in Washington and another set was flown around the world on commercial airplanes. On one trip, the plane flew west to east, in the direction of the earth's rotation, hence traveling faster than the clocks which remained in Washington. Due to their faster speeds, the clocks on the plane were predicted to slow down a total of 184 billionths of a second. However, due to the high altitude of the flight, the clocks were also in a weaker gravitational field, because they were farther from the earth's center. This altitude effect should have made the clocks gain 144 billionths of a second. The combined effects, therefore, should have resulted in the clocks on the plane being slower than the ones on the ground by 40 billionths of a second. At the conclusion of the flight, the two sets of clocks differed by the predicted amount, within the experimental error.

On a second flight, the clocks were carried westward around the world. Since the plane flew in the opposite direction to the earth's rotation, the "moving" clocks traveled slower than the ones on the ground and hence should have shown that time passes more rapidly. Combined with the faster time due to altitude, the westbound clocks were expected to gain 275 billionths of a second compared to the clocks on the ground. Again, the differences in the two sets of clocks agreed within the experimental error. The effect is real; time goes by at different rates for observers in different circumstances. For most purposes on earth, however, the effect is too small to be significant, and we can agree to use a standard time.

The special and general theories of relativity constitute two of the greatest intellectual achievements in the history of humanity. By now, many experiments in laboratories throughout the world have been conducted under conditions explained by the special theory. No real understanding of the nature of the smallest known particles, those which make up the nuclei of atoms, could have been attained without the special theory.

The general theory, on the other hand, has had much more limited application because of the usually weak nature of gravitation. In the last few years, astronomical objects such as neutron stars have been discovered. Gravitation is so strong in these objects that only calculations employing the general theory of relativity can be useful in describing their properties. Astronomers are openly speculating about the existence of black holes, objects of such strong gravitation that the spaces around them are curved into isolated or pinched-off regions of the universe. Alternately, one can picture a black hole as a region where gravity is so strong that its effect on time has reached the ultimate extreme, and time has absolutely stopped, as seen by an outside observer. Light in such a region can never reach another part of the universe, the path it follows being bounded. The entire universe itself involves dimensions so great that the curvature of space leads to incredible consequences. The point where the universe originated becomes the spherical boundary of an apparent three-dimensional universe.

Copernicus, Galileo, Kepler, Newton, and now Einstein: Is there yet another startling discovery to be made and another name to add to the list?

APPENDIX II
Powers of Ten

Critics of any particular item in the budget of the federal government usually will refer to the funding as "astronomical"—a reference to the large numbers used for the vast distances involved in astronomy. The nearest star to the sun, for instance, is 40,500,000,000,000 kilometers away, and reasonably distant objects are hundreds of millions of times still farther away. To avoid having the reader count the number of digits to the left of the decimal point, a convenient shorthand is used; 40,500,000,000,000 equals 4.05×10^{13}, where 13, the *exponent* or *power of ten,* counts the number of places the decimal point was moved to the left to arrive at the 4.05. At the other extreme, 0.0000000000218 (the energy in ergs required to ionize a hydrogen atom) is written 2.18×10^{-11}. This shows that the decimal point was moved eleven places to the right to give the value 2.18.

Not only is the power of ten much easier to read and write, but it makes computation much easier when dealing with large numbers. An object moving at 1 percent (or 10^{-2}) of the speed of light (3×10^{10} cm/sec) would move at 3×10^{8} cm/sec. That is, dividing by 100 is equivalent to moving the decimal point two places to the left or subtracting 2 from the exponent. Subtraction is so much easier than normal division that life is much simpler when we can perform a division by subtracting exponents. For example, the density of the planet Mercury is its mass divided by its volume. Mercury's mass is 3.2×10^{26} grams and its volume is 6.0×10^{25} cubic centimeters (written cm^3). Hence, the density is 3.2×10^{26} divided by 6.0×10^{25}, or $(3.2 \div 6.0) \times 10^{(26-25)}$, or 0.54×10^{1}, or 5.4 g/cm^3.

To multiply using powers of ten, simply add the exponents. If we know that the sun is 1.5×10^{8} km from the earth, and the next nearest star is 2.7×10^{5} times farther away, we find that this star is $(1.5 \times 2.7) \times 10^{(8+5)}$ km, or 4.05×10^{13} km away.

APPENDIX III
The Metric System

Anyone who ever wanted to know how many inches are in 100 miles, or how many teaspoons make one gallon, or how many ounces are in a ton (long or short) has discovered how cumbersome the traditional English system is. The two disadvantages of the system are the multitude of different names—some of them ambiguous—that must be learned; and the conversion factors are virtually never simple powers of ten, so manipulation is awkward with our decimal number system.

Discovering the nature of the universe and the laws governing it is most difficult, and the last thing a scientist needs is a confusing and cumbersome system of measurement. The far simpler metric system has thus been adopted by scientists and is coming into widespread usage throughout the United States. Metric units are used in this text.

The basic units of length and mass in the metric system are the meter and the gram. A meter (m) is a unit of length slightly longer than a yard (39.37 in), while a gram (g) is a measure of mass (28 g = 1 oz). All metric units of length and mass are some power of ten compared to the meter and gram. The prefix *kilo* designates a unit 1000 times larger, so one kilometer (or 1 km) equals 1000 m. A kilogram (kg) equals 1000 g. A prefix of *centi* designates a unit of 100 times smaller, so a centimeter (cm) is 0.01 m. The following table lists various metric units and their approximate equivalent in the English system.

Prefix	Meaning	Length	Mass
milli	10^{-3}	1 millimeter = 1/25 in	1 milligram = 1/28,000 oz
centi	10^{-2}	1 centimeter = 2/5 in	rarely used
	1	1 meter = 39.37 in	1 gram = 1/28 oz
kilo	10^{3}	1 kilometer = 5/8 mi	1 kilogram = 2.2 lb
mega	10^{6}	rarely used	rarely used

Finding the number of inches in 100 mi involves some computing ($12 \times 5280 \times 100 = 6.336 \times 10^6$). But the number of centimeters in 100 km is found much more easily ($10^2 \times 10^3 \times 10^2 = 10^7$).

People who use the metric system usually use Celsius temperatures instead of Fahrenheit. The freezing point of water is 0°C and 32°F, while the boiling point is 100°C and 212°F. Hence, a change in temperature of 1° Celsius corresponds to a change of 1.8° Fahrenheit. A third temperature scale is also used in the physical sciences—the absolute or Kelvin scale. A change of 1°C is the same as a change of 1K, but 0°C equals 273K. Zero Kelvin is called *absolute zero,* the point where black body radiation completely ceases. The Kelvin scale avoids using negative numbers, but its main advantage is that various formulae used in physics and astronomy are considerably simpler if temperature is expressed in the Kelvin scale.

APPENDIX IV
Astronomical and Physical Constants

Speed of light (c) = 2.9979×10^{10} cm/sec
Gravitation constant (G) = 6.670×10^{-8} dynes cm^2/g^2
Mass of hydrogen atom = 1.67×10^{-24} g
Mass of electron = 9.11×10^{-28} g
Stefan-Boltzmann constant (σ) = 5.67×10^{-5} erg cm^2/sec
Astronomical unit (AU) = 1.50×10^{13} cm = 1.50×10^{8} km
Parsec = 3.09×10^{18} cm = 3.09×10^{13} km
= 206,265 AU = 3.26 light-years
Light-year = 9.46×10^{17} cm = 9.46×10^{12} km
Radius of the sun = 6.96×10^{10} cm = 6.96×10^{5} km
Mass of the sun = 1.99×10^{33} g
Mass density of the sun = 1.41 g/cm^3
Luminosity of the sun = 3.90×10^{33} erg/sec
Radius of the moon = 1738 km
Mass of the moon = 7.35×10^{25} g
Mean density of the moon = 3.34 g/cm^3
Mean distance from the earth to the moon = 384,400 km

APPENDIX V
Planetary Data

Orbital Data

	Semi-major axis	Eccentricity	Inclination	Sidereal period	Synodic period
Mercury	5.79×10^7 km	0.206	7° 0′	88 days	116 days
Venus	1.08×10^8 km	0.007	3°23′	225 days	584 days
Earth	1.50×10^8 km	0.017	0° 0′	365 days	—
Mars	2.28×10^8 km	0.093	1°51′	687 days	780 days
Jupiter	7.78×10^8 km	0.048	1°18′	11.86 years	399 days
Saturn	1.43×10^9 km	0.056	2°29′	29.46 years	378 days
Uranus	2.87×10^9 km	0.047	0°46′	84.0 years	370 days
Neptune	4.50×10^9 km	0.009	1°46′	164.8 years	367 days
Pluto	5.91×10^9 km	0.249	17°10′	248.5 years	367 days

Physical Data

	Diameter at equator	Mass	Density	Rotation period	Gravitational acceleration at surface	Escape velocity
Mercury	4880 km	3.30×10^{26} g	5.4 g/cm³	58.6 days	363 cm/sec²	4.2 km/sec
Venus	12,112 km	4.87×10^{27} g	5.3 g/cm³	243 days retrograde	860 cm/sec²	10.3 km/sec
Earth	12,756 km	5.97×10^{27} g	5.5 g/cm³	1 day	980 cm/sec²	11.2 km/sec
Mars	6800 km	6.42×10^{26} g	3.9 g/cm³	24^h37^m	374 cm/sec²	5.0 km/sec
Jupiter	143,000 km	1.90×10^{30} g	1.3 g/cm³	9^h50^m to 9^h55^m	2590 cm/sec²	60 km/sec
Saturn	121,000 km	5.69×10^{29} g	0.7 g/cm³	10^h14^m to 10^h38^m	1130 cm/sec²	35 km/sec
Uranus	52,000 km	8.69×10^{28} g	1.2 g/cm³	14^h	850 cm/sec²	22 km/sec
Neptune	49,000 km	1.03×10^{29} g	1.7 g/cm³	18^h	1200 cm/sec²	25 km/sec
Pluto	<6000 km	$<1 \times 10^{26}$ g	?	6.4 days	?	?

APPENDIX VI
Primary Nuclear Reactions in Main-Sequence Stars

Proton-Proton Reaction

This process is most important for stars with less than one solar mass, that is, main-sequence stars having central temperatures less than 16 million K.

$$ {}_1H^1 + {}_1H^1 \rightarrow {}_1H^2 + e^+ + \nu $$

$$ {}_1H^2 + {}_1H^1 \rightarrow {}_2He^3 + \gamma $$

$$ {}_2He^3 + {}_2He^3 \rightarrow {}_2He^4 + 2{}_1H^1 $$

Carbon Cycle

This process is most important for stars with more than one solar mass, that is, main-sequence stars having central temperatures greater than 16 million K.

$$ {}_6C^{12} + {}_1H^1 \rightarrow {}_7N^{13} + \gamma $$

$$ {}_7N^{13} \rightarrow {}_6C^{13} + e^+ + \nu $$

$$ {}_6C^{13} + {}_1H^1 \rightarrow {}_7N^{14} + \gamma $$

$$ {}_7N^{14} + {}_1H^1 \rightarrow {}_8O^{15} + \gamma $$

$$ {}_8O \rightarrow {}_7N^{15} + e^+ + \nu $$

$$ {}_7N^{15} + {}_1H^1 \rightarrow {}_6C^{12} + {}_2He^4 $$

Note that in both reactions, four hydrogen nuclei eventually are consumed in forming one helium nucleus. Because more energetic neutrinos are released in the carbon cycle than in the proton-proton reaction, however, each completed carbon cycle only delivers 4.00×10^{-5} erg directly to the stellar material, while each proton-proton reaction delivers 4.19×10^{-5}. The less massive stars not only use their nuclear fuel more slowly, they also use it more efficiently.

Glossary

aberration (of starlight) The apparent shift in the direction to a star due to the orbital velocity of the earth.

absolute magnitude The apparent magnitude of a star if it were at a distance of 10 parsecs.

absolute zero The temperature at which matter does not radiate.

absorption spectrum A continuous spectrum with superposed dark lines due to absorption by a cool gas lying between the source of the continuous spectrum and the observer.

acceleration The rate of change of velocity.

accretion theory (of the moon) An explanation of the origin of the moon which states that the moon condensed from material orbiting the earth.

altitude An object's angular distance above or below the horizon.

amplitude The variation of a quantity (for example, the variation in light of a star or the variation in velocity of a component of a binary star).

angular diameter The angle subtended by the diameter of an object; also known as *apparent diameter.*

annular eclipse An eclipse of the sun in which the moon fails to cover the sun completely, leaving a ring or annulus of light.

Antarctic Circle The parallel of latitude at 66½°S within which the sun, at some time, can be seen as a circumpolar object.

aperture The diameter of the objective of a telescope.

aphelion That point of an object's orbit which is most distant from the sun.

apparent magnitude A measure of the brightness of a star as it appears in the sky.

apparent orbit The projection onto the sky of the orbit of the secondary component of a visual binary star relative to the primary.

Arctic Circle The parallel of latitude at 66½°N within which the sun, at some time, can be seen as a circumpolar object.

artificial satellite An object put into orbit around the earth.

asteroid One of many small objects with orbits lying predominantly between the orbits of Mars and Jupiter.

astrology A primitive religion which originated in ancient Mesopotamia and which taught that each person's destiny is fixed at the instant of his birth by the configurations of the sun, planets, and stars.

astrometric Having to do with astrometry, the branch of astronomy that deals with the measurements of the precise positions and motions of celestial objects.

astrometric binary A binary star in which only one star can be seen; but through its motions, the presence of a second star can be determined.

astronomical unit Semi-major axis of the earth's orbit about the sun; roughly 150 million km.

astronomy The science which deals with the nature of the universe and its constituents.

astrophysics The branch of astronomy which deals directly with the physical processes occurring in the various constituents of the universe.

atom The smallest unit of an element of matter that retains the usual chemical properties of the element.

aurora A glowing of the atmosphere resulting from light emitted by atoms and ions in the ionosphere, usually occurring in the polar regions. In the north it is called *aurora borealis,* in the south, *aurora australis.*

autumnal equinox The intersection of the ecliptic and the celestial equator such that a point moving eastward along the ecliptic would pass from the Northern to the Southern Hemisphere as defined by the celestial equator.

Balmer lines Emission or absorption lines in the spectrum of hydrogen occurring in visual wavelengths which result from transitions between the second level (or orbit, in the simple model) and higher levels (or larger orbits).

bands (in spectra) Emission or absorption lines in spectra of molecules so numerous and close together that they coalesce into broad bands.

barred spiral galaxy A spiral galaxy with arms starting from the ends of a bar passing through the center of the galaxy.

Be star A star of spectral type B with emission lines in its spectrum which are formed by material surrounding the star.

big-bang universe An evolutionary model of the universe which originates with a primeval explosion.

binary star A pair of stars gravitationally bound to each other.

black body A hypothetical perfect radiator that completely absorbs and re-emits all energy incident upon it.

black body radiation Radiation emitted by a black body.

black dwarf The presumed end product in the evolution of a white dwarf star after it has cooled to the point where it no longer emits radiation.

black hole A hypothetical object for which the velocity of escape exceeds the speed of light, and thus no light could escape it.

Bode's law An empirical relation between the order of increasing distance of the planets from the sun and their distances in astronomical units.

bolide An extremely bright meteor or fireball.

brightness A term used to describe the amount of radiant energy from an object falling on a unit area in a second of time at the earth.

butterfly diagram The plot of the latitude and number of sunspots versus time; the outline of the plotted points resembles the outspread wings of a butterfly.

capture theory (of the moon) An explanation of the origin of the moon which states that the moon formed in another part of the solar system and was later captured by the gravitational field of the earth.

carbon cycle A chain of nuclear reactions which involve carbon as a catalyst and result in the transmutation of hydrogen to helium.

Cassegrain focus The focus of a reflecting telescope which uses a secondary mirror to form an image behind the objective mirror.

cataclysmic variable A variable star that undergoes a tremendous and rapid change in luminosity.

celestial equator The intersection of the plane of the earth's equator and the celestial sphere.

celestial mechanics The branch of astronomy that deals with motions of celestial objects under the influence of gravitational forces.

celestial meridian The great circle on the celestial sphere which passes through the celestial poles and the observer's zenith.

celestial navigation The art of navigating by sightings of the sun, moon, planets, and stars.

celestial poles The intersections of the earth's axis of rotation and the celestial sphere.

celestial sphere The apparent sphere of the sky.

center of eccentric In the Ptolemaic system, the center of the circle on which a planet, the sun, or the moon moves about the earth, the earth being located at a point not at the center of this circle.

center of mass (two bodies) A point located such that the product of the mass of the first body and its distance from this point is equal to the product of the mass of the second body and its distance from the point.

cepheid variable A pulsating yellow supergiant.

Ceres The first asteroid to be discovered; also the largest asteroid.

Chandrasekhar limit The largest mass that a white dwarf star could possess.

chromosphere The part of the solar atmosphere that lies just above the photosphere.

chromospheric network A network of fine, bright lines which form a cell-like structure in an ionized calcium spectroheliogram.

circumpolar Any object having a diurnal circle always above or below an observer's horizon.

circumstellar Refers to material surrounding and relatively close to a star.

classical cepheid A star similar to the cepheid variables found in the Small Magellanic Cloud, also called Type I cepheids.

cluster of galaxies Any system comprised of several to thousands of galaxies more closely spaced than the average separation between galaxies.

cluster turn-off point The point that marks the upper end of the main sequence on a color-magnitude diagram of a star cluster.

cluster variable A name often used for RR Lyrae variable stars because many are found in globular clusters.

cluster variable gap A gap in the horizontal branch on the color-magnitude diagram of a globular cluster where no stars with constant luminosity are found.

color index The difference between the photographic (blue) magnitude and the photovisual (yellow) magnitude of a star.

color-magnitude diagram A plot of the apparent magnitudes of the stars in a cluster against their color indices.

comet A small object composed of ices and dust which orbits the sun, usually with great eccentricity; when near the sun, it vaporizes in part, forming a luminous head and long tail.

companion The fainter star in a visual binary.

comparison spectrum An emission spectrum of an element placed alongside the spectrum of a celestial object for wavelength determinations.

conduction The transport of energy by the exchange of energy from particle to particle. The particles can be electrons, atoms, or molecules.

configuration An arrangement of the sun, moon, planets, or stars in the sky.

conjunction The configuration of a planet when it is essentially in the direction of the sun; more generally, any two celestial objects in the same direction are said to be *in conjunction*.

constellation Originally a geometrical position or grouping of stars; now an area in the sky (including the original grouping) bounded by parts of hour circles and parallels of declination. There are now 88 constellations defined by action of the International Astronomical Union.

continuous spectrum An uninterrupted spectrum comprised of radiation at all wavelengths.

convection The transport of energy by moving groups of particles.

convergent point The point on the celestial sphere marking the direction toward which a moving cluster is traveling.

Copernican system A model universe centered on the sun.

core The central region in a star in which nuclear fusion is occurring or has occurred.

corona The outermost atmosphere of the sun.

cosmic rays Nuclei of atoms streaming through space with high energies.

cosmological model A model describing the structure, origin, and evolution of the universe.

cosmological principle The premise which states that the general state of the universe at any given time would be observed to be the same regardless of the location of the observer.

cosmology The study of the structure, origin, and evolution of the universe.

crater A more or less circular depression on the surface of a moon, the earth, or a planet.

crescent phase The phase of the moon or an inferior planet when it is seen less than half-illuminated.

daylight saving time The standard time of the next time zone to the east.

declination The angular distance of a celestial object from the celestial equator measured along the hour circle of the object.

deferent In the Ptolemaic system, a circle about the earth upon which the sun, the moon, or planetary epicycles move.

density The amount of mass per unit volume.

deuterium A form of hydrogen in which the atomic nucleus contains one proton and one neutron.

differential rotation Rotational motion of a system in which the period of rotation varies with distance from the axis of rotation.

diffraction The apparent bending of light as it passes by an obstacle.

diffraction grating A system of close, equidistant, and parallel lines cut into a polished surface which, by diffraction and interference, produces a spectrum.

diffuse nebula A reflection or emission nebula consisting of interstellar material.

direct motion The prevailing eastward motions of objects in the solar system.

disc (of Galaxy) The wheel-like structure formed by the preponderance of stars and interstellar matter in the Galaxy.

disc (of sun, moon, or planet) The apparent circle on the sky which is seen for a spherical object.

dispersion The separation of a single beam of radiation into beams of different wavelength, or the amount by which light is spread in a spectrum.

distance modulus The difference between an object's apparent magnitude and its absolute magnitude.

diurnal circle The apparent daily path of an object in the sky due to the earth's rotation.

diurnal motion The apparent daily motion of a celestial object due to the earth's rotation.

Doppler shift The change in wavelength or frequency of a wave due to a changing distance between the source of the wave and the observer.

double star Two stars which appear close together in the sky. (Unless otherwise qualified, this term is usually used as a synonym for *binary star.*)

dust particle An extremely small solid particle.

dust tail That component of a cometary tail which consists of dust particles, often separated from the gas tail.

dwarf A main-sequence star.

dwarf elliptical galaxy A small elliptical galaxy containing as few as a million stars.

dynamics The study of the motion of objects under the influence of forces.

earthshine Sunlight reflected by the earth onto the moon.

eccentric An off-center circle used in the Ptolemaic system to describe the motion of a planet as seen from the earth.

eccentricity A measure of the deviation of an ellipse from a circle; in particular, the distance between the foci of the ellipse divided by the length of the major axis.

eclipse The obscuration of light from one celestial body by another.

eclipse path The path followed by the moon's shadow as it sweeps across the earth during a solar eclipse.

eclipsing binary A binary star oriented so that the components of the binary periodically eclipse one another.

ecliptic The apparent path of the sun relative to the stars resulting from the earth's yearly revolution about the sun.

electromagnetic radiation Radiation associated with electricity and magnetism, including gamma rays, X-rays, ultraviolet, visible, infrared, and radio waves.

electron A negatively charged subatomic particle.

element A substance that cannot be reduced to simpler form by chemical means.

ellipse A curve such that any point on it has the same total distance to a pair of fixed points as any other point on the curve.

elliptical galaxy A galaxy with an ellipsoidal distribution of stars.

elongation The angular distance of a planet from the sun as seen from the earth.

emission nebula A nebula that shows an emission line spectrum usually with very weak continuous radiation.

emission spectrum A spectrum that consists of radiation at discrete wavelengths.

energy The capacity for doing work.

envelope The part of a star surrounding the core.

epicycle In early models of the universe, the circular path followed by a planet, the center of which itself followed a circular path about some reference point in the universe.

equant In the Ptolemaic system, the point opposite to the earth from the center of eccentric and at the same distance from it as the earth.

equinox One of the intersections of the celestial equator and the ecliptic.

erg A unit of energy; the energy of motion possessed by a two-gram mass moving with a speed of one centimeter per second.

evolutionary model A model of the universe which is constructed on the hypothesis that the universe changes with time, that is, evolves.

evolutionary track The theoretical path followed by the point representing a model star on the H-R diagram, as the star changes with time.

excited That state of an atom when one or more of its electrons are not in the lowest allowable levels.

expanding universe A concept of the universe based on the interpretation that the red shifts in the spectra of distant galaxies are due to velocities of recession.

exponent A symbol written above and to the right of another symbol denoting how many times the latter symbol is to be multiplied by itself. For example: $c^2 = c \times c$.

extragalactic nebula Originally a nebula seen far away from the plane of the Milky Way, but now recognized as any galaxy separate from our own Galaxy.

eyepiece A lens used to magnify the image formed by a telescope.

filaments Dark, threadlike markings seen in H-alpha spectroheliograms.

filar micrometer A device primarily used to measure the angular separations of components of binary stars.

filter An optical device which absorbs practically all light except in a particular band of wavelengths, so only light at those wavelengths is transmitted.

fireball An extremely bright, spectacular meteor.

first quarter The phase of the moon when it is half-illuminated and is moving from the new phase toward full phase.

fission Transmutation of an element through the splitting of its nucleus into nuclei of other elements, subatomic particles, and radiation.

fission theory (of the moon) An explanation of the origin of the moon which states that the moon was once part of the earth, but was thrown off when the earth was newly formed and rotating very rapidly.

flare A sudden and tremendous eruption in the solar chromosphere.

flash spectrum The emission spectrum of the chromosphere seen at the instant that the moon blocks the light from the photosphere during a solar eclipse.

focal length In a simple lens (or mirror), the distance from the lens (mirror) to its focus.

focus The plane in which the sharpest image of an object is formed by an optical system.

focus of an ellipse One of the two points associated with an ellipse from which the sum of the distances to any point on the ellipse is a constant.

following sunspot In a pair of sunspots, the one that follows in the direction of the solar rotation.

force That which can accelerate a body free to move; given by Newton's law, $F = ma$.

Foucault pendulum A pendulum used to demonstrate the rotation of the earth.

frequency The number of waves passing a given point each second.

full moon The phase of the moon when it is seen fully illuminated.

fusion The process of forming nuclei of heavier elements by the combination of lighter nuclei.

fusion shell The region in which nuclear fusion is occurring in an evolved star, about an inner core of different composition.

galactic cluster An open cluster of stars.

galactic disc The wheel-like structure formed by the preponderance of stars and interstellar matter in the Galaxy.

galactic halo A sparsely populated spheroidal distribution of stars about the center of the galaxy, for the most part defined by the distribution of globular clusters.

galactic nucleus A luminous concentration of stars and interstellar material at the center of a galaxy.

galaxy A large system of stars containing millions to hundreds of billions of stars and sometimes containing dust and gas.

Galaxy The system of stars, dust, and gas to which the sun belongs and which we see as the Milky Way.

gamma rays Radiation consisting of photons with higher energy (shorter wavelength) than X-rays.

gas tail That component of a cometary tail which consists of gas, often separated from the dust tail.

geocentric Centered on the earth.

giant A star of large size and great luminosity having a position to the right and above the main sequence in an H-R diagram.

gibbous The phase of the moon or a planet when it is seen more than half but less than fully illuminated.

globular cluster A tightly grouped, gravitationally bound spherical system of stars containing around 100,000 stars.

globule A dark, cold concentration of interstellar material of more or less spherical shape.

granulation The cellular structure of the solar photosphere which gives it a mottled appearance.

gravitational constant The constant in Newton's law of gravitation, usually designated by G; $F = GM_1M_2/r^2$.

gravitational red shift The increase in wavelength associated with the loss in energy of radiation (photons) escaping the gravitation of the radiating object.

gravity The force exerted by the gravitation of a material object.

greatest elongation The greatest angular separation between an inferior planet and the sun.

Great Red Spot A large, red oval feature in the atmosphere of Jupiter.

Great Rift A concentration of interstellar dust that obscures light from the Milky Way, dividing it into two bands extending from the constellation Cygnus to the constellation Scorpius.

Gregorian calendar A calendar introduced by Pope Gregory XIII in 1582 with the consultation of the astronomer Clavius.

H-alpha The strongest absorption or emission line in the visible spectrum of hydrogen resulting from transitions of electrons between the second and third levels of hydrogen atoms.

harmonic law One of Kepler's laws of planetary motion which states the squares of the sidereal periods of the planets are in proportion to the cubes of the semi-major axes of their orbits.

head (of a comet) The cloud of gas and dust surrounding the solid nucleus of a comet when the comet is near the sun.

heliocentric Centered on the sun.

helium flash The explosive ignition of helium fusion in the metallike core of a giant star.

Hertzsprung-Russell (H-R) diagram A plot of the absolute magnitudes of stars versus their spectral types.

H-minus ion A stable, negatively charged ion consisting of a proton and two electrons.

horizon The great circle on the celestial sphere 90° from the zenith.

hour circle (of a star) The great circle on the celestial sphere passing through the celestial poles and the star.

Hubble constant The constant of proportionality in Hubble's law of red shifts, usually designated by H.

Hubble's law The relation between the velocity of recession associated with the red shift in the spectral lines of a galaxy and its distance; $v = Hd$.

hydrogen The simplest element; an atom consisting of a single proton and a single electron.

hydrogen fusion The transmutation of hydrogen through fusion to form helium.

hyperbola An open curve that might be followed by an object as a result of a single gravitational encounter with another object; mathematically, the intersection of a plane and a right circular cone such that the plane does not slice the entire cone and is not parallel to a line on the surface of the cone.

image The optical picture of an object produced by a lens or mirror.

inclination The angle between the plane of an orbit and a reference plane, such as the plane of the ecliptic in the case of the objects in the solar system.

inertia The property of a material object which causes it to resist change in its motion.

inertial frame of reference A frame of reference in which Newton's laws of motion are valid.

inferior conjunction The configuration of an inferior planet in which the planet is in the direction of the sun and between the earth and the sun.

inferior planet A planet moving in an orbit smaller than the earth's orbit.

infrared Radiation consisting of photons with energy less than the energy of visible light but greater than that of radio waves.

infrared nebula A nebula that emits strongly in infrared radiation.

interference The partial or total reinforcing or cancelling of combined waves.

interferometer A device that combines light or radio waves through the process of interference.

interstellar dust Dust particles lying between the stars.

interstellar gas Gas atoms, ions, and molecules lying between the stars.

interstellar lines Absorption lines in the spectra of stars formed by intervening interstellar gas.

interstellar reddening The reddening of starlight by the scattering of light by interstellar dust.

intrinsic variable A star that varies in brightness as a result of changes in the structure and rate of energy flow in the star rather than as a result of external obscuration of its light.

ion An atom which does not have its normal number of electrons.

ionization The process resulting in the removal of one or more electrons from an atom.

ionosphere The outer regions of the earth's atmosphere where atoms are ionized.

irregular cluster of galaxies A cluster of galaxies having no symmetry or concentration in the distribution of its member galaxies.

irregular galaxy A galaxy with no definite form or structure.

irregular variable A pulsating star whose variation in brightness has no definite periodicity.

iron meteorite A meteorite containing predominantly iron and about 5 to 15 percent nickel.

isotropic Having the property of being the same in all directions.

Jovian planets Jupiter, Saturn, Uranus, and Neptune.

Julian calendar A calendar introduced by Julius Caesar in 45 B.C. with the consultation of the astronomer Sosigines.

Kepler's laws Three laws, established by Johannes Kepler, that describe the motion of the planets.

last quarter The phase of the moon when it appears half-illuminated and is moving from full phase to new phase.

latitude The angular distance of a point on the earth from the equator, measured along the meridian of the point.

law of equal areas One of Kepler's three laws, which states that the line joining the planet and the sun sweeps out equal areas in equal intervals of time.

law of force Newton's second law of motion, which states that the force on an object is equal to the product of the object's mass and its acceleration; $F = ma$.

law of inertia Newton's first law, which states that an object at rest or in uniform motion in a straight line continues at rest or in uniform motion in a straight line unless acted on by an unbalanced force.

law of red shifts Hubble's law.

law of universal gravitation Newton's law which states that between any two objects in the universe there is a force of attraction that is proportional to the product of the masses of the objects and inversely proportional to the square of the distance between them; $F = GM_1M_2/r^2$.

leading sunspot In a pair of sunspots, the one that leads in the direction of the solar rotation.

leap year A calendar year which is 366 days long.

light Electromagnetic radiation visible to the eye.

light curve A plot of the variation of the brightness of a star versus time.

light-gathering power The relative capability of a telescope to collect light; depends upon the area of the telescope objective.

light-year The distance traveled in one year by light moving at 300,000 kilometers per second; about 9½ trillion km.

limb The apparent edge of an object as seen in the sky.

limb darkening The dimming of the photosphere toward the limb.

line profile A plot of the intensity of radiation in a spectral line versus wavelength.

local apparent solar time The local hour angle of the apparent sun plus twelve hours.

local group of galaxies The irregular cluster of galaxies to which our Galaxy belongs.

local hour angle The angle between the celestial meridian and the hour circle of a celestial object or point, measured westward along the celestial equator.

local mean solar time The local hour angle of the mean sun plus twelve hours.

local sidereal time The local hour angle of the vernal equinox.

local standard of rest An arbitrary point in space in the vicinity of the sun which has a velocity equal to the average velocity of all stars, including the sun, lying within a distance of 100 parsecs from the sun.

longitude The angular distance of the meridian of a point on the earth's surface from the prime meridian measured along the equator.

long-period variable A pulsating star with a period between about three months and two years.

luminosity The rate at which a star radiates energy.

luminosity class The classification of a star of given spectral type according to its luminosity.

lunar Pertaining to the moon.

lunar eclipse The passing of the moon through the earth's shadow.

lunar highlands The older heavily cratered regions of the moon's surface.

Magellanic Clouds Two very close member galaxies of the local group, visible from southern latitudes.

magnetic field The region of space near a magnetic object or an electrical current in which magnetic forces due to the object or current can be detected.

magnetic storm A strong, temporary disturbance in the earth's magnetic field caused by ionized particles ejected from the region of a solar flare.

magnifying power The angular diameter of an object viewed with a telescope compared to its angular diameter viewed with the unaided eye.

magnitude A number used to indicate the brightness of a celestial object (usually means apparent magnitude unless qualified).

main sequence A concentration of points representing the majority of stars, running from the lower right to the upper left in the H-R diagram.

major axis The longest line across an ellipse; the line defined by the foci of an ellipse with end points on the ellipse.

maria The Latin word for *seas* applied to the dark, more or less smooth areas of the moon.

mass The amount of material in an object measured by its inertia or by its gravitational influence.

mass-luminosity relationship A relation between the masses of stars, principally main-sequence stars, and their luminosities.

mass transfer The process by which material can be transferred from one component of a binary star to the other component.

Maunder Minimum The extended interval from 1645 to 1715 when very few sunspots or other signs of solar activity were observed.

mean sun A fictitious sun which moves uniformly along the celestial equator, completing one revolution in the same time that the real sun takes for a complete trip along the ecliptic.

mechanics The branch of physics that deals with forces and their effects on objects.

megaparsec One million parsecs.

meridian (celestial) The great circle on the celestial sphere which passes through the celestial poles and the zenith.

meridian (terrestrial) A great circle on the earth which passes through the poles.

meridian transit An instrument used to determine the precise instant at which a particular star crosses the celestial meridian of the location of the instrument.

meteor The streak of light produced when a meteoroid enters the earth's atmosphere.

meteorite That part of a meteoroid which succeeds in passing through the earth's atmosphere and actually reaches the earth's surface.

meteoroid An object in orbit about the sun capable of producing a meteor.

meteor shower A display of many meteors appearing to originate from a common point in the sky.

microwave A very short-wavelength radio wave.

Milky Way The luminous band stretching the entire way around the sky due to the concentration of stars and interstellar material near the plane of the Galaxy.

minor planet One of an enormous number of small objects, ranging in size from several hundred kilometers to less than a kilometer in diameter, which orbit the sun principally in the region between the orbits of Mars and Jupiter; an asteroid.

Mira variable A long-period, pulsating red giant star having characteristics similar to the star Mira.

model A representation of the physical properties of an object or system of objects.

model star A quantitative, mathematical model of a star.

molecule Two or more atoms bound together by electrical and magnetic forces.

moving cluster method A method that employs the motion of a cluster of stars in the determination of the cluster's distance.

multiple star Three or more stars in a gravitationally bound system.

nadir The point on the celestial sphere lying directly beneath an observer; that is, the point 180° away from the zenith.

nebula A cloud of interstellar material.

nebular theory The concept that the solar system formed from a condensing nebula.

neutrino A subatomic particle with zero charge and extremely low mass.

neutron A subatomic particle with zero charge and a mass approximatley equal to the mass of the proton. Together with protons, neutrons are the basic constituents of atomic nuclei.

neutron star An extremely small, dense star composed of neutrons bound closely together by gravity.

new moon The phase of the moon when the moon is essentially in the same direction as the sun.

Newtonian focus The focus of a reflecting telescope in which a flat mirror is used to deflect the light beam to the side of the telescope.

Newton's laws The laws of mechanics formulated by Sir Isaac Newton.

node One of the two points at which an object crosses a designated reference plane during its orbital motion (The ecliptic is the reference plane usually used in the solar system.)

normal spiral galaxy A galaxy with a central concentration from which spiral arms extend.

northern lights A common name for the *aurora borealis*.

nova A cataclysmic variable that suddenly increases its light and energy output, typically by several ten thousandfold, then gradually subsides to its original brightness.

nuclear force The force that binds the constituents of atomic nuclei together.

nucleus of an atom The positively charged, heavy, centrally located part of an atom.

O-association A loose association of stars consisting mostly of O and B stars, usually located in spiral arms.

objective The main lens or mirror of a telescope.

objective prism A prism placed in front of the objective of a telescope to form simultaneous spectra of all the stars in the field of view of the telescope.

obliquity of the ecliptic The 23½°-angle between the planes of the ecliptic and the celestial equator.

Oort comet cloud A spherical shell with an average distance of about 50,000 AU from the sun containing the preponderance of comets.

open cluster A group of from several hundred to several thousand stars, usually located near the plane of the Galaxy.

opposition Configuration of a superior planet when it is essentially opposite in direction from the sun.

optical double Two unconnected stars which happen to appear in nearly the same direction.

orbit The path of a celestial object moving under the gravitational influence of another object or system of objects.

Orion arm The name of the spiral arm in which the sun is located.

oscillating universe A universe which periodically expands and contracts, beginning each new expansion with an explosion.

parabola A type of open curve followed by an object as a result of a single gravitational encounter; mathematically, the intersection between a right circular cone and a plane which is parallel to a straight line on the surface of the cone.

parallax An apparent change in the position of an object due to a change in position of the observer; in astronomy, the parallax resulting from a displacement of the observer by one AU.

parallax angle The total angular change in position of an object at the two positions from which it is seen.

parsec A unit of distance equal to 206,265 AU; the distance at which an object would have a parallax of one second of arc.

partial lunar eclipse An eclipse of the moon where only part of the moon is covered by the umbra of the earth's shadow.

partial solar eclipse An eclipse of the sun in which the moon covers only part of the sun.

peculiar velocity The velocity of a star relative to the local standard of rest.

penumbral eclipse The lunar eclipse in which the moon passes through the penumbra of the earth's shadow but not through the umbra.

penumbra (of a shadow) The part of a shadow cast by an object in which the light from the source is only partially excluded.

penumbra (of a sunspot) The relatively bright outer regions of a sunspot.

perihelion The point of closest approach to the sun on the orbit of an object revolving about the sun.

period The interval of time in which one rotation, one revolution, or one cycle takes place.

period-luminosity relation The relation between the pulsation periods and the luminosities of cepheid variables and RR Lyrae stars.

perturbation The deviations from the expected motion of a body calculated as if only two spherical bodies were present; caused by the influence of other bodies or nonspherical shapes.

phase A measure of the fraction of a cyclic process which has been completed. When referring to the moon or a planet, it is the fraction of the illuminated disc which is visible.

photoelectric effect The ejection of electrons from a metal being irradiated by photons.

photon A particle of electromagnetic radiation.

photosphere The apparent surface of the sun.

physical double A binary star.

plage A bright region of the solar chromosphere observed in photographs taken in the light of particular spectral lines.

Planck's constant The constant of proportionality in the relation between the energy and frequency of a photon; $E = h\nu$.

Planck's law A formula which relates the intensity to wavelength of radiation from a black body at a specified temperature.

plane of the Milky Way The plane which passes through the center of the Galaxy and divides the galactic disc into two more or less equivalent discs.

planet Any of the nine major objects orbiting the sun.

planetary nebula An expanding shell of gas, heated by a centrally located hot star from which the nebula presumably was ejected.

planetoid A minor planet.

polar axis The supporting axis of a telescope which is arranged to be parallel to the earth's axis of rotation.

population I Stars found predominantly in the disc of the galaxy, containing the youngest objects from an evolutionary viewpoint.

population II Stars found predominantly in the halo of the galaxy, containing the oldest objects from an evolutionary viewpoint.

position angle The orientation in the plane of the sky of the line joining the two components of a binary star.

positron A subatomic particle with positive charge and a mass equal to the mass of the electron.

precession A slow, conical rotation of the earth's axis due to the gravitational effects of the moon and sun on the equatorial bulge of the earth.

primary minimum The minimum in the light curve of an eclipsing binary that corresponds to the eclipse of the hotter star by the cooler star.

primary of eclipsing binary The hotter component.

primary of visual binary The brighter component.

prime focus The focus of a mirror when no auxiliary lenses or mirrors are used.

prime meridian The meridian of Greenwich, England.

primeval atom In the big-bang universe, the object formed by the concentration of the entire universe into a single point at the instant of creation.

primeval fireball The exploding primeval atom.

Principia Short name for *Philosophiae Naturalis Principia Mathematica,* in which Newton presented his laws of mechanics and law of universal gravitation.

principle of equivalence The principle which states that for a given noninertial frame of reference, there is an equivalent inertial frame with an appropriate local gravitational field.

principle of relativity A postulate defining the independence of the laws of physics from the observer's frame of reference.

prism In optics, a transparent wedge of glass used to disperse light.

progression of Titius More popularly known as Bode's law.

prominence Gas trapped in solar magnetic fields and seen in emission at the solar limb.

proper motion The annual change in the angular position of a star after removing the effect of trigonometric parallax, expressed in seconds of arc per year.

proton A subatomic particle with positive charge which, together with the neutron, is a basic constituent of the nuclei of atoms; also, the nucleus of a hydrogen atom.

proton-proton reaction A series of three successive processes by which helium is formed from hydrogen (named after the first process, which results in the combination of two protons).

protostar A condensation of interstellar material in the process of becoming a star.

Ptolemaic system A model universe centered on the earth.

pulsar A source of rapidly varying radio emission with a period in the range 0.03 to 5 seconds; presumably, a rapidly rotating neutron star.

pulsating variable star A star with variable brightness that pulsates in size.

quadrature The configuration of the moon or a superior planet when its elongation is 90°.

quasar Short name for quasi-stellar source.

quasi-stellar object A stellar-appearing object with a very large red shift presumed to be extremely far away and extremely luminous, but which differs from a quasi-stellar source in its lack of strong radio emission.

quasi-stellar source A strong radio source with a stellar appearance and a very large red shift presumed to be extremely far away and extremely luminous.

radar A technique for measuring the direction and distance to an object by transmitting radio waves to the object and analyzing the reflected signals.

radial velocity The rate at which the distance to an object changes.

radiation The process by which energy can be transmitted through a vacuum, either by electromagnetic waves or by particles.

radiation pressure The pressure exerted by electromagnetic radiation.

radio astronomy The study of celestial objects using the radio portion of the spectrum.

radio telescope A telescope capable of detecting radio waves.

reaction law Newton's third law, which states that to each and every action there is an equal and opposite reaction.

red giant A very large and luminous red star.

red shift A shift to longer wavelengths of light; can be due to a Doppler shift or a gravitational shift.

reflection The return of waves from a surface.

reflection nebula A nebula illuminated by the reflection of starlight from its component dust.

reflector Short name for a telescope employing a mirror as its objective; reflecting telescope.

refraction The bending of light rays in passing from one transparent medium to another.

refractor Short name for a telescope employing a lens as an objective; refracting telescope.

regression of the nodes The motion of the moon's nodes along the ecliptic due to the slow westward rotation of the moon's orbit.

regular cluster of galaxies A spherically symmetric cluster of galaxies with a concentration of galaxies toward the center of the cluster.

relative orbit The orbit of one body about another, as measured from one of the bodies.

resolution The ability of a telescope to separate two points of light very close together; the ability to detect fine detail.

resolving power A measure of a telescope's resolution.

retrograde motion The apparent westward motion of a planet relative to the stars.

revolution The cyclical motion of one object around another.

right ascension The angle between the vernal equinox and the hour circle of an object measured eastward along the celestial equator.

rill A long, narrow valley on the moon.

rotation Turning of an object about an axis through the object.

RR Lyrae star A pulsating giant star with a period less than one day and with characteristics like the star RR Lyrae.

satellite A small object in orbit around a larger object; a moon.

Schmidt telescope A reflecting telescope with a thin refracting corrector lens which gives a large field of view.

secondary minimum The minimum in the light curve of an eclipsing binary caused by the eclipse of the cooler component by the hotter component.

semi-major axis One-half the major axis.

separation of a binary star The observed angular distance between the components of a visual binary.

Seyfert galaxy A spiral galaxy with a very bright nucleus that has emission lines in its spectrum and is often a strong source of radio and infrared emission.

shooting star A meteor.

sidereal month The period of revolution of the moon about the earth relative to the stars.

sidereal period The period of revolution of a celestial object about another with respect to the stars.

sidereal time The local hour angle of the vernal equinox.

sidereal year The period of revolution of the earth around the sun relative to the stars.

solar activity Sunspots, flares, prominences, and other isolated phenomena in the atmosphere of the sun.

solar antapex The direction opposite to the solar apex.

solar apex The direction in which the sun is moving relative to the nearby stars.

solar eclipse The obscuration of the sun by the moon.

solar system The system consisting of the sun, planets, satellites, minor planets, comets, and all other objects orbiting the sun.

solar wind The outward streaming of nuclei of atoms (mostly protons) and electrons from the sun.

solstice One of two points on the ecliptic where the ecliptic reaches its greatest distance from the celestial equator.

space probe A device launched into space for the purpose of measuring properties of objects in the solar system.

space velocity The velocity of a star in space relative to the sun.

spectral type Classification of a star according to its spectral characteristics.

spectrograph An instrument designed for the photography of spectra.

spectroheliogram A photograph of the sun in the light of a particular spectral line.

spectroheliograph An instrument designed to take spectroheliograms.

spectroscopic binary A binary star that shows its duplicity by its spectrum.

spectrum An array of the various colors or wavelength components of light which is formed when light is dispersed.

speed The rate at which an object moves, irrespective of direction.

spicule A narrow spike of rising material in the chromosphere of the sun.

spiral galaxy A galaxy with arms of young stars and interstellar material spiraling outward from its central regions.

standard time The local mean solar time of a standard meridian, kept by all clocks in the time zone represented by that standard meridian.

steady-state model A model of the universe based on the assumption that the large-scale characteristics of the universe never change.

Stefan's law A law relating the energy emitted each second by a unit area of a black body to the temperature of the black body; $E = \sigma T^4$.

stony-iron meteorite A meteorite consisting in part of rock and in part of iron and nickel.

stony meteorite A meteorite consisting of rock.

summer solstice The point on the ecliptic where it reaches its greatest northern distance from the celestial equator.

sunspot A dark region on the surface of the sun.

sunspot cycle The cyclic variation in the number of sunspots, which has an average period of about eleven years.

supergiant A large star of very great luminosity.

superior conjunction The configuration of an inferior planet when it is essentially in the same direction as the sun but on the other side of the sun.

superior planet A planet with an orbit that is larger than the orbit of the earth.

supernova A cataclysmic variable that undergoes a sudden and tremendous increase in brightness, reaching a luminosity as great as 10 billion times the luminosity of the sun.

synodic month The period of the moon's cycle of phases.

synodic period The interval of time between two successive passages of a planet through the same phase or configuration.

tail (of a comet) Material ejected from a comet and forced away from the head by radiation pressure and the solar wind.

tangential velocity The component of the velocity of an object projected onto the sky.

T-association An association of T Tauri stars.

temperature (absolute or Kelvin) Temperature measured on the Celsius (or centigrade) scale, but with the zero point set at absolute zero.

temperature (Celsius or centigrade) Temperature measured on a scale with the boiling point of water set at 100° and the freezing point of water set at 0°; absolute zero equals −273°C.

temperature (effective) The temperature which a black body with the same dimensions as an observed body would have to have to radiate the same energy as the observed body.

terrestrial planet Any of the planets Mercury, Venus, Earth, or Mars.

theory of relativity The term used to describe Einstein's formulation of the laws of physics in accordance with a principle of relativity.

thermonuclear reaction A nuclear reaction resulting from the high velocities of the nuclei due to high temperature.

three-degree background radiation Isotropic radiation with an intensity-wavelength relation described by Planck's law for a black body at 3° above absolute zero; presumably originating from the primeval fireball.

tide The deformation of a body due to the differences in the gravitational force on its various parts resulting from the influence of a second body.

total lunar eclipse An eclipse of the moon in which the moon is entirely covered by the umbra of the shadow of the earth.

total solar eclipse An eclipse of the sun in which the moon totally obscures the sun.

transit The passage of a celestial object across the celestial meridian.

transition In a simple model of an atom, the moving of an electron from one orbit to another.

trigonometric parallax An astronomical term for the angular shift that would be observed in the apparent position of a star if it were viewed from the center of the sun and from a point exactly one AU away in a direction perpendicular to the star.

tropical year The interval of time between two successive passages of the sun through the vernal equinox.

true orbit The actual orbit of the secondary component of a binary star relative to the primary component; that is, the orbit as it would appear if viewed perpendicular to its plane.

T Tauri star An irregularly variable star usually found immersed in interstellar material.

tuning fork diagram A diagram representing Hubble's classification scheme for galaxies.

21-cm radiation Characteristic radio radiation from cold hydrogen gas.

Type I cepheid A classical cepheid variable.

Type II cepheid A pulsating variable star with characteristics similar to classical cepheid variable stars but with a somewhat lower luminosity.

ultraviolet radiation Radiation with wavelengths just shorter than violet light and not visible to the human eye.

umbra That part of a shadow cast by an object in which the light from the source is completely obscured; also the dark, central part of a sunspot.

universal time The local mean solar time at the prime meridian (Greenwich).

universe The entirety of matter and energy and the space occupied by it.

variable star A star that varies in brightness.

velocity The rate at which an object changes position, including the direction of change.

velocity curve A plot of the radial velocities of the components of a binary star versus time.

velocity of escape The minimum speed that an object must have in order to escape the gravitational attraction of another object.

vernal equinox The intersection of the ecliptic and the celestial equator such that a point moving eastward along the ecliptic would pass from the Southern Hemisphere to the Northern Hemisphere as defined by the celestial equator.

visual binary A binary star in which the two components can be resolved.

waning The phases passed through by the moon as it moves from full phase to new phase.

wavelength The distance between successive crests of a system of waves.

waxing The phases passed through by the moon as it moves from new phase to full phase.

white dwarf A small, dense star in which no new energy is being generated and which has contracted to the point where further contraction does not proceed.

winter solstice The point on the ecliptic where it reaches its greatest southern distance from the celestial equator.

wrinkled ridges Ridges on the relatively smooth surface of the maria.

X-rays Radiation composed of high-energy photons between those of gamma rays and ultraviolet light.

year Period of revolution of the earth around the sun.

Zeeman effect The splitting of a spectral line due to a magnetic field.

zenith The point in the celestial sphere directly above the observer, determined from the direction of gravity at the earth's surface.

zero-age main sequence The hypothetical main sequence formed by the points on the H-R diagram, representing stable stars with similar, homogeneous chemical compositions.

Zodiac The belt of constellations around the sky centered on the ecliptic.

zodiacal light A glow along the ecliptic caused by scattering of sunlight by interplanetary dust.

zone of avoidance A region centered on the plane of the Galaxy in which few or no galaxies can be seen, due to obscuration by interstellar material in our own Galaxy.

Bibliography

General

Abell, George A. *Exploration of the Universe*. 3rd ed. New York: Holt, Rinehart and Winston, 1975.

Frederick, Laurence W., and Baker, Robert H. *Astronomy*. 10th ed. New York: Van Nostrand Reinhold, 1976.

Smith, Elske V. P., and Jacobs, K. C. *Introductory Astronomy and Astrophysics*. Philadelphia: W. B. Saunders, 1973.

Historical

Cumont, F. V. M. *Astrology and Religion Among the Greeks and Romans*. New York: Dover, 1960.

Pannekoek, A. *A History of Astronomy*. London: Allen and Unwin, 1961.

Struve, O., and Zebergs, V. *Astronomy of the Twentieth Century*. New York: Macmillan, 1962.

Instruments and Light

Hey, J. S. *The Evolution of Radio Astronomy*. New York: Neale Watson Academic Publications, 1973.

Miczaika, G. R., and Sinton, W. M. *Tools of the Astronomer*. Cambridge, Mass.: Harvard University Press, 1961.

Minnaert, M. *The Nature of Light and Colour in the Open Air*. New York: Dover, 1954.

Solar System

Gibson, E. G. *The Quiet Sun*. Washington, D. C.: NASA, 1973.

Hartmann, W. K. *Moons and Planets*. Tarrytown-on-Hudson, N. Y.: Bodgen & Quigley, 1972.

Hartmann, W. K., and Raper, O. *The New Mars*. Washington, D. C.: NASA, 1974.

Scientific American, September 1975. (This issue is devoted completely to the solar system and contains many recent results obtained through the space program.)

Van de Kamp, P. *Elements of Astromechanics*. San Francisco: W. H. Freeman, 1964.

Whipple, F. *Earth, Moon and Planets*. 3rd ed. Cambridge, Mass.: Harvard University Press, 1968.

Wood, J. A. *Meteorites and the Origin of Planets*. New York: McGraw-Hill, 1968.

Stars and Stellar Systems

Bok, B. J., and Bok, P. E. *The Milky Way.* 4th ed. Cambridge, Mass.: Harvard University Press, 1974.

Shapely, H. *Galaxies*. 3rd ed. Cambridge, Mass.: Harvard University Press, 1972.

Cosmology and Relativity

Hodge, P. W. *The Physics and Astronomy of Galaxies and Cosmology.* New York: McGraw-Hill, 1966.

Kaufmann, W. J. *Relativity and Cosmology.* New York: Harper & Row, 1973.

Guides to the Sky

Norton, A. P. *Norton's Star Atlas*. 16th ed. Edinburgh: Gall & Inglis, 1973.

The Observer's Handbook. Toronto: Royal Astronomical Society of Canada, published annually.

Ottewell, G. *Astronomical Calendar.* Greenville, S. C.: Department of Physics, Furman University, published annually.

Periodicals

Scientific American. Scientific American, Inc., New York.

Sky and Telescope. Sky Publishing Corp., Cambridge, Mass.

Scientific American Articles of Special Interest

History of Astronomy:

Drake, Stillman. "Galileo and the First Mechanical Computing Device." April 1976.

Drake, Stillman. "The Role of Music in Galileo's Experiments." June 1975.

Drake, Stillman, and MacLachlan, James. "Galileo's Discovery of the Parabolic Trajectory." March 1975.

Gingerich, Owen. "Copernicus and Tycho." December 1973.

Lerner, Lawrence S., and Gosselin, Edward A. "Giordano Bruno." April 1973.

Ravetz, J. R. "The Origins of the Copernican Revolution." October 1966.

Wilson, Curtis. "How Did Kepler Discover His First Two Laws?" March 1972.

Relativity Theory:

Callahan, J. J. "The Curvature of Space in a Finite Universe." August 1976.

Van Flandern, Thomas C. "Is Gravity Getting Weaker?" February 1976.

Will, Clifford M. "Gravitation Theory." November 1974.

The Earth and Moon:

Dyal, Palmer, and Parkin, Curtis W. "The Magnetism of the Moon." August 1971.

Eglinton, Geoffrey; Maxwell, James R.; and Pillinger, Colin T. "The Carbon Chemistry of the Moon." October 1975.

Goldreich, Peter. "Tides and the Earth-Moon System." April 1972.

Mason, Brian. "The Lunar Rocks." October 1971.

Moorbath, Stephen. "The Oldest Rocks and the Growth of Continents." March 1977.

Siever, Raymond. "The Earth." September 1975.

Wood, John A. "The Lunar Soil." August 1970.

Wood, John A. "The Moon." September 1975.

Astronomical Instruments and Techniques:

Connes, Pierre. "How Light is Analyzed." September 1968.

Goldberg, Leo. "Ultraviolet Astronomy." June 1969.

Kellermann, K. I. "International Radio Astronomy." February 1972.

MacDougall, J. D. "Fission-Track Dating." December 1976.

Melbourne, William G. "Navigation Between the Planets." June 1976.

Neugebauer, G., and Leighton, Robert B. "The Infrared Sky." August 1968.

Shapiro, Irwin I. "Radar Observations of the Planets." July 1968.

Planets and Moons:

Carr, Michael H. "The Volcanoes of Mars." January 1976.

Cruikshank, Dale P., and Morrison, David. "The Galilean Satellites of Jupiter." May 1976.

Eshleman, Von R. "The Atmospheres of Mars and Venus." March 1969.

Hartmann, William K. "Cratering in the Solar System." January 1977.

Horowitz, Norman H. "The Search for Life on Mars." November 1977.

Hunten, Donald M. "The Outer Planets." September 1975.

Ingersoll, Andrew P. "The Meteorology of Jupiter." March 1976.

Leory, Conway B. "The Atmosphere of Mars." July 1977.

Murray, Bruce C. "Mercury." September 1975.

Pollack, James B. "Mars." September 1975.

Sagan, Carl. "The Solar System." September 1975.

Siever, Raymond. "The Earth." September 1975.

Veverka, Joseph. "Phobos and Deimos." February 1977.

Wolfe, John H. "Jupiter." September 1975.

Young, Andrew, and Young, Louise. "Venus." September 1975.

Interplanetary Matter and Origin of the Solar System:

Cameron, A. G. W. "The Origin and Evolution of the Solar System." September 1975.

Cameron, I. R. "Meteorites and Cosmic Radiation." July 1973.

Chapman, Clark R. "The Nature of Asteroids." January 1975.

Grossman, Lawrence. "The Most Primitive Objects in the Solar System." February 1975.

Hartmann, William K. "The Smaller Bodies of the Solar System." September 1975.

Lawless, James G.; Folsome, Clair E.; and Kvenvolden, Keith A. "Organic Matter in Meteorites." June 1972.

Lewis, John S. "The Chemistry of the Solar System." March 1974.

Van Allen, James A. "Interplanetary Particles and Fields." September 1975.

Whipple, Fred L. "The Nature of Comets." February 1974.

The Sun:

Eddy, John A. "The Case of the Missing Sunspots." May 1977.

Gosling, J. T., and Hundhausen, A. J. "Waves in the Solar Wind." March 1977.

Howard, Robert. "The Rotation of the Sun." April 1975.

Parker, E. N. "The Sun." September 1975.

Pasachoff, Jay M. "The Solar Corona." October 1973.

Structure and Evolution of Normal Stars:

Abt, Helmut, A. "The Companions of Sunlike Stars." April 1977.

Bahcall, John N. "Neutrinos from the Sun." July 1969.

Bok, Bart J. "The Birth of Stars." August 1972.

Dickman, Robert L. "Bok Globules." June 1977.

Neugebauer, G., and Becklin, Eric E. "The Brightest Infrared Sources." April 1973.

Percy, John R. "Pulsating Stars." June 1975.

Supernovae, X-ray Stars, Pulsars, and Black Holes:

Charles, Philip A., and Culhane, J. Leonard. "X-rays from Supernovae Remnants." December 1975.

Clark, George W. "X-Ray Stars in Globular Clusters." October 1977.

Gorenstein, Paul, and Tucker, Wallace. "Supernova Remnants." July 1971.

Gursky, Herbert, and van den Heuvel, Edward P. J. "X-ray-Emitting Double Stars." March 1975.

Hawking, S. W. "The Quantum Mechanics of Black Holes." January 1977.

Kirshner, Robert P. "Supernovas in Other Galaxies." December 1976.

Ostriker, Jeremiah P. "The Nature of Pulsars." January 1971.

Pacini, Franco, and Rees, Martin J. "Rotation in High-Energy Astrophysics." February 1973.

Penrose, Roger. "Black Holes." May 1972.

Ruderman, Malvin A. "Solid Stars." February 1971.

Schnopper, Herbert W. and Delvaille, John P. "The X-ray Sky." July 1972.

Stephenson, F. Richard, and Clark, David H. "Historical Supernovas." June 1976.

Strong, Ian B., and Klebesadel, Ray W. "Cosmic Gamma-Ray Bursts." October 1976.

Thorne, Kip S. "Gravitational Collapse." November 1967.

Thorne, Kip S. "The Search for Black Holes." December 1974.

Interstellar Matter:

Barrett, Alan H. "Radio Signals from Hydroxyl Radicals." December 1968.

Ginzburg, V. L. "The Astrophysics of Cosmic Rays." February 1969.

Greenberg, J. Mayo. "Interstellar Grains." October 1967.

Maran, Stephen P. "The Gum Nebula." December 1971.

Miller, Joseph S. "The Structure of Emission Nebulas." October 1974.

Sanders, R. H., and Wrixon, G. T. "The Center of the Galaxy." April 1974.

Turner, Barry E. "Interstellar Molecules." March 1973.

Galaxies:

Disney, Michael J., and Veron, Philippe. "BL Lacertae Objects." August 1977.

Groth, Edward J.: Pebbles, P. James E.; Seldner, Michael; and Soneira, Raymond M. "The Clustering of Galaxies." November 1977.

Rees, Martin J., and Silk, Joseph. "The Origin of Galaxies." June 1970.

Rubin, Vera C. "The Dynamics of the Andromeda Nebula." June 1973.

Strom, Richard G.; Miley, George K.; and Oort, Jan. "Giant Radio Galaxies." August 1975.

Toomre, Alar; and Toomre, Juri. "Violent Tides Between Galaxies." December 1973.

Weymann, Ray J. "Seyfert Galaxies." January 1969.

Quasi-Stellar Objects and Cosmology:

Alfven, Hannes. "Antimatter and Cosmology." April 1967.

Burbidge, E. Margaret, and Lynds, C. Roger. "The Absorption Lines of Quasi-Stellar Objects." December 1970.

Burbidge, Geoffrey, and Hoyle, Fred. "The Problem of the Quasi-Stellar Objects." December 1966.

Callahan, J. J. "The Curvature of Space in a Finite Universe." August 1976.

Gott, J. Richard, III; Gunn, James E.; Schramm, David N.; and Tinsley, Beatrice M. "Will the Universe Expand Forever?" March 1976.

Pasachoff, Jay M., and Fowler, William A. "Deuterium in the Universe." May 1974.
Peebles, P. J. E., and Wilkinson, David T. "The Primeval Fireball." June 1967.
Schmidt, Maarten, and Bello, Francis. "The Evolution of Quasars." May 1971.
Schramm, David N. "The Age of the Elements." January 1974.
Webster, Adrian. "The Cosmic Background Radiation." August 1974.

Index

Index